Defence Nuclear Waste Disposal in Russia: International Perspective

NATO ASI Series

Advanced Science Institute Series

A Series presenting the results of activities sponsored by the NATO Science Committee, which aims at the dissemination of advanced scientific and technological knowledge, with a view to strengthening links between scientific communities.

The Series is published by an international board of publishers in conjunction with the NATO Scientific Affairs Division

A Life Sciences **B Physics**	Plenum Publishing Corporation London and New York
C Mathematical and Physical Sciences **D Behavioural and Social Sciences** **E Applied Sciences**	Kluwer Academic Publishers Dordrecht, Boston and London
F Computer and Systems Sciences **G Ecological Sciences** **H Cell Biology** **I Global Environment Change**	Springer-Verlag Berlin, Heidelberg, New York, London, Paris and Tokyo

PARTNERSHIP SUB-SERIES

1. Disarmament Technologies	Kluwer Academic Publishers
2. Environment	Springer-Verlag / Kluwer Academic Publishers
3. High Technology	Kluwer Academic Publishers
4. Science and Technology Policy	Kluwer Academic Publishers
5. Computer Networking	Kluwer Academic Publishers

The Partnership Sub-Series incorporates activities undertaken in collaboration with NATO's Cooperation Partners, the countries of the CIS and Central and Eastern Europe, in Priority Areas of concern to those countries.

NATO-PCO-DATA BASE

The electronic index to the NATO ASI Series provides full bibliographical references (with keywords and/or abstracts) to about 50,000 contributions from international scientists published in all sections of the NATO ASI Series. Access to the NATO-PCO-DATA BASE is possible via a CD-ROM "NATO Science and Technology Disk" with user-friendly retrieval software in English, French, and German (©WTV GmbH and DATAWARE Technologies, Inc. 1989). The CD-ROM contains the AGARD Aerospace Database.

The CD-ROM can be ordered through any member of the Board of Publishers or through NATO-PCO, Overijse, Belgium.

Series 1: Disarmament Technologies – Vol. 18

Defence Nuclear Waste Disposal in Russia: International Perspective

edited by

M. J. Stenhouse
QuantiSci,
Melton Mowbray,
United Kingdom

and

V. I. Kirko
Research Institute of Physics
and Engineering (NIFTI),
Krasnoyarsk,
Russia

Kluwer Academic Publishers

Dordrecht / Boston / London

Published in cooperation with NATO Scientific Affairs Division

Proceedings of the NATO Advanced Research Workshop on
Defence Nuclear Waste Disposal in Russia: Implications for the Environment
Krasnoyarsk, Russia
June 24-27, 1996

A C.I.P. Catalogue record for this book is available from the Library of Congress.

ISBN 0-7923-5112-6

Published by Kluwer Academic Publishers,
P.O. Box 17, 3300 AA Dordrecht, The Netherlands.

Sold and distributed in North, Central and South America
by Kluwer Academic Publishers,
101 Philip Drive, Norwell, MA 02061, U.S.A.

In all other countries, sold and distributed
by Kluwer Academic Publishers,
P.O. Box 322, 3300 AH Dordrecht, The Netherlands.

Printed on acid-free paper

BACK ROW: Tim McEwen, Mark Gerchikov, Jaap Hoek, Jor-Shan Choi, German Kolonin, Andrey Rybalchenko, Mike Stenhouse, Boris Burakov, Stanislav Shabalev.
MIDDLE ROW: Etienne Trauwaert, Erich Merz, Manuel Martins, Harold Sullivan, Dimcho Evstatiev, Vladimir Komlev.
FRONT ROW: Kishor Mehta, Peter Booth, Leo Baetsle, Valentina Bogdanova, Alice Komogortseva, Nikolai Markov, Maryna Victorovna, Evgeniy Anderson, Yuri Dublyansky.
Photograph courtesy of Hans Engelmann.

TABLE OF CONTENTS

PREFACE

Large quantities of radioactive waste have arisen from more than 40 years of reprocessing activities associated with plutonium production and defence-related activities at Mayak, Tomsk and Krasnoyarsk, in central Siberia (see *Figure 1*). Given the additional volumes of waste to be generated, combined with a lack of adequate storage facilities at any of these sites, there is an urgent need to establish a waste management programme in terms of reliable waste treatment and conditioning methods, and to identify suitable sites for the disposal of conditioned waste.

To this end, an international group of specialists in the radioactive waste field attended a NATO Advanced Research Workshop (ARW) in Krasnoyarsk (June 24-27, 1996), focussing to some extent on the particular requirements of one of the above mentioned facilities: Krasnoyarsk-26.

The Krasnoyarsk-26 facility is situated 40 km from the city of Krasnoyarsk which hosted the ARW. The entire complex, operated by the Integrated Mining-Chemical Company, comprises 5 nuclear reactors, with associated reprocessing and plutonium production facilities, all of which were constructed underground in a large excavated volume, estimated to be comparable with that of the Moscow metro. Krasnoyarsk-26 is a town in its own right with some 100,000 inhabitants, populated mainly by workers at the complex and their families. Up until 1992, this town had been closed to the outside world, such that many of the inhabitants of nearby Krasnoyarsk were unaware of its existence.

The neighbouring city of Krasnoyarsk, with about 1 million inhabitants, lies on the Yenesei river and is the capital of the Krasnoyarsk Krai region, a massive area of Siberia about the size of Scandinavia, extending from the Russian/Mongolian border in the south to the Taymyr Peninsula close to the Arctic Circle. The Yenesei river extends the length of this region.

Since the end of the Cold War and *glasnost*, the local population around Krasnoyarsk-26 have become fully aware of the nuclear facility and are naturally concerned about the status of residual radioactivity from effluent and from the radioactive wastes and byproducts generated from this facility. A commitment was given in 1994 by President Yeltsin for a reactor to be constructed at Krasnoyarsk-26 for civilian purposes, *i.e.* civilian power generation, but this reactor will not be permitted to operate until a disposal site has been identified to accommodate the large volumes of radioactive waste from the nuclear complex. Since power reactor start-up was scheduled for 2010, there is limited time for site characterisation and selection, as well as the construction of a new (surface) reprocessing and conditioning facility (RT-2 Plant) to treat residual spent fuel prior to disposal. Work on the surface reprocessing facility had started but stopped before the plant

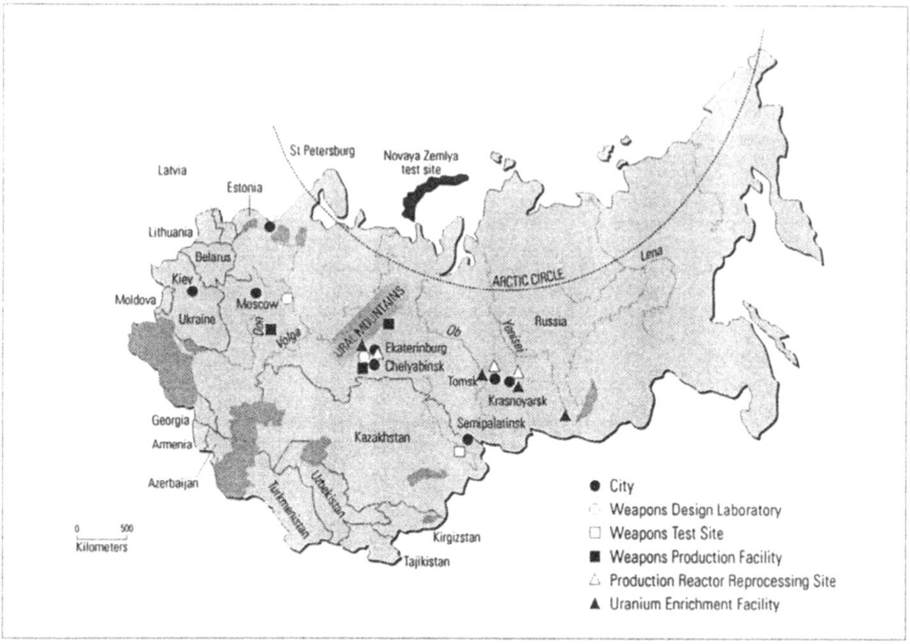

Figure 1. Location of Defence-Related Reprocessing Facilities in Russia

was completed, owing to lack of funds.

Previous NATO workshops on the theme of nuclear disarmament have been held in St. Petersburg and Moscow, but this workshop was the first of its kind to be held in central Siberia. We felt that a NATO Workshop held in the Krasnoyarsk Krai region would demonstrate to the local population a willingness on behalf of the international community to use funds, scientific knowledge and experience in pursuit of a solution to the major nuclear waste management problem in the area. In this context, the workshop was the culmination of a two-year planning effort involving staff of QuantiSci and the Research Institute of Physics and Engineering of Krasnoyarsk State University.

As of June, 1996, there were no direct flights to the Krasnoyarsk from outside Russia and, owing mainly to the time difference (Krasnoyarsk is 5 time zones from Moscow), it requires more than 24 hours to reach there from most European cities. This ARW was the first opportunity for scientists and the local population of the Krasnoyarsk Krai Territory of Siberia to discuss in detail key issues concerning radioactive waste disposal with an international group of scientists. The workshop also coincided with the Third International Radioecological Conference, allowing additional exchange of information between scientists from each meeting.

About 35 specialists from U.S.A., Canada, U.K., Belgium, Germany, Bulgaria and Russia (47 %) participated in the workshop, holding in-depth discussions on waste management options, waste conditioning methods, site characterisation techniques, and methodologies for site selection and performance assessment. Over 25 oral presentations provided a broad overview of the problems associated with the management and disposal of defence-related radioactive waste and ways in which similar problems in the civilan nuclear industry in Europe and North America are being tackled, to a large extent, successfully, although no repository has yet been constructed to contain high-activity, long-lived radioactive waste.

The workshop was divided into five major topical themes:

- Conditioning of Waste: Waste Forms
- Geological Aspects of Radioactive Waste Disposal
- Site Selection and Characterisation Methods
- Safety and Performance Assessment Modelling
- Special Issues Concerning Radioactive Waste Disposal (including Deep-Well Injection of Liquid Radioactive Waste)

A visit to the Krasnoyarsk-26 facility had been planned to take place at the end of the workshop. Unfortunately, owing to a change in the Russian administration prior to the meeting, authorisation was not possible. Nevertheless, the meeting was considered by all to be a great success.

Besides fulfilling its primary objective in providing a forum for discussion of state-of-the-art technologies, the ARW encouraged the creation of strong East-West scientific links for close operation in future R&D programmes. In this respect, the Workshop complemented efforts being made in other collaborative programmes involving Russia. For example, the U.S. Department of Energy and Russia's Minatom have already signed agreements for sharing waste treatment and conditioning technologies. Thus, encouragingly, the number of collaborative projects between Russian and European scientists continues to increase steadily.

Dr. Michael Stenhouse
Prof. Vladimir Kirko
Co-Directors, NATO Advanced
Research Workshop

ACKNOWLEDGEMENTS

The editors are most grateful to NATO, and especially to Mrs. Nancy Schulte, Director of the Disarmament Technologies Programme, for providing the majority of the financial support, without which the workshop could not have taken place. The editors are also grateful for additional funding from Krasnoyarsk State University and from the Administration of the Krasnoyarsk Krai Territory.

One of the editors (MJS) would like to thank the Krasnoyarsk State University, particularly Professor Vladimir Kirko and his staff of the Research Institute of Physics and Engineering, for their generous hospitality and for hosting the workshop so successfully. Special thanks are due to Dr. Nikolai Tsirulnichenko who devoted much time and effort to planning the meeting and to the compilation and translation of Russian abstracts.

The organising committee and the editors would like to express their thanks to the speakers and commentators, and to all others who attended the workshop and who took part in the many lively discussions. All of these people contributed to the success of the workshop and made this publication possible.

We are also grateful to Annelies Kersbergen of Kluwer Academic Publishers for her patience, for her editorial support and for general guidance in ensuring the efficient publication of the material herein.

Secretarial assistance (Mrs. Joy Ball and Mrs. Hilary Moore) and the preparation of graphics material (Mr. Paul Weston) by QuantiSci staff are acknowledged with thanks. This work and the substantial commitment of time associated with the organisation of the Workshop and preparation of its Proceedings, were undertaken with the enthusiastic support of Dr. David Hodgkinson (Managing Director, QuantiSci) and Professor Neil Chapman (Group Leader, Geosciences Group, QuantiSci).

Finally, MJS is particulary grateful to his wife, Margaret, for her continued patience and tolerance throughout a long editorial process.

DEDICATION

This volume is dedicated to the people of Siberia who greeted all visitors attending the Advanced Research Workshop with natural warmth and hospitality. We look forward to continued scientific collaboration and friendship.

LIST OF SPEAKERS AND PARTICIPANTS

ANDERSON, Evgeniy B.
Deputy Director General
V.G. Khlopin Radium Institute
194021 St. Petersburg
RUSSIA

ANIKOLENKO, V.A.
Institute of Experimental Geophysics
United Institute of Earth Physics
109004 Moscow
RUSSIA

BAETSLE, L.H.
SCK•CEN
Boeretang 200
B-2400 Mol
BELGIUM

BAKSHT, F.B.
Research Institute of Geology
and Mineral Resources
660036 Krasnoyarsk
RUSSIA

BOGDANOVA, V.I.
Institute of Mineralogy and Petrography
630090 Novosibirsk
RUSSIA

BOOTH, P.M.
EA Partnership
BNFL Engineering Group
Risley, Warrington
Cheshire WA3 6AS
U.K.

BURAKOV, B.E.
Mineralogical Group
V.G. Khlopin Radium Institute
194021 St. Petersburg
RUSSIA

CHOI, J.-S.
Science Fellow
Center for International Security
and Arms Control
Stanford University
Stanford CA94305-6165
U.S.A.

DUBLYANSKY, Y.V.
Associate, T.R.A.C.
Institute of Mineralogy and Petrography
630090 Novosibirsk
RUSSIA

ENGELMANN, H.J.
General Division Manager
Technology Development
D.B.E.
D-31201 Peine
GERMANY

EVSTATIEV, D.E.
Geological Institute of
Bulgarian Academy of Sciences
1113 Sofia
BULGARIA

GERCHIKOV, M.
Consultancy Services
EA Partnership
BNFL Engineering Group
Risley, Warrington
Cheshire WA3 6AS
U.K.

GUREVICH, Y.L.
Head
Laboratory of Ecological Biotechnology
Institute of Biophysics
660036 Krasnoyarsk
RUSSIA

HOEK, J.
Waste Environmental Group
AEA Technology
Didcot, OXON OX13 6JJ
U.K.

KIRKO, V.I.
Director
Research Institute of Physics (NIFTI)
and Engineering
660036 Krasnoyarsk
RUSSIA

KOLONIN, G.R.
Head of Laboratory of Experimental
Modelling of Ore Systems
Institute of Geology, Geophysics
and Mineral Resources
630090 Novosibirsk
RUSSIA

KOMLEV, V.N.
Director, Mining Institute
Kola Scientific Centre
184200 Apatity
Murmansk Region
RUSSIA

KRININ, V.A.
Research Institute of Physics (NIFTI)
and Engineering
660036 Krasnoyarsk
RUSSIA

MANSUROV, V.A.
Russian Engineering Academy
660073 Krasnoyarsk
RUSSIA

MARKOV, N.G.
Cybernetics Centre
Tomsk Polytechnical University
Tomsk
RUSSIA

MARTINS, M.F.R.
Ministry of Defence
Academia da Força Aérea
Granja-do-Marquês
2710 Sintra
PORTUGAL

McEWEN , T.J.
Principal Consultant
QuantiSci Ltd.
Melton Mowbray, Leics. LE13 1AF
U.K.

MEHTA, K.K. (AECL)
Dosimetry Section
Division of Human Health
IAEA
A-1400 Vienna
AUSTRIA

MERZ , E.
Research Center Jülich
KFA-ISR-3
D-52425 Jülich
GERMANY

MIRONENKO , V.A.
Mining Institute of St. Petersburg
199026 St. Petersburg
RUSSIA

POZDNIAKOV, V.A.
Head, Laboratory of Seismic
and Acoustic Methods
Research Institute of Physics
and Engineering (NIFTI)
660036 Krasnoyarsk
RUSSIA

RYBALCHENKO, A.I.
Head of Research Laboratory fo
Underground Disposal of Liquid Industrial
Waste
Research Institute of Technology
115409 Moscow
RUSSIA

SADOVSKY, M. G.
Senior Researcher
Laboratory of Ecological Biotechnology
Institute of Biophysics
660036 Krasnoyarsk
RUSSIA

SIGAEV, B.R.
Head Engineer
Mining Chemical Engineering Plant (K-26)
Zheleznogorsk
RUSSIA

SINCLAIR, J.
Waste Environmental Group
AEA Technology
Didcot
OXON OX13 6JJ
U.K.

STENHOUSE, M.J.
Principal Consultant
QuantiSci Ltd.
Melton Mowbray, Leics. LE13 1AF
U.K.

SULLIVAN, L.H.
Group Leader
Engineering&Safety Analysis Group (N-6)
Nuclear Engineering Technology
Los Alamos National Laboratory
Los Alamos, NM 87545
U.S.A.

TRAUWAERT, E.
Policy Research Corporation N.V.
Jan Moorkensstraat 68
B-2600 Antwerp
BELGIUM

TSIRULNICHENKO, N.M.
Research Institute of Physics
and Engineering (NIFTI)
P.O. Box 8678, Akademgorodok
660036 Krasnoyarsk
RUSSIA

TSYRCHENKO, N.M.
Geophysical Institute
Krasnodar
RUSSIA

ACRONYMS

AECL:	Atomic Energy of Canada Ltd.
AVM:	Atelier de Vitrification de Marcoule
BNFL:	British Nuclear Fuels Ltd.
CCDF:	complementary cumulative distribution function
DCRT:	Double-containment receiving tanks
DOE:	Department of Energy
DST:	double-shell tank
EC:	European Community
EDE:	effective dose equivalent
EIS:	Environmental Impact Statement
EPA:	Environmental Protection Agency
ESD:	event-sequence diagram
HEPA:	high-efficiency particulate air (filter)
HLW:	high-level radioactive waste
HTF:	Hanford Tank Farm
ICRP:	International Commission on Radiation ProteCtion
ILW:	Intermediate-level radioactive waste (also MLW)
LANL:	Los Alamos National Laboratory
LFL:	lower flammability limit
LLW:	low-level radioactive waste
MLD:	Master logic diagram
MLW:	medium-level radioactive waste (also ILW)
NAGRA:	Nationale Genossenschaft für die Lagerung radioactiver.Abfälle (Swiss Nuclear Waste Management Company)
NEA:	Nuclear Energy Agency
NIRAS-ONDRAF:	National Institute for Radioactive Waste and Fissile Materials (Belgium)
OECD:	Organisation for Economic Cooperation and Devlopment
PAG:	Protective Action Guideline
PSA:	probabilistic safety analysis
SCK•CEN:	Belgian Nuclear Research Centre
SKB:	Svensk Kärnbränslehantering AB (Swedish Nuclear Waste Management Company)
SST:	single-shell tank
TRU:	Transuranic
TVO	Teollisuuden Voima Oy (Finnish Nuclear Waste Management Company, now POSIVA)
U.K.	United Kingdom
U.S.A.	United States of America
U.S.S.R.	former Soviet Union
WHC:	Westinghouse Hanford Company

THE ROLE OF PERFORMANCE ASSESSMENT IN RADIOACTIVE WASTE DISPOSAL

M.J. STENHOUSE
QuantiSci
47 Burton Street
Melton Mowbray, LE13 1AF
United Kingdom

Abstract.
Performance assessment has many applications in the field of radioactive waste management, none more important than demonstrating the suitability of a particular repository system for waste disposal. The role of performance assessment in radioactive waste disposal is discussed with reference to assessments performed in civilian waste management programmes. The process is, however, relevant, and may be applied directly to the disposal of defence-related wastes. When used in an open and transparent manner, peformance assessment is a powerful methodology not only for convincing the authorities of the safety of a disposal concept, but also for gaining the wider acceptance of the general public for repository siting.

1. Introduction

Performance assessment (PA) is a formal and systematic process by which the safety performance of a system is quantified. In the context of radioactive waste management, ultimate performance is quantified in terms of the radiological impacts from the repository calculated both for its current state, *i.e.* at closure and, more importantly, for its predicted evolution. The long timescales over which repository performance is assessed, typically 10^4–10^6 a, distinguishes a deep disposal system for radioactive waste from other, more complex systems such as chemical and power plants [1].

PA has many different roles as summarised in *Figure 1*, *e.g.*

- to demonstrate the feasibility of a disposal concept;
- to assess near- and far-field contributions to the multibarrier system (see *Figure 2*);

M. J. Stenhouse and V. I. Kirko (eds.),
Defence Nuclear Waste Disposal in Russia: International Perspective, 1–16.

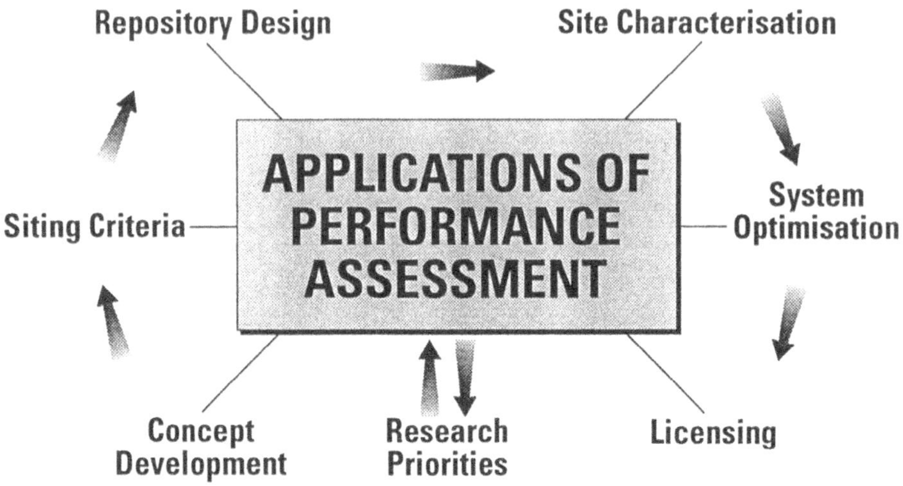

Figure 1. Examples of performance assessment applications (from [4]).

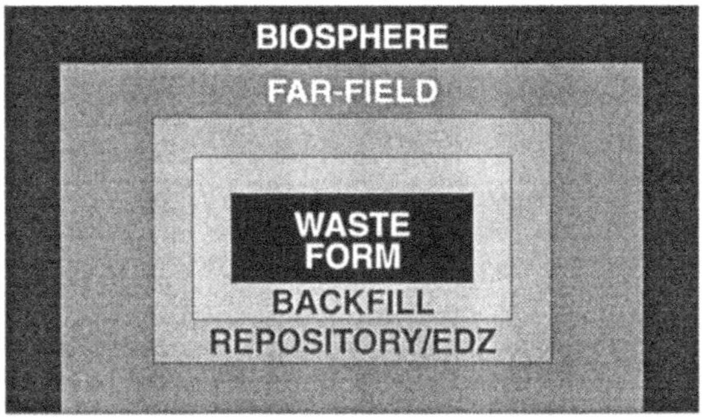

Figure 2. Multibarrier disposal system for the disposal of radioactive waste

The innermost barrier is the waste itself which has undergone some form of treatment and/or conditioning to create a 'stable' waste form, resistant to degradation, and normally contained within a metal canister of metal. The waste packages are surrounded by a backfill barrier, designed to slow the rate of ingress of groundwater and also the rate of migration of radionuclides away from the waste packages. The host rock surrounding the backfill (excavation damaged zone or EDZ) is the final component of the near-field barriers. Beyond the near-field are the far-field (rocks or sediments) and parts of the biosphere (soil, water, biota), which act as additional barriers to the release of those contaminants which eventually escape from the near-field. A full discussion of the various barriers of a radioactive waste disposal system can be found in [2] as well as in this volume [3].

- to help in the optimisation of a disposal system by highlighting key design parameters;
- to help direct site investigation programmes;
- to direct areas of research, *e.g.* by identifying strengths and weaknesses of individual components of an engineered barriered system (EBS);
- to demonstrate safety (integrated PA of disposal system);
- to provide a framework for presenting the assessment procedure and results in an open and transparent manner, for review by the public, thereby generating confidence in, and gaining acceptance for, a particular repository system.

In view of recent efforts involved to address public concerns over radioactive waste disposal, in particular the delays/difficulties experienced in obtaining authorisation to proceed with underground research laboratories (URLs), the last item above is assuming greater significance.

PA calculations are carried out at all stages of the design and development of a repository system and can focus on the whole, or only part, of the disposal system. At the early stages of repository design, PA may be used to aid decision makers in the selection of a suitable site or repository concept. In the case of site selection, the far field is being assessed. PA may also be applied to the near field to aid selection of the most suitable combination of engineered barriers, or used for a specific site to determine the critical parameter(s) (sensitivity studies). Ultimately, however, PA is carried out as an integrated assessment to demonstrate to the regulators and to the general public confidence in the final choice of disposal system.

2. Elements of PA Process

The key elements of an integrated PA of a radioactive waste disposal system are shown in *Figure 3.* The lower part of this diagram shows the various components of the disposal system to be studied as well as some of the key scientific disciplines involved, with information and/or data from all these sources feeding into the analysis depicted in the upper part of the diagram, *viz.*

- *Scenario analysis*, a systematic procedure which addresses initially the base case or predicted normal evolution of the repository system, by considering a comprehensive series of features, events and processes (FEPs) which describe different aspects of the disposal system [4]. Scenario analysis is normally extended to include one or more less likely but significant perturbations to the repository system which might affect radiological exposures [5].

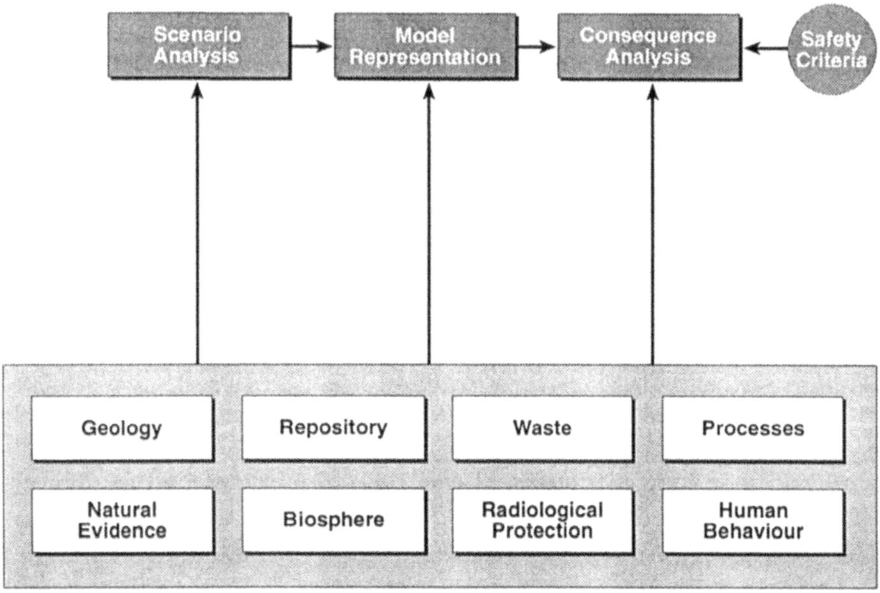

Figure 3. Key elements of integrated safety assessment of disposal system (from [4]).

- *Model representation*, whereby relatively complex features of each scenario are represented by one, or more often, a series of models which represent specific sub-systems (such as those shown in *Figure 4*) and which describe mathematically the processes therein. Within each sub-system, the models are, by necessity, simplified to include the key processes only, although additional, detailed modelling and calculations are carried out to provide supporting input and justification for the simplified models.
- *Consequence analysis* in which the consequences of individual scenarios are calculated using the models previously developed. The calculations may be deterministic, whereby specific values of parameters are used in calculations, or probabilistic, in which case distributions of values reflecting parameter value uncertainties are input to Monte Carlo simulations involving thousands of realisations in order to generate a corresponding distribution of results [6,7].

The output from modelling the behaviour of individual radionuclides within each component of the disposal system is the resultant release rates or fluxes from one component to the next. For example, if only the near-field is being studied, radionuclide fluxes are an appropriate end-point which becomes the source term for the next component (far-field). When the overall disposal system is being considered, a common end-point is the

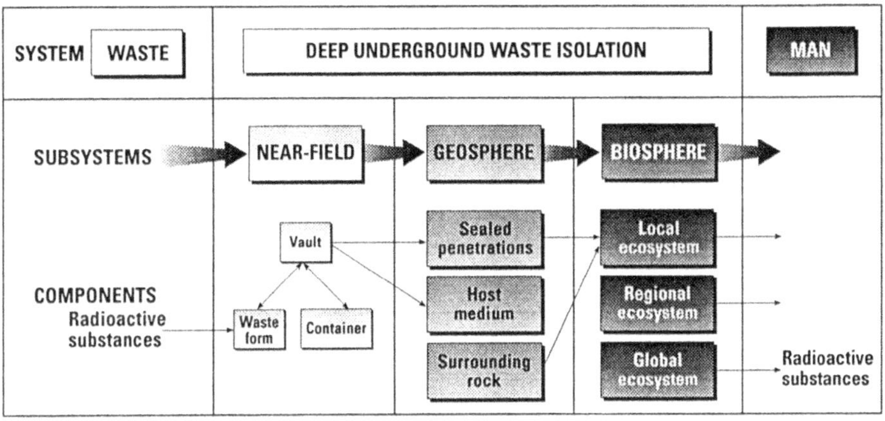

Figure 4. Typical treatment of disposal system in integrated safety assessment (from [8]).

maximum or peak dose, per radionuclide, to a member of the critical group (maximally exposed subset of the population).

In adhering to a compartmentalised approach, it is important that the interfaces between zones are described adequately. Thus, other important characteristics must be reviewed to assess their effect on radionuclide transport in subsequent zones; for example, chemical properties and the effects of an alkaline plume from a repository with cementitious engineered barriers on the host rock).

For the final part of an integrated assessment, the results obtained from the consequence analysis are compared with pre-established safety criteria, typically annual effective dose or associated risk to members of a critical group, in order to assess the overall safety of the system.

Due to computational constraints, conceptual models have to be a gross simplification of reality, in order to generate a mathematical representation which can be evaluated. Given this situation, it is imperative, for regulatory and public acceptance purposes, that the conceptual model, mathematical representation and parameter values are chosen to produce a robust solution. On the other hand, it is also important that the assessment is not 'over robust' in the sense that parameter (*e.g.* solubility) values are deliberately selected to be several orders of magnitude too conservative. In this way, the suitability of an 'acceptable' site may be overlooked.

3. Data Requirements of PA Calculations

The ultimate aim of PA of a potential repository site is to demonstrate by prediction, and with confidence, that radionuclides will be isolated from the biosphere for a time period sufficient to afford a level of protection to the environment which satisfies regulatory requirements.

The traditional approach to modelling solute (radionuclide) transport is to estimate, based on specific assumptions, groundwater velocity and to incorporate this velocity in a transport equation which accounts for advection, diffusion, linear equilibrium sorption, hydrodynamic dispersion as well as radioactive decay, *viz.*

$$\frac{\partial(\phi\ R\ C_L)}{\partial\ t} = -\nabla.(qC_L) + \nabla.(\phi\ D.\nabla C_L) - \lambda\ \phi\ R\ C_L \tag{1}$$

where $\phi =$ kinematic porosity ($m^3\ m^{-3}$),
$R =$ retardation factor due to linear equilibrium sorption (dimensionless),
$C_L =$ concentration of radionuclide in groundwater (moles m^{-3}),
$q =$ Darcy velocity ($m\ a^{-1}$),
$\lambda =$ radioactive decay constant (a^{-1}), and
$D =$ diffusion and dispersion tensor.

In an isotropic case, D has components:

$$D_{ij} = D_p\ \delta_{ij} + d_T\ v\delta_{ij} + (d_L - d_T)\frac{v_i v_j}{v} \tag{2}$$

where $D_p =$ pore water diffusion coefficient ($cm^2\ s^{-1}$),
$d_L =$ longitudinal dispersion length (m),
$d_T =$ transverse dispersion length (m), and
$v =$ transport velocity with components v_i, v_j,

Given equations 1 and 2 above, it might appear that the data requirements for PA calculations are relatively straightforward; for example, that values for the following parameters only are required:

- Darcy velocity;
- pore-water diffusion coefficient;
- retardation factor;
- source term (initial C_i);
- longitudinal and transverse path lengths, porosity.

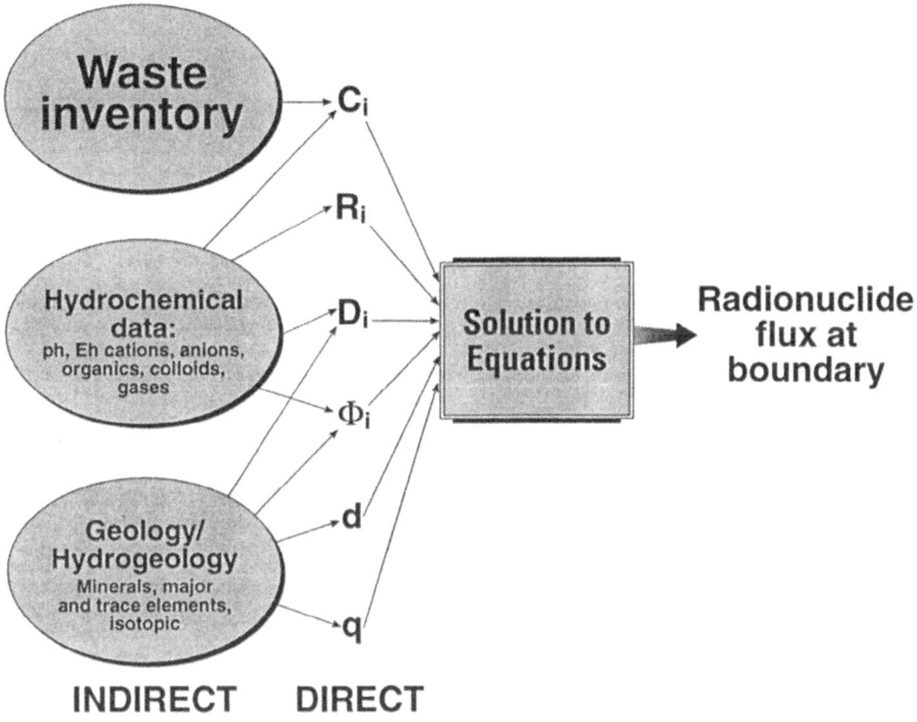

Figure 5. Data requirements for typical performance assessment of disposal system.

Values of source term are radionuclide and element dependent – controlled by radionuclide and total element inventory. Values for diffusion coefficient and retardation factor are element dependent only. In fact, for most radionuclides/elements, no single value applies to each of these parameters. Rather, PA calculations must take account of the evolution of the systems (repository and natural barriers) with time and the resultant changes to parameter values caused by the changing state of the system. Thus, other key parameters are required as *indirect input* to define the existing state of the system, in particular pH and *Eh*. *Figure 5* illustrates the general data requirements of a typical PA.

As suggested in this diagram, a proportionately larger database containing supporting information on the site is required as indirect input to PA, for example, in support of scenario development. This would include how the site evolved to its present state and a sound prediction of how it will evolve given the presence of a repository. Thus, as discussed in [9], both direct and indirect data are essential to a convincing PA.

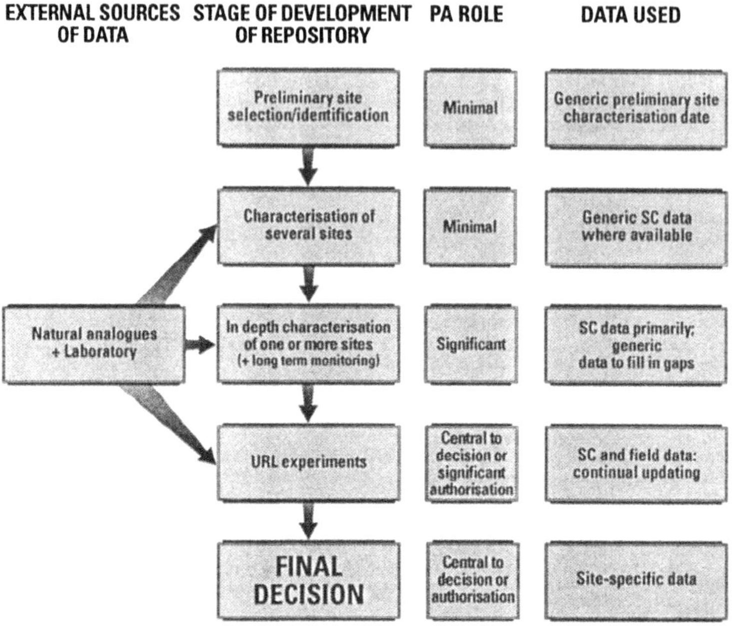

Figure 6. Relationships between repository development and PA requirements

3.1. METHODS OF DERIVING PA DATA

A variety of methods is available for deriving data for PA calculations, the method used depending to a large extent on the state of development of the site *Figure 6.* For example, PA may be performed at an early stage as a means of identifying suitable candidate sites for a repository using generic data applicable to the type of host rock, perhaps supported by preliminary site characterisation data. As site characterisation proceeds, perhaps at several sites and before underground work starts, more site-specific data (particularly geological, hydrogeological, hydrochemical and geochemical data) are generated from the site characterisation process, including borehole data, for input to PA calculations. However, reliance on generic data will still be required at this stage.

Underground construction, either for an underground research laboratory (URL) or for the repository itself, will take place only after detailed characterisation (surface and borehole) has taken place and only if the results of an in-depth assessment are favourable. The URL may be designed as a preliminary stage of the final repository or may be constructed in a body of rock which exhibits similar characteristics to those of the potential host rock. While field work proceeds, a comprehensive laboratory experimental programme should be taking place in parallel, also to provide data in support of PA. Throughout the whole process – site characterisation

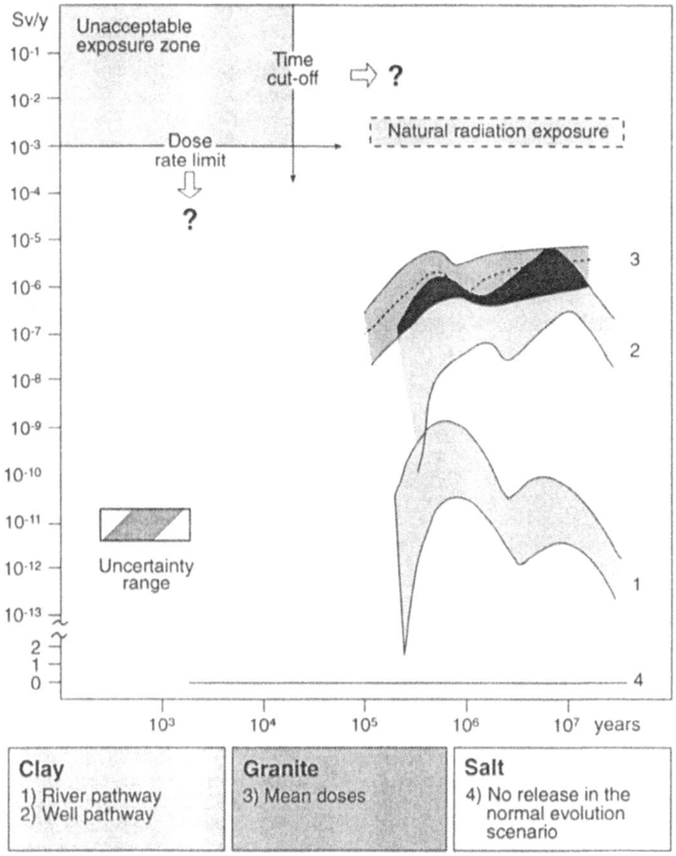

Figure 7. Results from safety assessments using generic data (from [11]).

Generic data were used in these assessments of the 3 main types of host formation. Consequently, results (annual doses) presented as a function of time appear as bands.

and URL, opportunities for studying natural analogues [10] will normally be available, providing data to complement existing data from site characterisation, laboratory experiments, and URL.

As an example of generic data, *Figure 7* shows the results of PA for the three main types of geological environment considered for radioactive waste disposal: clay, granite (hard rock), and salt. The results are from the PAGIS project (Performance Assessment of Geological Isolation Systems for Radioactive Waste; [11]) which made use of generic data. The uncertainty bands reflect the variability in geological data.

4. Examples of Integrated PA

The following fully-integrated PAs have been completed in the past five years:

- SKB-91 performed by SKB [12], Sweden, for the disposal of spent fuel;
- TVO-92 performed by TVO [13], Finland, for the disposal of spent fuel;
- Kristallin I performed by NAGRA [14], Switzerland, for the disposal of vitrified high-level waste (HLW);
- H-3 performed by PNC [15], Japan, for the disposal of HLW;
- AECL 94 by AECL [16], Canada, for the disposal of spent fuel;
- SITE 94 performed by SKI [17], Sweden, for the disposal of spent fuel.

As one example of the above studies, *Figure 8* shows the PA results obtained in Kristallin I, with the regulatory guideline included for comparison. In this particular example, the role of the geosphere, or far-field, in the safety assessment is shown by comparing *Figures 8a* and *8b*, the latter being the results where releases from the near field entered the biosphere directly, by-passing the far field. In both cases, the maximum doses were several orders of magnitude below the authorisation guidelines.

It is important to make results of assessment calculations available for peer review and to compare results with those in other countries In *Figure 9*, the results of six PAs are compared: SKB 91, TVO 92, AECL 94, H-3, Kristallin I [12-16], and Project Gewähr, the original assessment carried out by Nagra in 1985 [20]. These assessments are suitable for comparison since they all feature the disposal of either HLW or spent fuel. Interestingly, although a range of disposal concepts is considered in these assessments, the results are relatively similar (except for Project Gewähr) in terms of maximum annual individual dose.

All these assessments also highlighted similar conclusions, *viz.*

- a few key radionuclides dominate the potential radiological impact, due mainly to a high release rate into the biosphere and/or a high radiotoxicity
    - chlorine-36, selenium-79, carbon-14, caesium-135, iodine-129;
- for certain actinides (protactinium-231, plutonium-239, plutonium-242, radium-226, neptunium-27), *decay chains are important*, and should be included, despite the associated increase in calculational complexity.

Clearly, one important challenge in nuclear waste management is to aim for a more balanced public perspective on nuclear issues. Thus, conveying to

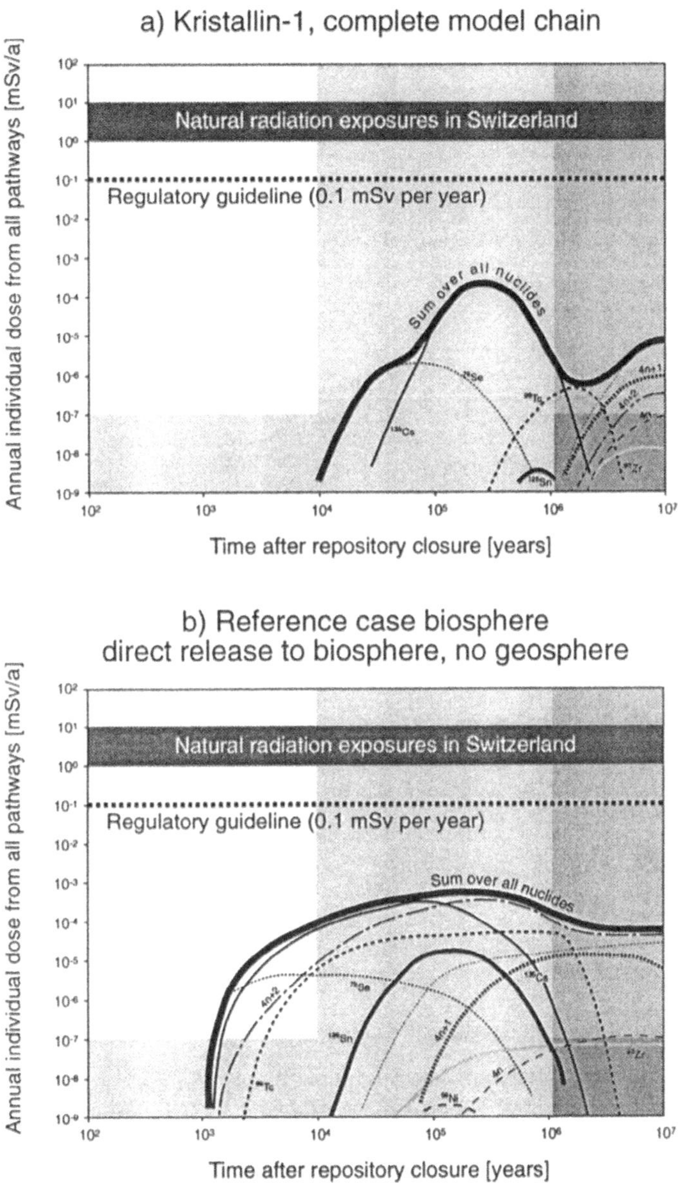

Figure 8. PA results from Kristallin I (from [18]).

Total annual dose (mSv a^{-1}) as a function of time, showing individual contributions of key radionuclides. (a) = Base case or Reference Scenario; (b) = Robust Scenario, with no credit taken for retardation and decay in the geosphere. Contributions from decay chains $4n$, $4n+1$, $4n+2$, and $4n+3$ are also included.

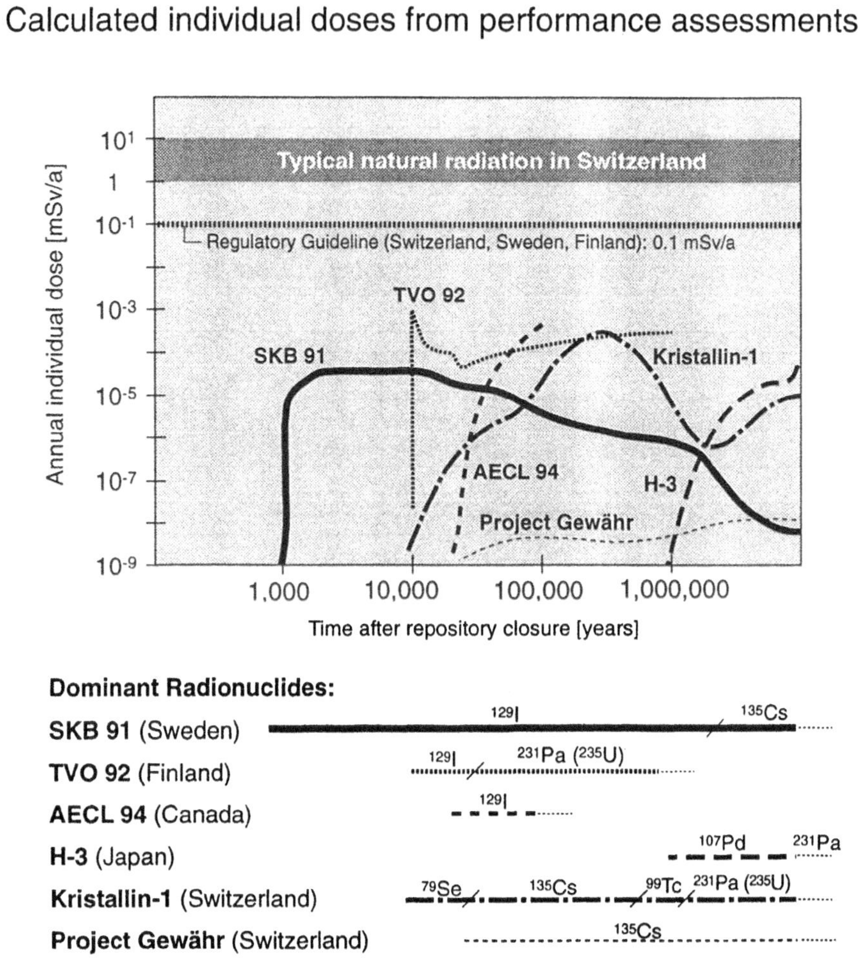

Figure 9. Comparison of results from six different PAs (from [19]).

Results presented as total annual individual dose (mSv a^{-1}) as a function of time. Dominant radionuclides, in terms of contributions to annual dose, are shown in the lower part of the diagram.

the public the impacts of radioactive waste disposal on future generations, is an important aspect of radioactive waste management, reflected in the need to transmit PA results in ways which are readily understood by both the authorities and the general population. For example, in *Figure 10*, the results of PA are compared with a range of natural sources of radioactivity including drinking water and mineral water.

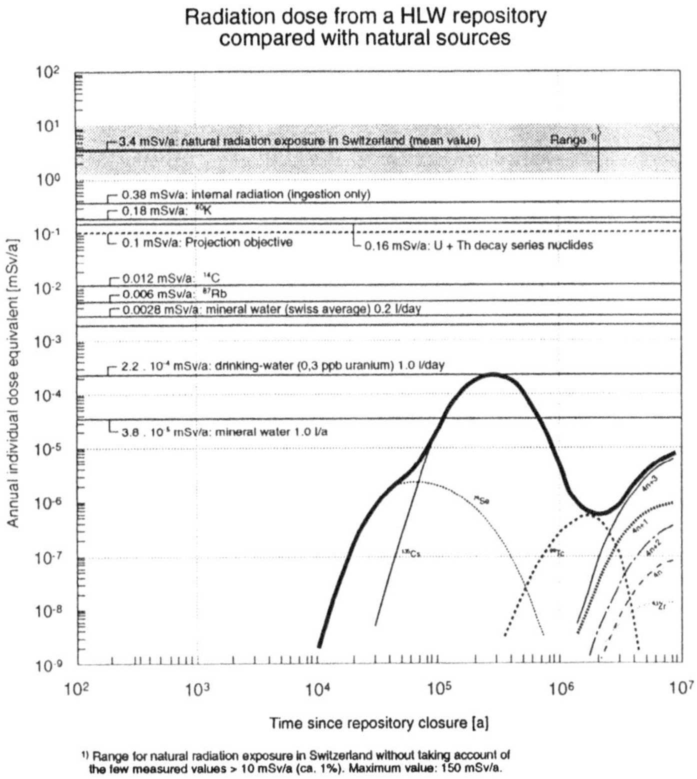

Figure 10. Comparison of PA results (individual radiation doses) for Swiss HLW disposal concept with doses from natural sources (from [21]).

5. Defence-Related Waste

So far, most of the discussion has focussed on the disposal of HLW or spent fuel from civilian nuclear programmes. However, the problems associated with the disposal of defence-related wastes will be similar in may ways, and PA methods may be applied directly to defence waste disposal. One important difference is the large quantity of weapons plutonium generated in the defence programme, the fate and treatment of which is discussed in depth in [22]. In this context, criticality issues are significant and additional precautions, in terms of ensuring the long-term presence of a material for absorbing neutrons, are likely to be necessary [23].

Besides highly-active waste, large quantities of long-lived radioactive waste (ILW) have been generated from reprocessing during defence-related nuclear programmes, once again similar to the wastes in those countries which have adopted reprocessing. Several comments are worth noting in connection with ILW:

- Since ILW contains significant quantities of long-lived radionuclides (including actinides), the disposal requirements in terms of isolation period, are similar to those for HLW or spent fuel, although, owing to the large volumes involved, the waste conditioning processes are different.
- Given the larger volumes of waste arisings, disposal concepts for ILW are likely to be demanding and it is interesting to note that few detailed integrated safety assessments have been performed on ILW (exceptions are [24,25]. Additional complicating issues such as the generation and release of large volumes of gases (H_2 from the anaerobic corrosion of metals, possibly CO_2 and CH_4 from the microbial degradation of any organic materials present), must be taken into consideration.
- The repository geometry and environment for ILW in most disposal concepts is entirely different from that for HLW disposal alone. Differences include repository layout, EBS design and a cementitious near-field with associated highly alkaline conditions which has the potential to affect the properties of the surrounding host rock adversely [26].

6. Summary

PA, including supporting data acquisition, has a wide application in civilian radioactive waste management programmes and is also relevant, and can be applied directly, to the disposal of defence-related wastes. Provided that the PA process is carried out in an open and transparent manner, this methodology can be used effectively to demonstrate the long-term safety of a repository system for both the authorities and the general population.

References

1. McCombie, C. (1985) Present status and current challenges in system performance assessments, in *System Performance Assessments for Radioactive Waste Disposal*, Proc. Nuclear Energy Agency Workshop, Paris, October 1985, pp. 29-40, NEA, Paris.
2. Savage, D. (1995) *The Scientific and Regulatory Basis for the Geological Disposal of Radioactive Waste*, John Wiley & Sons, Chichester, England.
3. Merz, E.R. (1998) Multibarrier effectiveness as the expedient measure for selcting the appropriate stabilisation and immobilisation procedure for the various wase categories, in M.J. Stenhouse and V.I. Kirko (eds.) *Defence Nuclear Waste Disposal in Russia: Implications for the Environment*, this volume, Kluwer Academic Publishers, Dordrecht.
4. NEA (1992) *Systematic Approaches to Scenario Development*, Report of Nuclear Performance Asessment Advisory Group of the Radioactive Waste Management Committee, OECD/NEA, Paris.

5. CHAPMAN, N.A., ANDERSSON, J., ROBINSON, P., SKAGIUS, K., WENE, C-O. W., WIBORGH, M., AND WINGEFORS, S. (1995) Devising scenarios for future repository evolution: a rigorous methodology. *Proceedings Materials Research Society Symposium* **Vol. 353**, pp. 495-502, Materials Research Society, Pittsburgh.
6. THOMPSON, B.G.J. (1988) A method of overcoming the limitations of conventional scenario-based risk assessments by using Monte Carlo simulation of possible future environmental changes, Her Majesty's Inspectorate of Pollution Technical Report HMIP/TR-DoE-12, Department of the Environment, London.
7. SINCLAIR, J.E. AND HICKFORD, G.E. (1998) Experience of safety assessment for the deep disposal of intermediate-level waste in the U.K., in M.J. Stenhouse and V.I. Kirko (eds.) *Defence Nuclear Waste Disposal in Russia: Implications for the Environment*, this volume, Kluwer Academic Publishers, Dordrecht.
8. NEA (1991) *Disposal of Radioactive Waste: Review of Safety Assessment Methodologies*, Report of Performance Asessment Advisory Group of the Radioactive Waste Management Committee, OECD/NEA, Paris.
9. SAVAGE, D. AND STENHOUSE, M. J. (1997) Geochemical data requirements for performance assessment, U.K. Environment Agency Technical Report P101, Bristol.
10. MILLER, W.M., ALEXANDER, R., CHAPMAN, N.A., MCKINLEY, I.G., AND SMELLIE, J. (1994) *Natural Analogue Studies in the Geological Disposal of Radioactive Wastes*, Studies in Environmental Science **57**, Elsevier, Amsterdam.
11. CEC (1988) Performance Assessment of Geological Isolation Systems for Radioactive Waste – Summary, Commission of European Community Nuclear and Science Report Series PAGIS, *EUR 11775EN*, Brussels.
12. SKB (1992) SKB-91: Final disposal of spent nuclear fuel: Importance of the bedrock for safety, Technical Report SKB TR 92-20, Swedish Nuclear Fuel and Waste Management Co. (SKB), Stockholm.
13. VIENO, T., HAUTOJÄRVI, L., AND NORDMAN, H. (1992) TVO-92 safety analysis of spent fuel disposal, YJT Report YJT-92-33E, Nuclear Waste Commission of Finnish Power Companies (YJT), Helsinki.
14. NAGRA (1994) Kristallin I: Safety assessment report, Nagra Technical Report NTB 93-22, Swiss National Cooperative for the Disposal of Radioactive Waste (Nagra), Wettingen.
15. PNC (1992) Research and development on geological disposal of high level radioactive waste, First progress report PNC TN 1410 93-059, Power Reactor and Nuclear Fuel Development Corporation (PNC), Japan.
16. GOODWIN, B.W., MCCONNELL, D.B., ANDRES, T.H., HAJAS, W.C., LENEVEU, D.M., MELNYK, T.W., SHERMAN, G.R., STEPHENS, M.E., SZEKELY, J.G., BERA, P.C., COSGROVE, C.M., DOUGAN, K.D., KEELING, S.B., KITSON, C.I., KUMMEN, B.C., OLIVER, S.E., WITZKE, K., WOJCIECHOWSKI, L., AND WIKJORD, A.G. (1994) The disposal of Canada's nuclear fuel waste: postclosure assessment of a reference system, AECL Report AECL-10717, COG-93-97, Atomic Energy of Canada Limited (AECL), Pinawa.
17. SKI (1994) SITE-94, Deep repository performance assessment project, SKI Report 96:36, Swedish Nuclear Inspectorate, Stockholm.
18. SMITH P., ZUIDEMA, P., AND MCKINLEY, I.G. (1995) Assessing the performance of the Nagra HLW disposal concept, *Nagra Bulletin* **25**, 34-46.
19. NEALL, F., SMITH, P., SUMERLING, T., AND UMEKI, H. (1995) Putting HLW performance assessment results in perspective, *Nagra Bulletin* **25**, 47-55.
20. NAGRA (1985) Project Gewähr, Feasibility Study, Nagra Project Gewähr Technical Reports NGB 85-01 to NGB 85-09, Baden.
21. BAERTSCHI, P. (1995) Radiation risk from a HLW repository compared with other risks, *Nagra Bulletin* **25**, 56-69.
22. MERZ, E.R. AND WALTER, C.E. (1996) eds., *Disposal of Weapon Plutonium: Approaches and Prospects*, Proc. NATO Advanced Research Workshop, St. Petersburg, Russia, May 1995, Kluwer Academic Publishers, Dordrecht, The Netherlands.

23. Choi, J.-S. (1998) Underground autocatalytic-criticality potential and its implications to Weapons Fissile-Material Disposition, in M.J. Stenhouse and V.I. Kirko (eds.) *Defence Nuclear Waste Disposal in Russia: Implications for the Environment*, this volume, Kluwer Academic Publishers, Dordrecht.
24. Wiborgh, M. (1995) (Ed.) Prestudy of final disposal of long-lived low and intermediate-level waste, Technical Report SKB TR 95-03, Swedish Nuclear Fuel and Waste Management Co. (SKB), Stockholm.
25. Nykyri, M. (1991) A digest of the safety assessment of TVO's VLJ repository. Technical Report TVO/VLJ-Repository Work Report 91-11, Nuclear Waste Commission of Finnish Power Companies (YJT), Helsinki.
26. Savage, D. and Rochelle, C.A. (1993) Modelling reactions between cement pore fluids and rock: implications for porosity changes, *Journal Contaminant Hydrology* **13**, 365-378.

MULTIBARRIER EFFECTIVENESS AS THE EXPEDIENT MEASURE FOR SELECTING THE APPROPRIATE STABILIZATION AND IMMOBILIZATION PROCEDURE FOR THE VARIOUS WASTE CATEGORIES

ERICH P. MERZ
Research Center Jülich KFA-ISR-3
D-52425 Jülich
GERMANY

1. Introduction

The management of radioactive wastes has become a major concern particularly with regard to the release of radioactive material to the environment and possible risks of contamination. The development of rational and acceptable options for radioactive waste disposal requires a clear understanding of radiation protection objectives and their application in planning, regulation and licensing.

Considerable progress has been made over the past three decades within many countries utilising nuclear power to develop strategies for the management of nuclear wastes. All wastes should be managed in such a way that high standards of conditioning are maintained and that potential hazards originating from their disposal are reduced to levels that are as low as reasonable and well below admissible levels. However, deficiencies are evident in some areas of nuclear weapon fabrication.

The nuclear fuel cycle is associated in the military weapon fabrication sector as well as in the civilian energy production field with two rather similar types of risk:

1. the risk due to the operation of the nuclear reactors and the appertaining fuel facilities, and
2. the risk contribution originating from the generation of radioactive wastes.

M. J. Stenhouse and V. I. Kirko (eds.),
Defence Nuclear Waste Disposal in Russia: International Perspective, 17–31.

The difference between these two categories of risk is that the first one has only a short time factor associated with it, since the lifetime of the plants is relatively short and drops to zero after plant shutdown. The second category is, more or less, a permanent kind of risk which will be inherited by future generations.

Actual health effects of waste on people and populations, particularly over long periods of time, are not necessarily related to the level of radioactivity. If intensely radioactive waste is effectively isolated, then the radiation dose it causes can be much less than that accumulating from widely-dispersed but low-activity waste, particularly if this includes long-lived radioisotopes.

By far the most important producers of nuclear wastes are nuclear power plants and their supporting fuel cycle facilities. Compared to nuclear power and reprocessing plants, only very small waste contributions come from research establishments, industry and medical applications. An additional source of radioactive waste arises from nuclear weapons production and related testing activities. The problems associated with disarmament requirements, however, are not dealt with in this paper.

2. Systems Approach in Waste Conditioning, Storage and Disposal

In evaluating an engineering project, it is advantageous to use a systems approach to interconnect the subtasks to an optimum functional entirety. An integrated system for the management of radioactive wastes refers to the complete spectrum of background policy, safety, environmental protection, and actual practices which define the classification, control, movement, conditioning, quality assurance, storage and disposal of wastes. Emphasis is placed upon demonstrating that all radioactive wastes can be safely isolated from the biosphere for the required time.

A deductive approach as depicted in *Figure 1* is recommended, according to which in a first step, pre-determined safety features of the overall system are laid down and the inferences are drawn related to each single step.

Four domains dictate the requirements concerning the properties and quality of the wastes to be disposed of:

- handling and transport;
- conditioning procedures;
- interim storage;
- final disposal in a geologic repository.

Thus, waste product and canister quality assurance measures must be oriented towards criteria derived from their overall safety assessments. A basic element is the effective combination of the so-called multibarriers

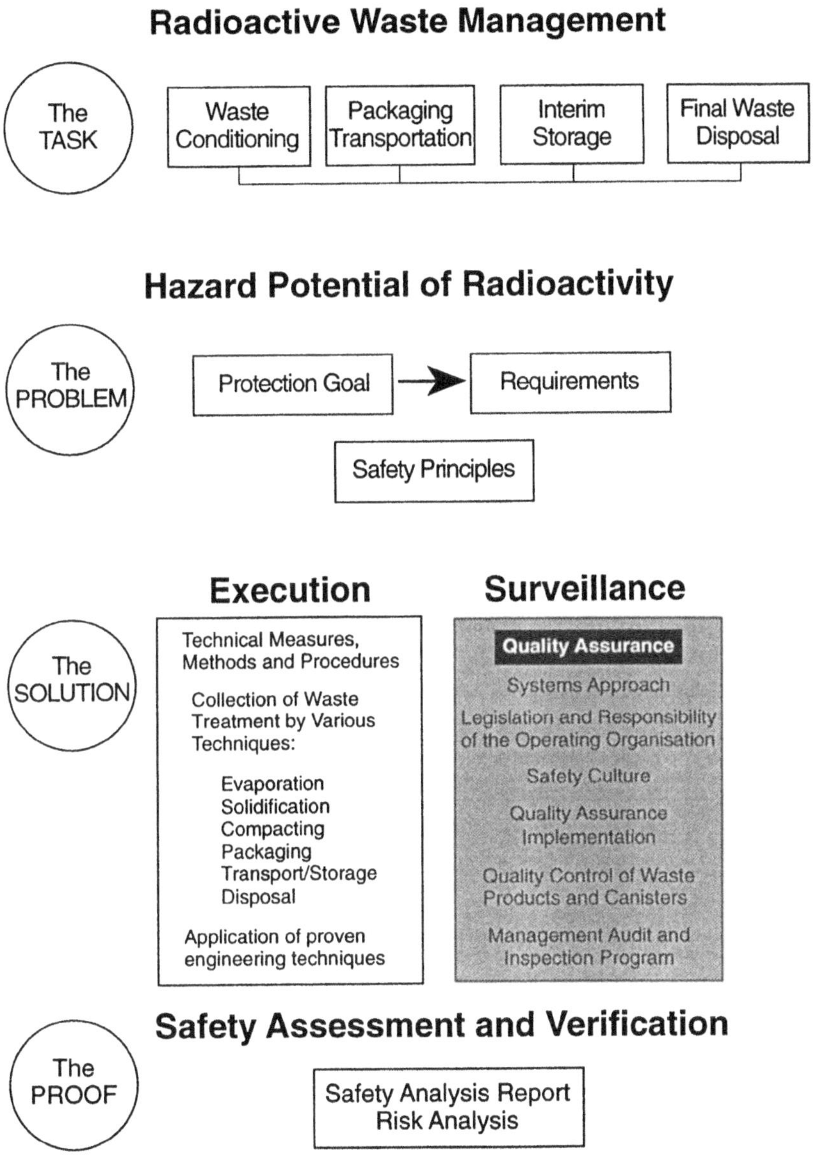

Figure 1. Outline of radioactive waste management structural elements.

providing long-term isolation of the radioactive materials. The waste form and package have to be designed in close coordination with the repository characteristics. The various engineered and natural barriers are capable of counteracting or minimising the natural processes which could lead to release of the waste. Thus, it is opportune to base safety assessments and comparisons of options on the assumption that the barriers act in concert, and that the behaviour of each can be quantified and predicted.

The final barrier, and the most important in many respects, is the host rock and geological environment in which the repository is constructed. This isolates the waste from the surface and controls water movement and geochemistry on both large and small effects. The first three barriers (waste form, container, and backfill) are within the engineer's control, and can be tailored to suit a particular concept or geological environment. In contrast to the "engineered barriers", the rock barrier is outside the engineer's control and can only be modified to a limited extent.

3. Treatment and Conditioning of Waste

After collection and sorting into appropriate categories, wastes generally have to be treated and conditioned in order to convert them into a form suitable for safe handling, storage and disposal. Various suitable techniques are available today. The selection of the favoured process depends on factors such as transport, storage and disposal requirements, the overall costs of the individual steps, and other licensing and acceptance requirements. Of foremost importance is the potential of waste volume minimisation and long-term safety behaviour. In principle, only a few techniques are of interest. Many of the otherwise proposed methods have not found widespread acceptance.

Some of the environmentally unsatisfactory practices that have been employed in the past comprise:

- direct disposal of unconditioned raw wastes in liquid or other inadmissible forms;
- discharge to waterways or sewer systems before evaluating the potential danger and harmful environmental effects;
- deep injection of liquid wastes into porous strata without proven barrier efficiency of the surrounding strata;
- disposal of sludges and slurries in landfills where leachate may contaminate groundwater;
- disposal on-site without proper precautions to avoid groundwater and surface water contamination;
- incineration without controlled emission of toxic or corrosive gases to the environment;
- gas generation in radioactive wastes and gas releases from these wastes due to corrosion and microbial activities if appreciable amounts of organics and moisture are present in the waste;
- swelling and pressure build-up and subsequent disintegration of the waste matrix barrier.

In general, the processing of solid and liquid waste streams aims at producing a fraction that has a smaller volume and higher concentration of radioactivity than the original raw waste, and a second fraction of sufficiently lower radionuclide concentration which permits release or re-use.

Radionuclides are removed from liquid and gas streams using established industrial technology. The solutions, resins, concentrates, sludges and filter materials in which the waste radionuclides are trapped can be reduced considerably by:

- evaporation of solutions,
- incineration of combustible constituents, and
- high-pressure compaction.

The most effective way to put a containment and isolation approach into practice is to contain the radioactive waste in a stable waste form, and to isolate the waste form from the biosphere by additional barriers.

Decisive criteria for the final waste product including the waste matrix are:

- chemical inertness;
- near-zero residual organic carbon content;
- absence of free water and humidity.

3.1. EVAPORATION/DRYING

Evaporation is a widely applied concentration method used in nuclear facilities to reduce volume, remove water from a solution, produce relatively pure water from impure water, and remove soluble and insoluble materials from liquid radwaste streams. Evaporation can be used on a wide variety of liquid wastes and slurries, particularly those with high dissolved solids content and requiring a high degree of decontamination. Since energy requirements for evaporation are high, other liquid waste treatment techniques are preferred if possible when large volumes of liquids have to be processed.

Wet solids require treatment and conditioning to reduce the free water content to less than 0.5 %, thereby making them acceptable for disposal. The selection of the appropriate technology for processing of semi-solids is dependent upon a number of waste characteristics.

3.2. COMPACTION

Compaction is a concentration technology in which a press is used to compress the dry solid waste into the final disposal container (see *Figure 2*). The volume reduction factor achieved during compaction is a function of

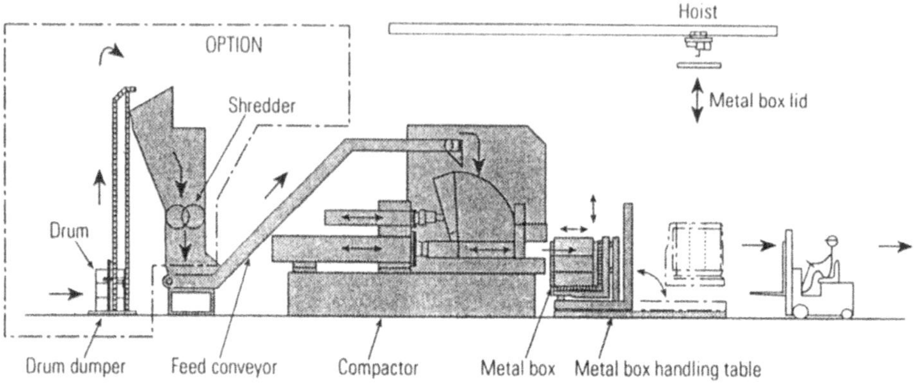

Figure 2. Compaction of dry solid radioactive waste.

the void space in the waste, the force applied by the press, the bulk density of the material, and its spring-back characteristics.

Compactors vary in size, design, and capacity and are often custom designed according to special requirements. Ordinary compactors apply pressures between 150 to 500 kN, supercompactors use pressures of up to 15,000 kN.

3.3. INCINERATION

The combustion process of incineration has long been used as a volume-reducing technique for low-level radioactive waste. However, the high costs have discouraged widespread utilisation. Incinerators designed for the treatment of all kinds of radioactive wastes today use a variety of combustion approaches either singly or in combination. The basic flowsheet is shown in *Figure 3.*

Incineration exhibits several advantages. It converts the combustible waste into radioactive ashes, residues that are chemically inert, non-flammable, and homogeneous, and gaseous effluents containing entrapped radioactive particulates. It is, therefore, essential to equip the incineration system with highly efficient, multistage off-gas control systems to reduce the radioactivity below environmental release standards before emission of the cleansed gases. Thus, in addition to the ash and other solid residues generated, the off-gas controls will produce a secondary waste stream (*e.g.* filters) that requires disposal.

Incinerators designed for the treatment of radioactive wastes use a variety of combustion approaches. The principal combustion techniques, categorised by process temperature ranges, are:

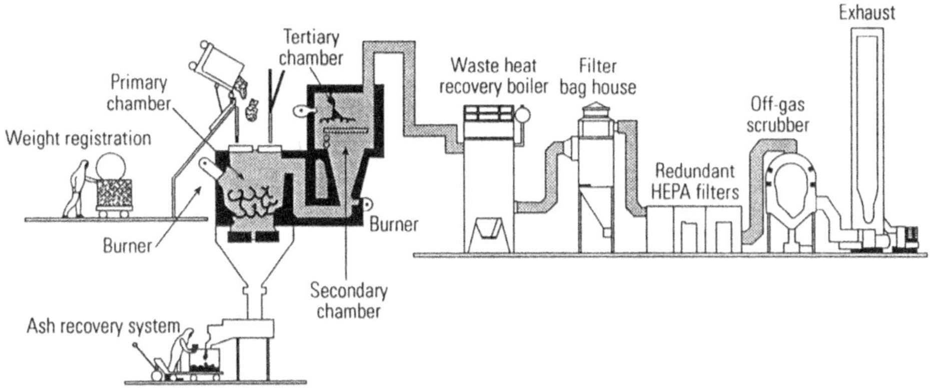

Figure 3. Incineration of combustible radioactive waste.

- *Controlled-air incineration*, which limits the air supply in the primary combustion phase and requires a secondary combustion phase to achieve complete combustion (800–1100 °C),
- *Excess-air incineration*, in which an excess of oxygen is fed into the primary combustion-phase to permit both the solid and gaseous components to burn directly (800–1100 °C).
- *Pyrolytic or thermal decomposition*, in which the organic waste material is essentially distilled in a highly reducing atmosphere (absence of oxidation) generating combustible liquids and ases, A carbon residue (char) remains after pyrolysis (500–600 °C).
- *Fluidized-bed incineration*, in which an inert bed of particles is suspended by air flow through the bed and the pre-shredded waste is burned upon introduction into the self-sustaining bed (800 °C).
- *Slagging incineration* at high-temperature (1400–1600 °C) to burn the organic material releasing sufficient heat energy to convert the non-combustibles to a molten slag residue.
- *Electromelt incineration*, in which the waste is burned and sorbed into a glass melt maintained by electromelt Joule heating (1200 °C).

3.4. SOLIDIFICATION

Solidification is used to immobilise sludges, semi-solids and dry solids. This is a technique by means of which these materials are immobilised within an inert matrix or binder. The optimum solid incorporating radioactive waste should:

- not exhibit microbial activity;

- have high mechanical strength and not fracture under load;
- be resistant to radiation damage for a long time period;
- have good thermal resistance;
- resist leaching from water infiltrating the package;
- demonstrate chemical compatibility between binder and the incorporated waste.

Glass and ceramics, cementitious materials and so-called geopolymers have evolved as the most suitable materials. Utilisation of bitumen and organic polymers favoured at many nuclear installations in the past has meanwhile become less popular due to several drawbacks (*e.g.* fire hazard, chemical reactivity, swelling, radiolytic gas generation), not least because of its susceptibility to microbial activity. There are many examples where bitumen and plastics covering materials have been destroyed in a damp environment.

3.5. VITRIFICATION

The established method for solidification of high-level liquid waste is vitrification. Borosilicate glass is the most comprehensively studied matrix at present and is very flexible in terms of waste content. Industrial scale vitrification plants are now operational in France (COGEMA) and the U.K. (BNFL), both utilising the French continuous two-stage AVM design (AVM = Atelier de Vitrification de Marcoule), and in Mol, Belgium, the German/Belgian joint PAMELA facility using the continuous one-step Joule-heated ceramic-lined furnace melting process. The same glass melting process will be used in the U.S.A. and Japan [1].

Thus, it is commonly accepted that borosilicate glass is a satisfactory waste form for high-level radioactive wastes (HLW). It should be recalled, however, that the first attempts to incorporate HLW in a glass matrix were based on using phosphate glasses which offer several obvious advantages relative to silicate-based glasses, including a lower processing temperature and a higher solubility for sulphates. The disadvantages, which are typical of sodium phosphate glass and which led to the abandonment of its further consideration as a suitable waste form, are the high corrosiveness of sodium phosphate glass melts and the higher tendency of this glass to devitrification in a certain temperature range compared to borosilicate glass.

The question arises whether the condemnation of phosphate glasses at that time was wise and justified. Further development of phosphate glasses continued in the former Soviet Union and some other countries. It can be shown that an improved type of phosphate glass composition counteracts nearly all of the weak points, thereby outstripping borosilicate glass.

Typically, lead-iron phosphate glasses appear to be a promising new storage medium for HLW [2]. They can be prepared by melting in either zirconia, alumina, or platinum crucibles. When properly prepared, all of the waste oxides and silicates, as well as many other chemical compounds, can be dissolved in the lead-iron glass melt at 1050 °C.

The results of corrosion tests on this category of phosphate glasses indicate that they are extremely resistant to groundwater corrosion when the solution pH is between about 5 and 9 and the solution temperature is below about 150 °C. It appears worthwhile to further explore the potential prospects of advanced phosphate glass compositions in combination with tailored repository concepts. A challenging option may arise from the disposition of phosphate waste glass blocks in deep boreholes (see Section 5).

3.6. CERAMIC HIGH-LEVEL WASTE FORMS

In addition to borosilicate and phosphate glasses, there are a large number of crystalline and ceramic waste forms that have been proposed, and many are being tested at present. Mineral assemblages can be chosen which are chemically very stable under all conditions likely to be encountered in a disposal repository. In particular, the flexibility of various aluminate and titanate matrix types with respect to variable waste compositions and loadings are being investigated. Intrinsically, such materials as SYNROC, zirconolite, alumina [3, 4] are more stable to leaching than glasses since the waste elements which can be successfully incorporated are strongly bound in the lattice structure of the host minerals on an atomic scale.

However, it should be noted that the importance of simple processing techniques has been greatly undervalued until now. A technical scale plant for manufacturing "hot" ceramic waste products has not yet been operable so far. Difficulties are obvious and may exclude practical application. Arguments in favour of further development needs for advanced waste forms in order to improve public acceptance of waste disposal or just to provide another waste form in addition to glass, do not justify additional efforts under the prevailing circumstances [5].

3.7. CEMENTATION

Cementation is a well-established conditioning method. Depending on the composition of the raw waste, Portland cement, high alumina cement or blast furnace slag cement may be employed. Additives are used to improve product quality.

The basic advantages of cementation are:

- simplicity of the process,

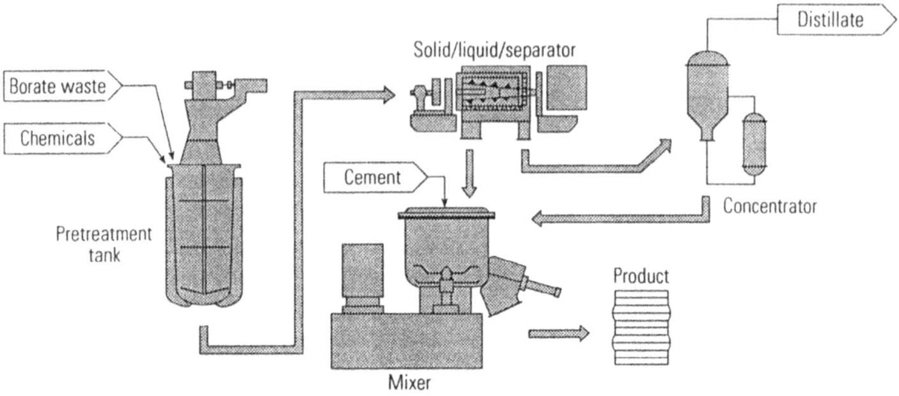

Figure 4. Solidification of liquid radioactive waste with cement.

- non-inflammability of the matrix,
- low processing temperatures, and
- high process and matrix flexibility.

The basic flowsheet of cementation is shown in *Figure 4*.

A negative effect is generally the relatively large waste form volume, especially after cementation of liquid waste. Another drawback can be the chemical reactivity of the cementitious material with metals, for example the reaction of aluminium metal, which gives rise to hydrogen production.

3.8. GEOPOLYMERS

The family of mineral polymer binders, already known since the 1940s [6] represents a second generation of solidification matrices for intermediate-level waste [7]. Geopolymers are aluminosilicate network binders that extended in three dimensions exhibiting amorphous to semi-crystalline structures. Alkali-activated aluminosilicate binders are a set of materials that are formed by the reaction of alkali solution with fine grain solid that is rich in silica and alumina. Poly-condensation reactions take place at low temperature similar to organic polymers [8]. The products are superior to cementitious ones with regard to their chemical properties, and water contents may be less than 1 % when heated to 100 °C [9].

The elementary compounds which form the basic building stones of the cross-linked inorganic polymers are:

- *aluminium silicate*:

$$2\ Al_2Si_2O_4(OH)_4 \xrightarrow{Heat} Al_2Si_4O_{11} + Al_2O_3 + 4\ H_2O$$

- *sodium aluminate*:

$$NaAl(OH)_4 \xrightarrow{Heat} NaAlO_2 + H_2O$$

Condensation reactions take place leading to high molecular weight oxo-compounds:

$$Si(OH)_3\text{-}O\text{-}Si(OH)_3 \quad \text{resp.} \quad Al(OH)_3\text{-}O\text{-}Al(OH)_3^{2-}$$

A typical example of a natural compound of this type is *montmorillonite*:

$$Al_2Si_4O_{10}(OH)_2.nH_2O$$

in which silicon atoms are partly replaced by aluminium atoms.

4. Stabilisation and Immobilisation of Plutonium Scrap and Residues

Plutonium scrap and residues are of major national and international concern because:

- significant environmental, safety, and health problems have been identified with their storage,
- all plutonium seized from the black market in Europe is apparently in this category,
- storage costs are high, and
- safeguards are difficult.

These difficulties are a direct result of the characteristics of these materials. Plutonium scrap and residue normally consist of only a few weight percent of plutonium, but the total volume and mass exceed that of clean plutonium. The materials have highly variable chemical and nuclear characteristics. Some of the chemical forms are hazardous and corrode their containers.

In principle, scrap plutonium could be recovered by applying costly chemical reclamation and purification procedures. However, the necessary technical expenditure does not justify the benefit. In some ways, the plutonium in scrap and residues is now even more of a security, environmental, and economical liability than metal plutonium pits from dismantled warheads. This is because such plutonium is far less well accounted for, often stored in deteriorating containers, and sometimes mixed with flammable substances.

Stabilising these forms of plutonium poses one of the more urgent safety and health problems in the nuclear weapons complex. These left-over materials should be given priority treatment. Interim safe storage criteria call for a standardised package that meets both safe off-site transportation requirements, as well as remote handling requirements for future storage facilities.

Among the different scrap conditioning procedures which yield a suitable disposal form are:

- various forms of glass; especially borosilicate or phosphate glass;

- sintered ceramics;
- a number of cement-related materials, *e.g.* cementitious matrices, cement glasses, and geopolymers.

Most of these waste forms have not been thoroughly evaluated for scrap plutonium stabilisation and immobilisation. Manufacturing technologies for many of them are not fully developed.

It is worthwhile mentioning the endeavours to classify plutonium waste. Stabilised plutonium is destined for disposition or geological disposal depending on its concentration. The material requires safeguarding within the concentration range 1000 g Pu kg^{-1} to 0.5 g Pu kg^{-1}.

Material with plutonium content between 0.5 g Pu kg^{-1} and 1.6 mg Pu kg^{-1} is classified as transuranic (TRU) waste and should only be admitted for deep underground geological disposal. Material with plutonium content below 1.6 mg Pu kg^{-1} (3.7 kBq g^{-1}) is called low-level waste (LLW) and may be admitted for near-surface disposal. TRU and low-level waste are exempt from safeguards.

5. Repository Design and Engineering

The objective of geological disposal is to achieve a sufficient degree of isolation of the waste so that any return of radionuclides to the biosphere will be at least sufficiently low so that any resulting radiation exposure to people would not represent an unacceptable health hazard.

The site and the engineered containment barriers must play a dual role:

1. protect the waste by restricting both the circulation of water in contact with the waste and the possibility of human intrusion;
2. limit any delay in the time necessary for sufficient decay of the radionuclides involved in transfer to the biosphere of any radioactive substances released by the waste.

There is international consensus that the disposal of HLW or long-lived radioactive waste by injection of liquids into deep geological (sandstone) formations is not an acceptable approach since the dispersal of radionuclides via uncontrolled migration cannot be prevented. The strict pursuit of the multibarrier principle requires prior solidification and incorporation of the liquids into a stable product form.

HLW repositories are of two broadly different designs, both involving the placement of the packaged waste in the repository in alternative ways:

- in shallow drill-holes in the floor of large repositories mined at depths of several hundred metres underground, or
- in very deep single boreholes.

The two different approaches are compared in *Figure 5*.

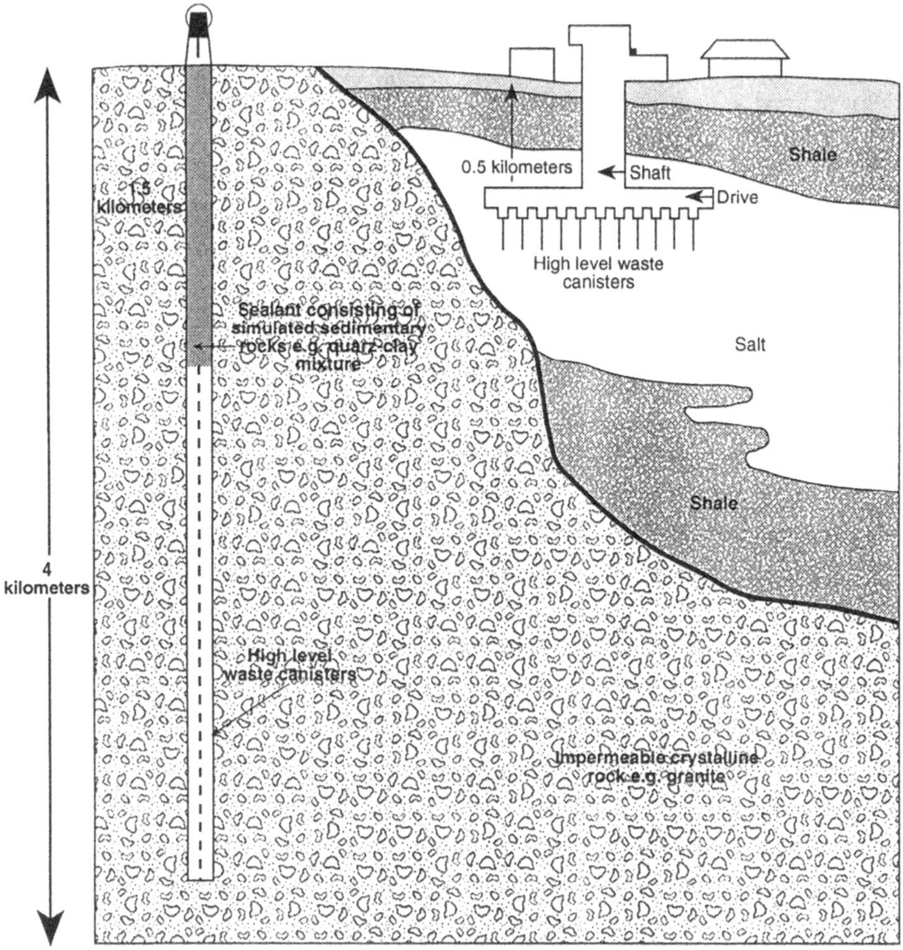

Figure 5. Two different approaches to geological disposal of high-level waste.

A major difference between the two is the long time required to fill and, consequently, to operate the mined repository in contrast to the short time to fill, seal and close a deep borehole.

Today, underground disposal in mined repositories is receiving the greatest attention. The excavations are planned to be at depths of 500 to 1000 m in crystalline rock and salt, but are limited to 200 to 300 m in clay formations because of their mechanical properties. The required technology is available and the civil and mining engineering industries in many countries have the ability to construct and operate such underground facilities.

Deep borehole disposition has been considered in recent decades for disposal of both hazardous and radioactive wastes [10], but its realisation was rejected for several reasons, mostly because of unsatisfactory maturity and the fear of licensing constraints and public opposition. It seems worthwhile to reconsider this disposal option as a better substitute for the hitherto poor method of liquid injection.

Desirable site characteristics could probably be revealed at the different nuclear sites, for example *Sneshinsk* (formerly Chelyabinsk 46), *Seversk* (formerly Tomsk 7), and *Zelesnogorsk* (formerly Krasnoyarsk 26). Desirable site characteristics are:

- a thick layer of ancient rock with good mechanical properties for drilling and sealing,
- a demonstrated history of geologic stability,
- no evidence of exploitable resources,
- little free void space, little water content, any water present should be very old, immobile and unusable ,
- large vertical separation distance from usable water.

It is believed that a suitable host rock can probably be found in the three areas identified above.

The deep borehole concept should be suitable for the disposal of phosphate glasses, a vitrification technique well advanced in Russia. The concept appears technically feasible, and has certain advantages with regard to the situation in Russia. It may be the concept capable of being realised most rapidly, and the most economic alternative under the prevailing circumstances.

6. Concluding Remarks

The criteria to be established for any kind of radioactive waste disposal must always be put in perspective:

1. What are the waste characteristics?
2. What time period for safe isolation is of interest?
3. Which geological disposal alternatives exist?

Different approaches may be used in the short- and long-term perspective. In either case, a general procedure is recommended which involves concentrating, containing and isolating the source of radiotoxicity as far as practicable.

Four domains dictate the requirements concerning properties and quality of the wastes to be disposed of:

- handling and transport;
- conditioning;
- interim storage;
- final disposal.

References

1. LUTZE, W. (1988) Silicate glasses, in W. Lutze and R.C. Ewing (eds.) *Radioactive Waste Forms for the Future*, Elsevier Science Publishers B.V., North-Holland, Amsterdam, pp. 193-231.
2. SALES, B.C. AND BOATNER, L.A. (1988) Lead-iron phosphate glass, in W. Lutze and R.C. Ewing (eds.), *Radioactive Waste Forms for the Future*, Elsevier Publishers B.V., North-Holland, Amsterdam, pp. 131-192.
3. HARKER, A.B. (1988) Tailored ceramics, in W. Lutze and R.C. Ewing (eds.), *Radioactive Waste Forms for the Future*, Elsevier Publishers B.V., North-Holland, Amsterdam, pp. 335-392.
4. SCHMIDT, S.O. (1990) Herstellung und Charakterisierung keramischer Produkte als Abfallverwahrform unter Einsatz der Sol-Gel Technik. *KFA Jülich Research Center Report* JÜL-2394, Jülich, Germany.
5. MERZ, E.R. (1991) Need for an advanced HLLW fixation matrix?, in R.G. Post (ed.) *Working Towards a Cleaner Environment*, Proc. Waste Management '91, Tucson, Arizona, Vol. I, pp. 789-794.
6. BERCK, D.W. (1975) *Synthetic Zeolites: Properties and Application in Industrial Minerals and Rocks*, 4th Ed., Portcity Press, Baltimore, U.S.A.
7. DAVIDOVITS, J. (1988) Soft mineralogy and geopolymers, *Geopolymer '88*, Vol. 1.
8. MALONE, P.G., RANDAL, C.A., AND KIRKPATRICK, T. (1985) Potential for use of alkali-activated aluminosilicate binders in military applications. *U.S. Corps of Engineering Report* WES/MP/GL-85-15.
9. MERZ, E.R. AND KHALIL, M.Y. (1992) Geopolymer and cement glass: promising fixation and solidification matrices for nuclear and hazardous wastes, in *SPECTRUM '92*, Proc. Am. Nuclear Society Inc. Meeting, 1992, Vol. 2, pp. 914-919, Am. Nuclear Society Inc., La Grange Park, Illinois, U.S.A.
10. O'BRIEN, M.T., COHEN, L.H., NARASIMHAM, T.N., SIMKIN, T.L., AND WOLLENBERG, H.A. (1979) The very deep borehole concept. Evaluation of an alternative for nuclear waste disposal. *Lawrence Berkeley Laboratory Report* LBL-7089, LBL, San Francisco.

DISPOSAL OF CONDITIONED HLW MATRICES AND INTERACTION WITH A DEEP CLAY LAYER AS HOST FORMATION

L.H. BAETSLE
Belgian Nuclear Research Center, SCK•CEN
Boeretang 200, B-2400 Mol, BELGIUM

Abstract.
About two decades ago the Belgian Nuclear Research Center, SCK•CEN, started the construction of an underground laboratory at a depth of 250 m in the Boom Clay formation extending from 180 to 280 m depth. The shaft was dug under freezing conditions because large aquifers surmount the clay layer. The underground galleries were constructed with conventional semi-automatic tunneling machines.

An extensive R&D program has been undertaken within the framework of the European Community's Research and Development Programme on the management and storage of radioactive waste. The main scientific fields of investigation are: local and regional hydrogeologic characterisation, migration of radionuclides in clay, interaction of waste forms with clay, gas migration in compact clay, corrosion of canister materials, thermal load and heat dissipation in clay, study and performance analysis of backfill materials, and safety analysis of an underground repository.

The underground laboratory at Mol has been used by many partners within the European Union to carry out *in situ* experiments in fully representative conditions.

1. Introduction

Belgium has a nuclear-electric generating capacity of 5400 MWe delivered by 7 PWRs which produce 60 % of the national electricity output. Two large fuel fabrication plants for UO_2- and MOX-fuel are situated at Mol, and produce for the whole European market.

The National Institute for Radioactive Waste and Fissile Materials (NIRAS-ONDRAF) has overall responsibility for radioactive waste treatment, carried out by its subsidiary Belgoprocess, and for waste disposal which is carried out in conjunction with the Belgian Nuclear Research Center (SCK•CEN) for R&D.

M. J. Stenhouse and V. I. Kirko (eds.),
Defence Nuclear Waste Disposal in Russia: International Perspective, 33–43.

Belgium has existing contracts with the reprocessing plants of COGEMA to reprocess a substantial part of the spent fuel arisings. Consequently, the basis of all former studies was the interaction, with the clay host formation, of high-level waste (HLW) glass, cementitious materials, and bituminous waste matrices as well as canister materials. Recently, attention has been given to the possibility of disposing of spent fuel directly into the clay repository. The Belgian parliament will review its position in 1998 with regard to reprocessing and fuel cycle policies.

About two decades ago, SCK•CEN started the construction of an underground laboratory at a depth of 250 m in the Boom Clay formation extending from 180 to 280 m depth. The shaft was dug under freezing conditions because large aquifers surmount the clay layer. The underground galleries were constructed with conventional semi-automatic tunneling machines. Much geotechnical work has been performed in order to determine the parameters for building industrial-sized galleries in the framework of a future repository for HLW and transuranic (TRU) waste.

An extensive R&D programme has been undertaken within the framework of the European Community's (EC) Research and Development Programme on the management and storage of radioactive waste. The main scientific fields of investigation are:

- Local and regional hydrogeological characterisation.
- Migration of radionuclides in clay.
- Interaction of waste forms with clay.
- Gas migration in compact clay.
- Corrosion of canister materials.
- Thermal load and heat dissipation in clay.
- Study and performance analysis of backfill materials.
- Safety analysis of an underground repository.

The underground laboratory at Mol has been used by many partners within the EU to carry out *in situ* experiments in fully representative conditions.

2. Local and Regional Hydrogeology

A very large and widespread net of piezometers has been installed in a region of roughly 2,000 km^2 surrounding the potential repository site of Mol in Belgium. The permeability of the clay layer is of the order of 3.10^{-10} m s^{-1} on a regional scale and about 2.10^{-12} m s^{-1} at the site of the underground laboratory. The interstitial clay water is alkaline (pH = 9) and this pH value is determined by the carbonate-bicarbonate buffer. The sulphides which are present in the clay body, determine the redox value at a mean

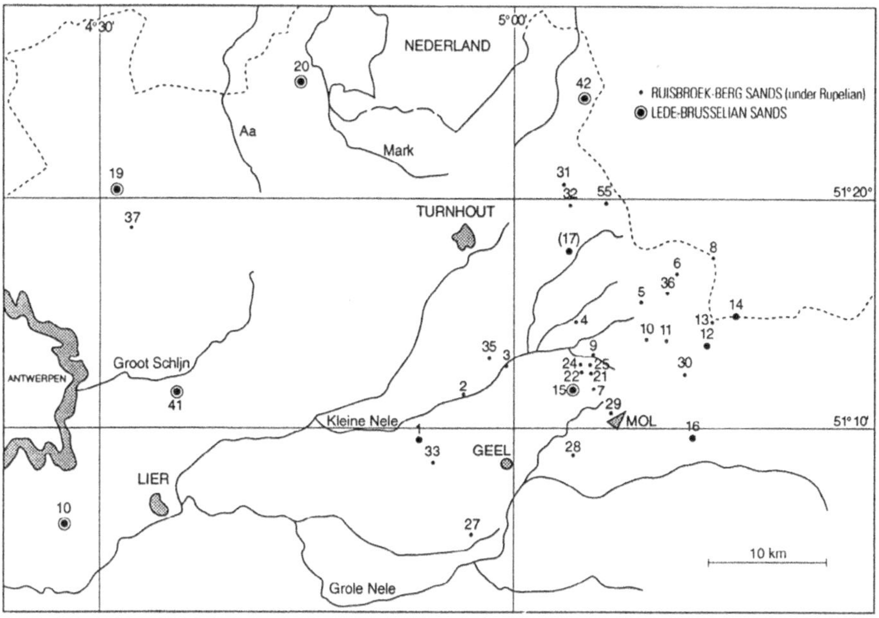

Figure 1. Location of the piezometric wells of SCK•CEN network

value of -315 mV. Water above the clay layer is neutral and constitutes a single "drinking water aquifer".

The boundary of the hydrogeologic region in the north east of Belgium, is the Scheldt river on the west side, and the Campine plateau on the east side. The rainfall of about 750 mm a^{-1} is partially evaporated, and partially drained by the surrounding rivers (Grote Nete, Kleine Nete) and discharged into the Scheldt river at about 40 km from the experimental repository site. Forty-two wells were drilled at depths varying from 200 to 600 m, and instrumented with piezometer wells for each of the aquifers above and below the clay layer. In this way, it was possible to monitor the water levels and piezometric heads of each aquifer over a relatively long time period. The drillings started in 1972 and all the wells are, and continue to be, monitored by automatic sensing devices (from 1982 onwards). *Figure 1* shows the location of the wells in the region.

A schematic view of the successive sedimentary layers is shown in *Figure 2*, which shows the general direction of flow and the leakage directions within the aquifers. From a radiological point of view, it is important to note that the leakage direction within the Boom Clay layer is directed downwards at the repository site. However, closer to the discharge zone (near Antwerp), the leakage direction is inverted.

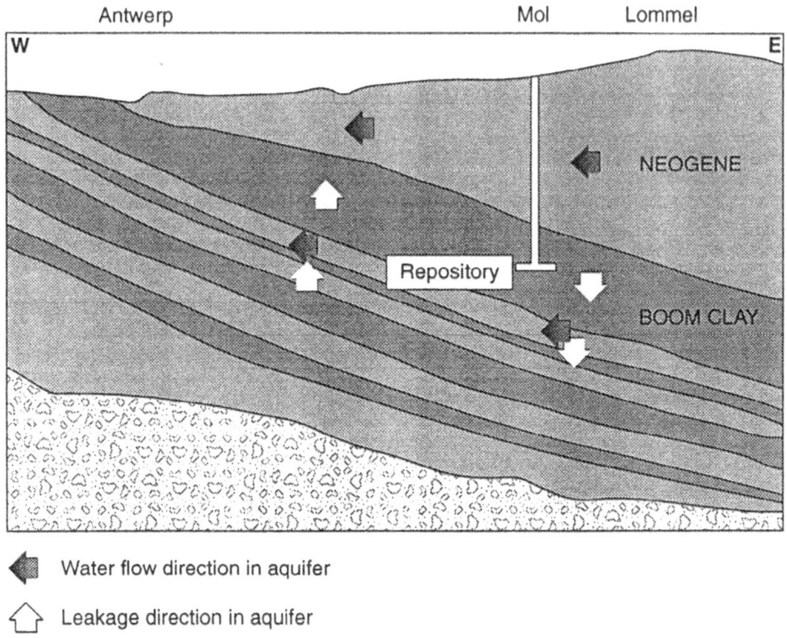

Figure 2. Geologic cross section of N.E. Belgium

Much work has been devoted to the interpretation of the complex hydrogeology of the region. The main objective was the development of a computer code capable of predicting the water flow patterns and their magnitudes in the present climatic conditions, in order to use the model for extrapolating these data to geoprospective simulations including: glaciation and a lowering of the oceanic water level.

Most of the codes for the simulation of transport in aquifers can solve both transport and water flow problems. However, the analysis of water flow transport of radionuclides for a specific site cannot be solved by using one single computer code. In many cases, as a first step, the water flow is studied on a large, so-called, regional scale. In a second step, a site model is applied at the local scale in order to perform detailed water flow simulations for boundary conditions derived from the results of the regional scale model. In a third step, detailed transport simulations are carried out on the local scale for abnormal hydraulic conditions.

Calculations of the water flow on a regional scale are performed with the quasi-three dimensional (2D plus Layers) finite difference model **NEWSAM**, which allowed the definition of a discretization mesh with reasonable detail in the area of interest, without over-exerting the computer calculations. In addition, a certain flexibility in the boundary conditions allows its application in a geoprospective simulation. Water flow and transport calculations on a local scale are performed mainly with the two-dimensional (2D) finite element code **AQUA** and in some cases with the three-dimensional (3D) finite elements code **WATERFLOW**. An advantage of these finite elements codes is that they allow the borders of the modelled area to be adapted to the physical borders of the aquifer. The geometry of the different layers and the variability of their characteristics can also be better accommodated. The 2D or 3D analytical code **AT123** is also applied for transport simulation in aquifers, using Darcy velocities calculated with one of the previous water flow codes. It has, however, the disadvantage that no meteorological recharge can be taken into account and, therefore, no dilution.

For the verification of computer codes, it is important to make use of at least two computer codes that can simulate the essential phenomena of a given flow transport problem. The validation of aquifer models is very difficult but necessary to accomplish a convincing long-term geoprospective simulation.

3. Migration of Radionuclides

Extensive studies were undertaken in the underground laboratory by means of injection cylinders and solid suspensions of clay loaded with the appropriate radionuclides. The migration phenomenon is essentially based on molecular diffusion within the interstitial clay water.

Diffusion coefficients of 10^{-9} down to 10^{-16} m^2 s^{-1} have been measured *in situ* for a series of nuclides (H-3, C-14, I-125, Se-75, Tc-99, Cs-137, Eu-152, U-233 and Am-241). Due to the high pH and low Eh values, the lanthanides and actinides have very low solubilities and occur mostly in their lower oxidation form. The migration of radionuclides in clay is determined by the physico-chemical characteristics of the clay body itself, its permeability, and the conditioning matrix which has been used to solidify the nuclides.

Due to the very compact stucture of the Boom-clay at 250 m depth, the hydraulic conductivity is about 10^{-12} m s^{-1} and in such a medium, molecular diffusion is the dominant phenomenon. Both the redox potential and alkalinity determine the chemical state of the nuclides during their migration. The presence of organic macromolecules (humic acids) and inorganic complexing agents, especially carbonates, influences the mobility of the dissolved species. The migration ability of a radionuclide is determined by the diffusion coefficient expressed in a unit m^2 s^{-1}. *Figure 3* shows a series of experimental data obtained in laboratory conditions and during *in situ* experiments in the underground laboratory. All these data have been validated in extensive R&D activities from the end of the 1970's up to the

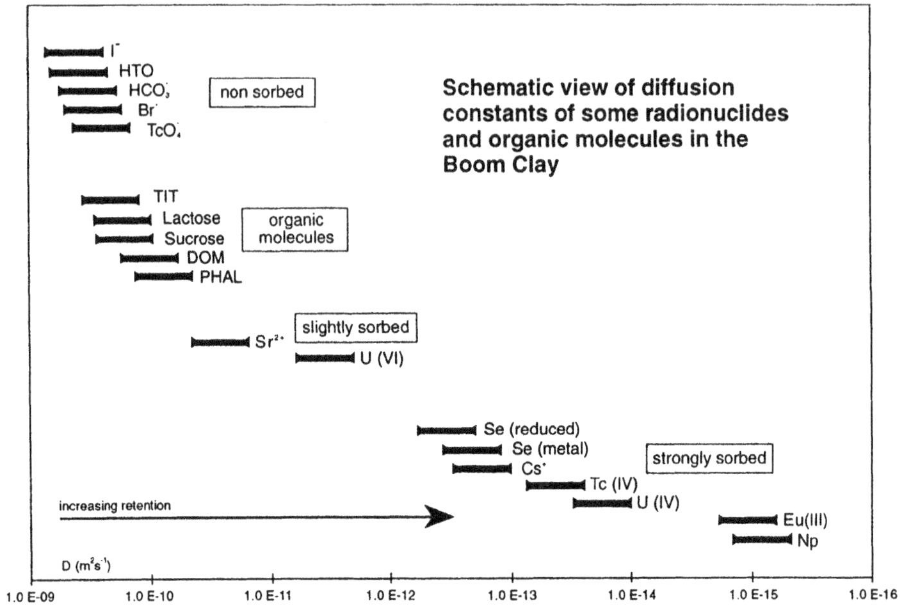

Figure 3. Overview of the diffusion constants of radionuclides and organic molecules studied in the Boom Clay (decreasing order of mobility)

present day.

Non–"sorbed" species are tritiated water, iodine compounds and oxidized technetium complexes which move, in principle, as fast as the conveying water flux. More complex organic molecules are slightly delayed because of their surface sorption on clay minerals. Cationic species, *e.g.* Cs, Sr, are generally sorbed through cation-exchange mechanisms on the negatively-charged surface of the aluminosilicates. Their retardation is inversely proportional to their distribution coefficient and may vary by a factor ranging from 2 to 100 or more. Some ions undergo complexation and move in the interstitial water at a relative velocity which depends on the net charge of the complex. This is the case with, for example, uranium species.

Finally, a series of nuclides do not move because they are insoluble in the interstitial water at the pH values of 8 to 9 which prevail in the clay body due to the presence of macroscopic quantities of carbonates. Most of the actinides and lanthanides fall in this category, their "relative velocity" being 100 to 1000 times slower than the water movement.

All these characteristics are expressed by the diffusion coefficient which is the fundamental parameter for assessing the retention potential of a geologic clay formation.

The redox potential in deep clay deposits is, in general, very reducing due to its pyrite (FeS_2) content. If anaerobic conditions remain predominant, the redox potential of deep clay is around −315 mV which is sufficient to keep most of the radionuclides in their lowest oxidation state, *e.g.* Pu(III), Am(III), U(IV), Np(IV) *etc.*

Finally, one has to investigate the complexation potential of the organic species which are present in clay deposits. *Humic acids* with their high molecular weight are sorption "sinks" for heavy nuclides and will ultimately determine the mobility of the lanthanides and actinides.

Laboratory experiments to determine the "apparent" diffusion coefficients are a first step in the investigation of the sorption characteristics of nuclides, but large *in situ* experiments are indispensable for validating the overall behaviour of radionuclides in the geological strata.

An analytic migration model **MICOF** has been developed and has been validated in many benchmark experiments carried out within the EC programme. Among the numerical "umbrella" codes, the **LISA** and **LISGAN** codes have yielded good, reliable results.

3.1. INTERACTION OF CANISTER MATERIALS WITH CLAY

Typical types of stainless steel which are used for vitrification canisters were studied in the laboratory and *in situ.* It was concluded that overpack materials were necessary to guarantee a canister lifetime of 1000 years.

Any corrodable material will produce an amount of hydrogen gas proportional to the weight of the metal transformed into oxide. The consequences of a macroscopic corrosion phenomenon should not be underestimated: if 1 tonne of iron corrodes, a total amount of 400 to 600 m^3 of hydrogen (H_2) gas will be produced. The timescale for such phenomena to occur is very long, such that the quantity of H_2 produced per unit time should be dissipated within the mostly impermeable clay body.

Two options for the selection of overpack materials were investigated:

- a thick overpack made of inexpensive materials (approximately 10 cm thick carbon steel), and
- a thin overpack based on "noble" materials. (Hastelloy C4, Inconel 625, Ti/02Pd alloys, AISI316).

Both options have pros and cons and the final choice will depend on the overall repository concept. However, the integrated gas production in the case of carbon steel may become the critical factor in the overall safety assessment. A thick carbon steel overpack is useful for reducing radiolysis phenomena in near-field materials. However, such an overpack will, after a several centuries, be oxidised unless a corrosion-resistant film builds up at the surface, or is deposited initially. A combination of a thin noble mate-

Figure 4. In situ corrosion experiment (direct contact)

rial coating on a thick carbon steel overpack is, without doubt, the most commendable, but also the most expensive, practice.

In situ experiments using all these materials have been performed in the underground laboratory over a 7.5 year period. *Figure 4* shows an "artist's view" of the undergound set up. Corrosion studies were made on the recovered samples and showed that carbon steel corrodes at a rate of about 2 to 8 μm a^{-1}, depending on the temperature (16, 90 and 170 °C) with a mean value of 4 μm a^{-1}. Pit-corrosion is the main phenomenon and maximum corrosion depths of 140 μm were observed.

High-nickel stainless steel samples, and *a priori* noble alloys, did not show any sign of corrosion.

4. Interaction of Waste Type Materials with Clay

The interactions of waste matrices and, in general, all nuclear waste type materials, have been extensively studied in the laboratory and *in situ*. The main attention was devoted to glass types used for the vitrification of HLW. A comprehensive experimental investigation was also conducted on medium-level waste (MLW) types (cement-, and bitumen-based matrices).

Borosilicate glass is by far the most widely studied waste material since it provides a very stable and inert waste form for liquid HLW which is produced during the reprocessing of spent fuel.

The most widely publicized vitrification method was developed in France and consists of a combined calcination and vitrification process. The final stage in the French R&D programme was the production of SON68 glass which has been adopted by many countries as the HLW standard. Belgium, which has current reprocessing contracts at the La Hague Reprocessing Plant, will receive this type of HLW for final disposal. For this reason, an important test programme was undertaken to investigate the compatibility between SON68 glass and the clay body in which it will eventually be disposed.

In the aftermath of the EUROCHEMIC project carried out in Belgium, a vitrification plant was built by the German reprocessing authority DWK. The project, called PAMELA, was based on the direct vitrification of liquid HLW by the so-called Joule-heating method. In the course of the previous decade, a total of 600 vitrified HLW drums were produced. The glass SM527 has a different composition according to the waste being treated. In some cases, where high alumina concentrations are present, the composition is quite different from SON68 glass.

Samples of the different types of waste glasses have been tested in the laboratory and in the underground test facility. Corrosion loops remained for 5 to 7.5 years in contact with the natural clay formation at different temperatures (16, 90, 170 °C). The two main types of glass samples investigated were the COGEMA glass SON68 and its precursors, and the PAMELA glass types, especially SM527. Subsequently, the corrosion loops were excavated from the clay formation according to the "overcoring" method which allows extraction of the entire experimental loop without disturbance of the glass-clay interface.

All the glass samples retrieved from the undergound tests were studied by electron microscopy (SEM-EDXA, and EPMA) and secondary-ion mass spectrometry (SIMS). Based on the comparison between the thickness of dissolved glass and the leached depth, it appeared that the corrosion is homogeneous but that selective substitutions of elements from the glass matrix occur at the surface. Some elements are depleted in the surface layer and adsorbed by the surrounding clay. All the radionuclides dissolved in the glass (Pu plus fission products) were found at distances of 1–4 mm from the glass surface.

As a first important conclusion, we would suggest that vitrification as borosilicate glass is an excellent method to condition HLW for final disposal in a clay-type repository.

Phosphate glass, which has been advocated by Russian scientists, is perhaps somewhat less stable than borosilicate glass, but it is worthwhile for further investigation because its production technology is simpler and perhaps cheaper. The studies on the interaction of glass doped with high amounts of TRU elements is projected in the next R&D progamme of

SCK•CEN within the framework of the 4th R&D progamme of the Commission of the European Union. The underground laboratory is available for any scientific partner who wants to study the interaction of its waste types with a clay body. A specific agreement between the partners has to be drafted in order to protect the intellectual property of the test runs and the underlying technology.

Acknowledgments

The author thanks his former collaborators for their contributions to this summary paper. Many thanks to Mr. B. Neerdael, the present project leader, for logistical support. Particular thanks to Dr. J. Marivoet, Mr. M. Put and Mr. P. Van Iseghem who provided a wealth of recent results.

References

1. BONNE, A. *et al.* (1986) Status of nuclear waste disposal programme in Belgium, in H.C. Burkholder (ed.) *High-Level Nuclear Waste Disposal*, Battelle Memorial Institute Press, ISBN 0-935470-29-8, pp. 1-9.
2. MANFROY, P. *et al.* (1986) Ten years of site-specific research and its relevance to the concept of a waste disposal system in clay, in H.C. Burkholder (ed.) *High-Level Nuclear Waste Disposal*, Battelle Memorial Institute Press, ISBN 0-935470-29-8, pp. 233-246.
3. BAETSLÉ, L.H. *et al.* (1996) Study of the Boom Clay layer as a geochemical barrier for long-lived radionuclides, in H.C. Burkholder (ed.) *High-Level Nuclear Waste Disposal*, Battelle Memorial Institute Press, ISBN 0-935470-29-8, pp. 313-324.
4. DE BATIST, R. *et al.* (1986) Waste form and waste package characterization studies related to disposal in a geological clay formation, in H.C. Burkholder (ed.) *High-Level Nuclear Waste Disposal*, Battelle Memorial Institute Press, ISBN 0-935470-29-8, pp. 585-596.
5. BAETSLÉ, L.H. AND BONNE, A. (1990) Disposal of radioactive waste into clay layers, the most natural option, in R.G. Post and M.E. Wacks (eds.) *Waste Management '90*, Proc. Symposium on Waste Management, Tucson, Arizona, Feb.–March 1990, Library of Congress No. 86-644175, pp. 907-913.

HYDROGEOLOGY

6. BEAUFAYS, R. *et al.* (1994) Characterization of the Boom Clay and its multilayered hydrogeological environment, Final Report, *CEC Nuclear Science and Technology Report* EUR 14961 EN, ISBN 92-826-7697-8, Brussels-Luxembourg.

MIGRATION OF RADIONUCLIDES

7. PUT, M. AND HENRION, P. (1992) Modelling of radionuclide migration and heat transport from a HLW-repository in Boom Clay, Final Report, *CEC-Nuclear Science and Technology Report* EUR 14156 EN, ISBN 92-826-4062-0, Brussels-Luxembourg.

PERFORMANCE ASSESSMENT

8. MARIVOET, J. AND BONNE, A. (1988) PAGIS Performance Assessment of Geological Isolation Systems for radioactive waste disposal in clay formations, *CEC Nuclear Science and Technology Report* EUR 11776, ISBN 92-825-9254-5, Brussels-Luxembourg.

9. Marivoet, J. and Zeevaert, T. (1991) PACOMA – Performance assessment of the geological disposal of medium-level waste and alpha waste in a clay formation in Belgium, Final Report, *CEC Nuclear Science and Technology Report* EUR 13042 EN.
10. Marivoet, J. (1992) PACOMA – UPDATING 1990. Updating of the performance assessments of the geological disposal of high level and medium level wastes in the Boom Clay, *SCK•CEN Report* BLG 634.

CHARACTERIZATION OF HLW

11. Conradt, R., Roggendorf, H., and Ostertag (1986) The basic corrosion mechanisms of HLW, Final Report, Fraunhofer Institut München, *CEC Nuclear Science and Technology Report* EUR 10680 EN, ISBN 92-825-6560-2 Brussels-Luxembourg.
12. Hall, A.R. (ed.) (1987) *Testing and Evaluation of Solidified High-Level Waste*, EUR10852 EN, Radioactive Waste Management Series, ISBN 0-86010-893-7, Graham & Trotman Co., London.
13. Van Iseghem, P. *et al.* (1996) Completion of the corrosion programme in Boom Clay – in situ experiments, Final Report for the period 1991-1994, NIROND 96-05, June 1996, to be published as EUR report.
14. Van Iseghem, P. *et al.* (1996) The European network for quality checking of waste packages: objectives and status, *Proc. 4th Conference of the European Commission on the Management and Disposal of Radioactive Wastes*, Luxembourg 25-29-03-1996.

WASTE CONDITIONING METHODS: ECONOMICAL AND ECOLOGICAL IMPACT

E. TRAUWAERT
Straatsburglaan 33,
B-2400 Mol, BELGIUM

1. Introduction

Between the collection of nuclear waste material, no longer useful for practical applications, and the final disposal of the material until complete radioactive decay, a great number of operations have to be performed. Omitting all transport and storage operations, which fundamentally leave the material unchanged, two major transformation categories can be distinguished. The first category aims at the chemical stabilisation and volumetric minimisation of the waste mass and is commonly called waste treatment. Frequently adopted treatment operations include sorting, cutting, incinerating, compacting, *etc.* These operations are usually followed by complementary operations aimed at giving the waste the required properties necessary to withstand transport manipulations, intermediate storage for some limited time (10 to 50 years), further transportation and final disposal, presumably to a shallow disposal site or a geological repository. This second category of operations is commonly defined as *conditioning* and it will be the subject of the present paper.

2. Immobilisation Goals

Apart from the chemical stability, which is supposed to be basically obtained by the waste treatment operations, specific goals are required for the immobilisation step. These goals result from the remaining stages in the lifecycle of the waste, *viz.* in general, transportation to a storage room for intermediate storage, intermediate storage and final disposal after further transportation. Each of these lifecycle stages generates its own require-

M. J. Stenhouse and V. I. Kirko (eds.),
Defence Nuclear Waste Disposal in Russia: International Perspective, 45–57.

ments, which normally are not uniquely different but are, nevertheless, distinguishable.

2.1. LONG-TERM STORAGE OR FINAL DISPOSAL

Depending on whether one considers low level waste (LLW), medium level waste (MLW) or high level waste (HLW), the long-term isolation requirements will be quite different, ranging from some 300 years to millions of years. Also the place of disposal may differ: shallow disposal for LLW and geological burial for MLW and HLW. The general requirements are, however, similar even if their intensities may differ.

The main requirement of burial is that the material remains in place with as little interference as possible with the surrounding environment. Therefore, the waste should be made dense enough to avoid 'floating' to the surface under the influence of geological movements. Ideally, the density should be comparable to that of the surrounding rocks or earth layers. The waste form should also be solid rather than liquid, in order to have a lower diffusion coefficient. Even if solid, the waste can still migrate as a result of leaching by groundwater. Therefore, the basic leachability of the conditioned waste material should be as low as possible.

Other aspects of conditioning, also affecting the leached mass, should be taken into account. The geometric form should be compact (an isoform cylinder rather than a set of individual needles), the mass should have no cracks and, if possible, the surface of the cylinder should even have a protective layer. Since in many instances, the radioactive waste will still produce some heat, the heat conductivity of the conditioned waste should be high enough to avoid too high temperatures being established in the core of the waste which could cause physical instabilities and violent reactions. Finally, the conditioned waste mass should also exhibit some radiation resistance to avoid deterioration of the characteristics of the waste form over time.

It should be noted here that not all the containment power is expected from the conditioned waste package alone. In fact, different levels of surrounding barriers will be developed and will greatly enhance the capability to contain the contamination for a long period of time. This concept is known as the multibarrier approach, which is addressed in this workshop [12].

2.2. INTERMEDIATE STORAGE

The requirements for intermediate storage are in many aspects similar to those of long-term storage. Some of the specifications are, however, quantitatively different.

The integrity of the waste package must be assured during the entire duration of intermediate storage in order to guarantee waste transportability to the final disposal site. This time period may vary from some 10 years for LLW to typically 50 years for MLW and HLW. For all practical reasons, such integrity will most often be realised by surrounding the conditioned waste mass with a metallic envelope or container.

Furthermore, the requirements on heat conductivity will normally be more stringent for intermediate storage than for long-term storage. This results from the fact that heat production decreases from the initial generation of isotopes to their final decay. However, the difference in heat conductivity requirement will normally be taken into account by the intermediate storage infrastructure rather than by the conditioned waste form itself.

As the intermediate storage infrastructure is expected to have a much shorter useful life than the waste itself, it is important that this infrastructure does not become contaminated. Therefore, integrity of waste packages is very important, and the consequences of this requirement will be quite different for LLW than for HLW.

2.3. TRANSPORTATION

Transportation also places its own requirements on the waste packages. Most stringent are the requirements for transport between intermediate and final storage sites. Indeed, after the period of intermediate storage (10 to 50 years), the packages must still be sufficiently resistant and intact to allow manipulations without putting the operators at risk or endangering the intermediate storage infrastructure or transport equipment.

This specific requirement most probably reduces to the quality characteristics of the containment of the waste, most importantly, the avoidance of corrosion of the metallic container of the waste. Two aspects have to be distinguished:

- corrosion from the outside due to interactions between the intermediate storage atmosphere and the waste container;
- corrosion from the inside of the container due to interactions between the conditioned material and its container.

This latter requirement is most difficult to satisfy as the corrosive action remains generally unnoticed until it is too late.

Less demanding with regard to material resistance, but important for its psychological impact, is the safe transport of waste between the conditioning plant and the intermediate storage facility. This is particularly true when the mode of transport involves the use of public routes, and especially when the waste is spent fuel or HLW. Given the difficulties that such transportation creates when using road or rail, the possibility of inland navigation should be considered. In practice, this form of transport is ideal for heavy loads, has very good safety records and is less exposed to disturbance from activist groups. Provided that an adequate transport cask is available, this mode of transport could satisfy the highest safety standards and help to reduce congestion in roads.

2.4. RADIATION PROTECTION

It should be stressed that at least one feature is generally not part of the requirements for waste packages: radiation shielding. Although some additive materials may be mixed with the matrix (haematite or barite), this important aspect of the waste disposal concept is taken into account mainly by the relevant infrastructure. For example, during transport, an adequate transport cask has to be used; in the storage facility, manipulation tools together with the distance factor, are used to assure the necessary radiation protection for the operators as well as for the population at large. Thus, this aspect will not be taken into account in the subsequent analysis.

3. Available Conditioning Methods

The main aim of conditioning radioactive waste after waste treatment is to enhance its resistance to leaching and to improve its mechanical strength with a view to transport operations, intermediate storage and final disposal [3]. It is also of the utmost importance that the conditioning matrix is compatible with both the waste materials and storage conditions. Further requirements are a good fire resistance and a long-term stability.

The current practice in Belgium, as in most nuclear countries, with respect to the various types of radioactive waste, is described below.

3.1. CEMENTATION

Cementation has the great advantage of being a well developed and proven process, achieved in cold conditions and using relatively simple equipment. This makes the process very attractive, certainly when taking into account the fact that the raw materials are widely available and inexpensive [18]. Furthermore, cementation assures the chemical stabilisation of many inorganic compounds as well as organic wastes.

This conditioning process can be applied under *inactive conditions* whereby liquid cement is poured around the solid or solidified radioactive waste that has to be conditioned. The cement is usually made fluid enough to fill all accessible voids within the waste package. In this way, good mechanical strength is obtained. Using this conditioning method, it is often the aim to have cement all around the waste, which can be achieved using centring devices in the drums. In this way, direct contact between the waste and the container is avoided, resulting in better protection of the waste by introducing, even at the level of an individual package, a multibarrier concept [12].

Cement can also be used under *active conditions*, whereby some radioactive elements are introduced during the preparation of the cement. These elements can either be introduced as solutes (dissolved) in the water that is added during the process, or as a component of the solidification materials such as sand. This sand can come from contaminated soils or from the dismantling of contaminated concrete buildings. In both cases, the sand has to be treated in order to satisfy certain process requirements; tests have shown, however, that these approaches are applicable generally.

The advantage of the active process is to incorporate waste material in an efficient way, thereby avoiding the contamination of materials that were originally inactive. An overall reduction of the waste volume will result. The drawback, however, is that all the cement is contaminated immediately on preparation and that any cement that is ruined has to be treated as contaminated waste. The same is true for contaminated cementation equipment.

Another potential problem associated with cementation is the retardation role that the waste material may play with regard to the setting cement phase. High waste loadings of some types can be expected to affect the cement chemistry and waste form structure; however, such an effect does not necessarily result in waste-form failure, although severe retardation could result in incomplete immobilisation and hence, in an unacceptable physical form for the conditioned waste. Considerable research and laboratory studies on these interaction mechanisms have been performed and are widely reported [10,16,18,21].

A less positive characteristic of cement is that its leachability resistance is not very high, resulting in the solubilisation of the radionuclides from the waste mass. This situation is true at least for some radioactive elements such as Cs and I, whereas for others (*e.g.* C, Sr, Co) that are strongly adsorbed by cement hydrates, the rate of release to the surrounding water will remain limited [9].

A critical problem with all cementation processes is to ensure that all residual water is removed, otherwise residual free water may react with waste materials, thereby initiating different possible corrosion reactions. This problem should be avoided by careful quality control of the cementation process and by assuring that the cement sets without undue delay.

3.2. BITUMINISATION

Waste conditioning by bituminisation is a well-established process characterised by a moderate operation temperature (160–190 C) and by low construction and operating costs. Its technological features make it particularly well suited for treatment of waste in aqueous solutions such as concentrates, sludges and resins. The main reason for the choice of bitumen as conditioning matrix is its very good leaching resistance even in moist environment conditions.

Since bitumen is not chemically inert and burns at high temperature, care has to be taken not to mix the bitumen with elements of such a type and quantity that an exothermic reaction could start, initiating a reaction that would rapidly grow out of control. This phenomenon has already been the cause of a number of fire incidents.

The methodology used to monitor the inflammability of the waste mixture is generally based on laboratory tests. Under normalised conditions, it is possible to measure if the waste becomes exothermic and to determine the limiting temperature and waste loading factor which should not be exceeded.

The advantage of this feature is, however, that bituminised waste can easily be re-treated, providing a reversible conditioning process. Two major bituminisation techniques have been developed. In the first technique, evaporation of the water from the waste precedes mixing with bitumen. To achieve this, the (radioactive) sludge is laid down as a thin film on a filter before being mixed with the bitumen at a suitable temperature. An alternative method uses extrusion to add the waste at a constant rate to the bitumen bath [5,7], thereby assuring the evaporation of redundant water. Due to bitumen's inflammability, bituminisation should preferably be undertaken in an inert atmosphere such as nitrogen.

Another drawback of bitumen is its poor mechanical resistance. If not properly supported in a container, the bitumen will collapse under its own weight after a period of time. Even in metallic containers, bitumen often spills out after a number of years.

3.3. VITRIFICATION

Glass is well known for its high leachability resistance, which is the reason for many of its applications. Different types of glasses have been tested with regard to their capability as a conditioning matrix, and borosilicate glass has proven capable of incorporating a significant amount of waste without losing its main characteristics.

The main properties on which the choice of glass is based are chemical inertness, a relatively high mechanical resistance and a reasonable thermal conductivity. The good performance of glass with regard to these features explains the choice of glass as a conditioning matrix for the most demanding type of waste, *i.e.* HLW. The vitrification technique is, in principle, also applicable for fissile material from weapons origin. However, limitations on the waste incorporated, related to the type of fissile isotope, have to be taken into account [19].

The price to pay for these advantages is a high-temperature process. Two alternative technologies have been developed:

- a two-stage process, with a calcination step at a lower temperature and the actual vitrification step at higher temperature;
- a process whereby the two steps are combined using an appropriate feeding mechanism.

Typical of the first approach is the French AVM (Atelier de Vitrification de Marcoule) whereas the second technique has been demonstrated successfully by the Belgian PAMELA installation, constructed in Dessel, in Germany. In the latter installation, liquid waste, possibly mixed with a certain amount of uranium or plutonium nitrate, is first denitrated. The concentrate is then mixed with borosilicate glass frit and fed into the ceramic melter at a rate compatible with the rest of the process. In the melter, the glass frit is heated to melting and homogenised with additions of waste.

The fused and mixed glass is poured into stainless steel canisters, each of which is covered with a lid after the necessary cooling period; the lid is subsequently sealed with a circular weld bead, thereby assuring the leak tightness of the container.

3.4. CONDITIONING BY POLYMERS AND CERAMICS

Additional forms of conditioning nuclear waste are under investigation and development. Examples are polymers, either used alone (Japan) or in combination with cement (France). However, these forms are presently not proven adequately and are the subject of many studies. The same is true for ceramic materials, such as SYNROC, developed in Australia. Although these materials may be very promising, it is too early to fully endorse them as valid conditioning methods capable of superseding one or more of the older techniques. Therefore, these alternative methods will not be discussed further in this paper.

4. Economics as a Function of the Waste Activity Level and Final Disposal Type

The cost of the different installations described so far is related more to the activity level of the waste to be treated than to the various types of process equipment. Indeed, LLW can be treated under relatively simple confinement using manual tools; for α-contaminated waste and MLW, however, glove-box conditions are necessary, whereas for HLW, heavily-shielded cells and remote handling are required. This difference in confinement requirements determines the major portion of investment costs.

In addition, the final disposal concept, which again is related to waste activity level, influences significantly the unit cost of different waste categories. Presently, essentially two broad concepts exist: shallow disposal and geological burial. Shallow disposal is being undertaken for LLW, containing radioactive isotopes with a decay half-life of 30 years or less. All other types of waste should be disposed of by deep (geological) burial. Whether such geological disposal sites are in granite, salt or clay is still the subject of many investigations [1], and acceptance of these sites is likely to depend on their respective availability in the countries concerned.

Given the importance of waste activity and the final storage concept on the overall cost of the waste evaluated from the waste conditioning stage to final burial, two main categories of cost can be distinguished. In order to analyse the costs within these broad classes, a representative type of waste is considered for each class, *viz.*

- LLW, conditioned in metal containers filled with cement and designated for shallow burial, and
- vitrified HLW, intended for a deep (geological) repository.

4.1. COST OF LLW

Considering the necessary equipment for the cementation of LLW, an order of magnitude cost per unit quantity of waste can be estimated at 3,000 \$ m^{-3} or about 1 \$ kg^{-1}. Adding the cost for intermediate storage and final disposal and for the necessary transportation of package between the sites, the total cost should amount to about 10,000 \$ m^{-3} or 3.5 \$ kg^{-1} [14,15]. This is an actualised cost for 1996.

In terms of radioactivity, LLW is generally characterised by the limit of 370 kBq g^{-1} for β,γ–radiation and by 3.7 kBq g^{-1} for α–radiation [13,19], although some variations on these limits may exist (*e.g.* 12 kBq g^{-1} for β, γ–activity in the U.K. [8]).

Accordingly, it is also interesting to express the cost versus the amount of radioactivity. Referring to the above figures, one obtains for β,γ–activity a cost of about 10 \$ GBq^{-1} and for α–activity 1,000 \$ GBq^{-1}. Since most of the LLW is not produced close to its activity limit, an average concentration ten times lower is probably more representative of actual waste packages [11]; this provides a unit cost of about 100 \$ GBq^{-1} for β,γ–activity and of 10,000 \$ GBq^{-1} for α–activity. *Table 1* gives a summary of these results.

TABLE 1. Typical cost figures for LLW

Cost per unit of mass		3.5 \$ kg^{-1}	
Type of radiation		β,γ	α
Radioactivity (kBq g^{-1})	limit	370	3.7
	average	37	0.37
Cost per unit of activity (\$ GBq^{-1})	limit	10	1,000
	average	100	10,000

TABLE 2. Typical cost figures for HLW

Cost per unit of waste mass	5,000 \$ kg^{-1}
Radioactivity	20,000 – 200,000 GBq kg^{-1}
Cost per unit of radioactivity	0.25 – 0.025 \$ GBq^{-1}

4.2. COST OF HLW

Similarly as for LLW, one can estimate the cost of conditioning HLW (vitrification) at $0.7 \cdot 10^6$ \$ m^{-3} or about 280 \$ kg^{-1} of glass [17]. However, since only 10–20 % of the glass is actual radioactive waste, the cost per unit quantity of waste is some 5 to 10 times higher: $7 \cdot 10^6$ \$ m^{-3} of waste and 2,800 \$ kg^{-1} of waste. Considering the additional cost for intermediate storage and final disposal and the necessary manipulations and transportation, the total cost amounts to $12 \cdot 10^6$ \$ m^{-3} or 5,000 \$ kg^{-1} of waste.

With regard to the radioactivity content, HLW has typically a specific activity in the range 50,000–500,000 TBq m^{-3}, or 20,000–200,000 GBq kg^{-1}, corresponding to a thermal power of 2–20 kW m^{-3} [6]. Expressing the cost versus the amount of activity, one obtains for the lowest activity level 0.25 \$ GBq^{-1}, which is considerably lower than for LLW. These results are summarised in *Table 2.*

This result confirms the good practice of reducing waste volumes and concentrating radioactive isotopes as much as possible: HLW is cheaper than LLW when expressed as cost per Bq.

5. Ecological Impact

The ecological impact of nuclear waste is a 'diamond with many faces'. In comparing the conditioning and subsequent burial methods for LLW and

HLW, attention is focused on two aspects of their ecological impact: volumetric and visual damage to the environment, and the potential radioactive burden caused by the solubility of the conditioned waste in groundwater.

5.1. VOLUMETRIC AND VISUAL IMPACT

The comparison between shallow burial for LLW and a geological repository for HLW is not straightforward. Firstly, it is probably trivial to note that shallow burial will leave some traces on the surface landscape depending on the kind of tumulus that will be built. In contrast, almost no indication of geological disposal will be seen on the earth's surface, once the site is closed.

With regard to the dimensions of a shallow burial site, a reference design considers modular bunkers of 10m x 10m x 8m or 800 m^3 internal dimensions to house 400 m^3 of LLW. Taking account of the external dimensions increases this volume to about 1,500 m^3, which, with an additional 20 % for the covering and the final closing, results in 1,800 m^3 of external volume for a useful (waste) volume of 400 m^3, *i.e.* the disposal volume is a factor of 4.5 more [13]. Thus, the 120,000 m^3 LLW expected from the Belgian nuclear programme until the year 2050, will produce a volume of about 540,000 m^3 potentially impacting the environment.

On the other hand, MLW and HLW will require a subsurface 'hole' approximately twice the size of the waste volume. Hence, for the 25,000 m^3 of MLW and the 5,000 m^3 of HLW from the Belgian nuclear programme until the year 2050, about 60,000 m^3 of subsurface volume will be needed. In this context, MLW and HLW interfere with a volume of environment about 10 times less than that for LLW.

Considering further the difference between LLW and HLW in terms of activity, the difference between the two burial concepts relative to 1 GBq is even more significant. With an average activity of some 37 kBq g^{-1} or about 100 GBq m^{-3} of conditioned LLW, one needs some $4.5{\cdot}10^{-2}$ m^3 of external volume to contain 1 GBq of LLW. For conditioned HLW, with an average activity of some 50,000 TBq m^{-3}, 1 GBq will require a burial volume of about $4{\cdot}10^{-8}$ m^3 which is about 1 million times less. If one considers that a geological repository will be used for both MLW and HLW, and taking into account their respective quantities and the negligible amount of radioactivity in MLW compared with HLW, this volume should be reduced by a factor of 6, which is still 150,000 times less volume per GBq than for LLW.

This conclusion should be tempered to some extent by the fact that LLW remains radioactive for a period of 300 years, whereas HLW emits radiation for millions of years, particularly if the transuranic elements (TRU) are included in the waste. This difference in timescale of concern is a good reason to try to eliminate TRU from HLW, leaving only radioactive fission products, with an activity level that will disappear much more rapidly [2].

5.2. LEACHING POTENTIAL OF THE WASTE FORMS

The ecological impact of the radioactivity enclosed in the waste packages is a fairly complicated function of the leaching dynamics and the waste content of the individual packages. The leaching dynamic is itself a function of the chemistry, physics and thermal characteristics of the conditioned waste packages, the backfilling materials and the host layers. In all ongoing reactions, reaction rate is an important factor not only because many retardation mechanisms operate but also because radioactivity itself is time dependent through radioactive decay.

The basic parameter affecting a comparison of the behaviour of cement and glass with respect to release of radioactivity is the leach rate. Leach rates for cement packages at room temperatures vary between 10 and 1 $g\ m^{-2}\ d^{-1}$ [18]. For conditioned glass, this characteristic under the same conditions reaches 10^{-2}–10^{-4} $g\ m^{-2}\ d^{-1}$ [4,20]. Hence, glass retains the radioactivity some 1,000 times better than cement. This is reasonable since the radioactivity concentration in glass is generally much higher than in cement. Indeed, it was seen that LLW encapsulated in cement has an activity of about 10^{-4} GBq cm^{-3} and HLW in glass, an activity of 50 GBq cm^{-3}, *i.e.* some 500,000 times greater, which, together with the 1,000 times lower leach rate, still leaves a factor 500 in favour of glass.

This reasoning fully justifies geological disposal for HLW conditioned in glass, as the geological barriers can assure some time delay of 1,000 to 100,000 years before radioactivity returns to the biosphere. In the case of HLW without a TRU component, the combined decay of isotopes will assure a reduction in radioactivity of 10^5 after 10,000 years which appears adequate [2]. If, however, TRU elements are not removed from the waste, the reduction factor during the same period will only be 10, which makes it less favourable than shallow burial of LLW and, consequently, potentially insufficient.

6. Conclusions

When dismantling and cleaning-up nuclear weapons as well as the industrial installations used to produce them, a range of nuclear waste types will be generated similar to that from the civilian nuclear industry.

Different acceptable conditioning methods have been demonstrated and are readily available for application. The techniques are derived from classical industrial technology, but adapted to the nuclear environment and the associated safety requirements.

Approximate cost estimates show that HLW is less expensive than LLW, when related to radioactive content. This observation confirms the common sense approach not to disperse, but rather to concentrate, radioactive material. Such an attitude is not only appropriate for the safety of the population and the operators, but also from a pure economics perspective.

From an ecological perspective, it was also seen that HLW requires a smaller volume of repository than LLW, not only in terms of overall volume but also per specific amount of radioactivity. The only problem and drawback for HLW is its very long 'active' life, which places it in a less favourable position than LLW in terms of a long-term concern. Whether the leachability characteristics of HLW will be acceptable will depend greatly on the performance of all the other barriers of the multibarrier concept for this waste. One certain way to relax to some extent the constraints for this type of waste would be to remove as much as possible the TRU content of this waste stream. This operation would result in a reduction in activity of HLW by a factor of 100 after only 300 years, and by 100,000 after 10,000 years; such a result could easily be obtained if plutonium was burnt as fuel in reactors rather than considered as waste.

Acknowledgment

The author wishes to thank M. Demonie for his helpful advice and his patient corrections to this paper.

References

1. BAETSLÉ L.H. (1998) Interactions of conditioned high-level-waste matrices with a deep clay layer as host formation, in M.J. Stenhouse and V.I. Kirko (eds.) *Defence Nuclear Waste Disposal in Russia*, Proc. NATO Advanced Research Workshop, Krasnoyarsk, June 24-26, 1996 (this volume), Kluwer Academic Publishers, Dordrecht.
2. BARRÉ J.Y. (1992) The future of the back end of the fuel cycle, *Scientific Basis for Nuclear Waste Management XV*, **Vol.257**, 1992.
3. COURTOIS C. AND SAAS A. (1992) Conditionnement des déchets, *Revue Générale Nucléaire* Nr.5.
4. DEJONGHE P., BAETSLÉ L., ET AL. (1982) General perspectives in radioactive waste management in Belgium, ANS topical meeting, Richland, WA, April 1982.
5. DELGRANDE J., DELRIEU J.C., MORAVCOVA Z., AND CECH I. (1994) Czechs check out and choose bitumenisation, *Nuclear Engineering International* **39** (483), 58-59.
6. FLOWERS R.H. (1992) Management of high level and intermediate level radioactive wastes, Summer School on Radioactive Waste and Decommissioning, Jesus College, Cambridge, June 29-July 2, 1992.
7. HILD W., DETILLEUX E., ESCHRICH H., LEFILLATRE G., AND TITS E. (1970) Development work on the homogeneous incorporation into bitumen of intermediate-level

waste from the Eurochemic reprocessing plant; management of low- and intermediate-level radioactive wastes; *IAEA Document SM-137/54*, IAEA, Vienna.
8. JOHNSON L. (1992) Low level waste disposal, Summer School on Radioactive Waste and Decommissioning, Jesus College, Cambridge, June 29-July 2, 1992.
9. KRISHNAMOORTHY T.M., JOSHI S.N., DOSHI G.R., AND NAIR R.N. (1993) Desorption kinetics of radionuclides fixed in cement matrix, *Nuclear Technology* **104** (12), 351-357.
10. LEA F.M. (1970) *Chemistry of Cement and Concrete*, Chemical Publishing Co., New York.
11. MARQUE Y. (1994) Low-level radioactive waste management in France: two decades of experience, BNS Annual Meeting, June 2-3, 1994.
12. MERZ E.R. (1998) Multibarrier effectiveness as the expedient measure for selecting the appropriate stabilization and immobilization procedure for the various waste categories, in M.J. Stenhouse and V.I. Kirko (eds.) *Defence Nuclear Waste Disposal in Russia*, Proc. NATO Advanced Research Workshop, Krasnoyarsk, June 24-26, 1996 (this volume), Kluwer Academic Publishers, Dordrecht.
13. MINON J.P. AND VAN MIEGROET J. (1994) Surface disposal on Belgian territory of low-level and short-lived radioactive waste, BNS Annual Meeting, June 2-3, 1994.
14. SALOMON S.N. (1994) Compilation of costs for low level waste disposal facilities, *Waste Management '94*, Proc. International Conf., Tucson, Arizona, Vol. 1, pp. 99-106.
15. J.G., GIANTELLI A.S., AND KENNEY J. (1994) Economic analysis of a centralized LLW storage facility in New York State, *Waste Management '94*, Proc. International Conf., Tucson, Arizona, Vol.3, pp. 1749-1755.
16. TAYLOR H.F. (1990) *Cement Chemistry*, Academic Press, San Diego.
17. TRAUWAERT E. AND DEMONIE M. (1996) Plutonium handling and vitrification: main process steps and their cost evaluation, in E.R. Merz and C.E. Walter (eds.) *Disposal of Weapon Plutonium – Approaches and Prospects*, Proc. NATO Advanced Research Workshop, St.Petersburg, Russia, May 14-17, 1995, pp. 51-58, Kluwer Academic Publishers, Dordrecht.
18. TRUSSELL S. AND SPENCE R.D. (1994) A review of solidification/stabilization interferences, *Waste Management* **14** (6), 507-519.
19. ULIBARRI A. AND ZULOAGA P. (1994) El Cabril: a near surface disposal facility of I&LLW in Spain; BNS Annual Meeting, June 2-3, 1994.
20. VERNAZ E.Y. AND GODON N. (1992) Leaching of actinides from nuclear waste glass: French experience, *Scientific Basis for Nuclear Waste Management XV* **257**, 1992.
21. YOUNG J.F. (1972) A review of mechanisms of set-retardation in Portland cement pastes containing organic admixtures, *Cem. Concrete Res.* **2**, 415-433.

A SEARCH FOR OPTIMAL FORMS FOR SOLIDIFYING HIGH-LEVEL RADIOACTIVE WASTE WHICH ARE GEOLOGICALLY COMPATIBLE WITH GRANITIC HOST ROCKS

B.E. BURAKOV, E.B. ANDERSON & S.I. SHABALEV
V.G. Khlopin Radium Institute
28, 2nd Murinskiy Ave.
St. Petersburg, 194021, RUSSIA

Abstract.
Various wasteforms for immobilising high-level radioactive waste (HLW) are considered in this paper with respect to their geochemical compatibility with granites of the Nizhnekansky massif, in the Krasnoyarsk region of Russia. The geochemical system of this granitic rock is assumed to be at, or close to, equilibrium, having existed in this state for billions of years. Construction of a shaft- or well-type HLW disposal facility within the massif will disturb existing geochemical conditions, however, although such a perturbation may be insignificant or relatively quickly cancelled out if materials with chemical and thermodynamic behaviour similar to that of the host rock (granite) are used as wasteforms for HLW immobilisation.

The thermodynamic characteristics of vitrified 'raw' (no separation of radionuclides) HLW are shown to differ sharply from those of the crystalline host rocks, thereby casting doubt on the reliability of the Nizhnekansky granite massif for isolating vitrified HLW over long time periods. On the other hand, the vitrification technology for the solidification of the caesium-strontium fraction of HLW is quite efficient, allowing the subsequent emplacement of glass blocks in different types of host rock.

For the reliable, long-term immobilisation of transuranic elements (TRU) and rare-earth elements (REE), as well as iodine-129, it is proposed that crystalline host phases in the form of ceramics, resistant to prolonged irradiation and environmental disturbance, are utilised.

The search for suitable host phases is based on a study of the most stable natural accessory minerals characteristic of the particular rock being considered. It is assumed that crystalline ceramics based on mineral host phases similar to the natural accessory minerals of certain types of rock, provide the most efficient material for the long-term isolation of HLW in these rocks. In addition, the reliability of the long-term immobilisation of HLW is assured not only by the physico-chemical strength of the crystalline host phases as engineered barrier, but also by preservation of the geochemical equilibrium in the bulk of the rock mass. In fact, a HLW disposal facility in the form of shafts or wells filled with crystalline ceramic blocks

M. J. Stenhouse and V. I. Kirko (eds.),
Defence Nuclear Waste Disposal in Russia: International Perspective, 59–68.

could be considered as an analogue of natural assemblages of accessory minerals, for example in pegmatite veins.

Two groups of stable minerals are considered as promising crystalline host phases:

1. those which concentrate REE, and
2. those which concentrate halogens.

It is shown that for the Nizhnekansky granite massif, the optimal host phases for 'fixing' TRU and REE are:

- zircon ($ZrSiO_4$),
- various modifications of zirconium dioxide,
- zirconolite ($CaZrTi_2O_7$), and
- monazite ($CePO_4$).

For iodine-129, immobilisation is best achieved using minerals of the tourmaline group. The problems concerning the synthesis of crystalline ceramics based on the above minerals are discussed.

1. Introduction

The problem of high-level radioactive waste (HLW) management is the main obstacle to the further development of nuclear power generation. At the present time, the most realistic of the different approaches to solving this problem is disposal of HLW in deep geological formations. It is assumed that application of the multibarrier protection principle will ensure the safe isolation of HLW for the necessary time period. The creation of a multibarrier protection system includes:

- transformation of HLW into a stable solid form;
- encapsulation of solidified HLW in corrosion-resistant metal canisters;
- application of additional sorbing materials, *e.g.* clays or zeolites, in the near field around the encapsulated HLW; and
- selection of the 'most' monolithic (least fractured) and least permeable sections of a rock mass for emplacement of HLW.

However, it should be noted that underground disposal of HLW has not yet been accomplished in any country of the world. Nevertheless, all existing storage facilities of vitrified HLW should be considered only as temporary, owing to the difficulty in predicting the behaviour of the long-lived radionuclide component of HLW. Such radionuclides have life-times comparable with the duration of geological processes ($\geq \sim 10^5$ years). In addition, different fractions of HLW, *e.g.* the caesium-strontium, actinide and iodine-129 components, exhibit different geochemical behaviour.

The V. G. Khlopin Radium Institute (VGKRI), in St. Petersburg, is currently carrying out investigations on the problem of HLW immobilisation based on the following considerations:

- An industrial process for separating HLW into fractions has been developed in Russia [1]. With this technology, each fraction should be immobilised in a form which ensures efficient isolation of its radionuclide component over a time period long enough to allow for radioactive decay.
- Existing vitrification immobilisation technologies may be applied successfully to the caesium-strontium fraction of HLW, the lifetime of which is comparatively small ($\sim$ 300 years).
- For the immobilisation of the long-lived fractions of HLW, it is necessary to use stable crystalline materials capable of ensuring the safe isolation of a given radionuclide under conditions which are characteristic of the geological formation over a long time period ($\approx 10^5$ years).
- The selection of crystalline materials for immobilising HLW should be carried out taking into account not only economic criteria, but also the practicality of using these materials within existing technologies. It is desirable that the crystalline host phases selected should also be suitable for actinide transmutation, as well as for the creation of mineral-like fuel in order to utilise excess weapons plutonium.
- An additional requirement imposed on the host phases is their geological compatibility with the host rocks in which the HLW burial facility is placed.

This work addresses the most suitable forms for solidifying HLW, which are geochemically compatible with granites of the Nizhnekansky massif in the Krasnoyarsk region of Siberia, a potential area in which to construct a deep disposal facility for radioactive wastes.

2. General Criteria for the Geochemical Compatibility of Solidified HLW and Granitic Rocks

Traditionally, in choosing a site for the location of a HLW burial facility in granitic rocks, complex geological information is evaluated to identify a stable 'block' of rock. The selection process takes into account the geochemical integrity of the block, the presence of fractures, the proximity to faults, *etc.* Even for the optimal choice in these circumstances, however, one should not ignore the fact that the selected block of rock is only a small part of the entire granite rock mass and as such, corresponding geochemical properties are intimately related.

The "lifetime" of granite massifs, from formation to complete destruction, spans hundreds of millions of years. This time interval is not compatible with the comparatively low resistance of granites to environmental disturbance. Such an apparent contradiction is readily explained, however, by considering the granite massif as an essentially closed system geochemically. An igneous rock mass is not monolithic, but consists of fractures containing considerable amounts of groundwater. Consequently, within the total volume of rock, millions of years after its formation, a natural equilibrium is established between the granite rock and groundwater in fractures, which determines the geochemical stability of the massif as a whole.

The creation of a shaft- or well-type HLW disposal facility will disturb the initial geochemical equilibrium conditions within the rock mass. The negative impact may be insignificant or equilibrium conditions quickly re-established if materials with chemical and thermodynamical properties similar to those of the host granite are used as wasteforms for HLW solidification.

Thus, the HLW wasteform prepared for emplacement in granite should fulfil the following requirements:

- high chemical and mechanical stability;
- chemical composition compatible with that of the host rock(s);
- crystalline structure of the matrix;
- monolithic matrix.

3. Vitrified HLW

HLW vitrification is currently performed in several "nuclear" countries, including France, Russia, the U.K. and Japan. The main disadvantage of this technology is the low resistance of the glass matrix to α-irradiation. Even according to the most optimistic calculations, the physico-chemical integrity of borosilicate glass doped with 'raw' HLW (no separation of radionuclides prior to vitrification) cannot be guaranteed to last more than 1,000 years.

Thus, vitrification ensures the safe immobilisation of only the caesium-strontium fraction of HLW, with a lifetime of about 300 years. Such a timescale is not significant in the context of a disturbance of the geochemical equilibrium within different rocks. The disposal in deep geological formations of glass blocks doped with the HLW actinide fraction does not appear to be acceptable, however, since such blocks do not fulfil the long-term safety (stability) requirements identified above. In fact, 1,000 years after burial, the long-lived transuranic elements (TRU) of vitrified HLW are expected to have formed separate, dispersed phases.

4. Crystalline Ceramics for HLW Immobilisation

Crystalline ceramics are currently under consideration for the immobilisation of the long-lived components of HLW. It is assumed that ceramics based on host phases similar to the natural accessory minerals of certain igneous rocks are the most efficient materials for the long-term ($\sim 10^5$ years) isolation of HLW buried in these rocks. In fact, a HLW disposal facility in the form of wells or shafts filled with crystalline ceramic blocks, could be considered as an analogue of natural assemblages of accessory minerals, for example in pegmatite veins.

5. The Choice of Host Phases for the Immobilisation of the Actinide Fraction of HLW

Natural minerals which concentrate rare earth elements (REE) represent potential crystalline host phases for TRU. In order to select actinide host phases which are geochemically compatible with the granite of the Nizhnekansky massif (near Krasnoyarsk), we considered three groups of minerals that have been discussed widely by different researchers [2-9]:

- *Zirconium minerals*:
  - $ZrSiO_4$, zircon,
  - ZrO_2 (monoclinic), baddeleyite,
  - ZrO_2 (cubic), tazheranite;
- *Titanium minerals*:
  - $CaTiSiO_5$, titanite (sphene),
  - $CaTiO_3$, perovskite,
  - $CaZrTi_2O_7$, zirkelite (zirconolite),
  - TiO_2, rutile;
- *Phosphate minerals*:
  - $CePO_4$, monazite,
  - $Ca_5(PO_4)_3(F,Cl,OH)$, apatite,
  - YPO_4, xenotime.

We have found representatives of these groups – zircon, rutile, apatite – in samples of Nizhnekansky granites. However it would be wrong to focus only on these minerals for the following reasons:

- Zircon is the natural phase most resistant to chemical perturbations and irradiation effects, and capable of incorporating considerable quantities of tetravalent actinides (U, Pu, Np) into its crystalline lattice; however, its isomorphic capacity with regard to trivalent elements (Am, Cm) has not been studied.

- Rutile is characterised by both chemical stability and radiation resistance, but its isomorphic capacity with regard to actinides is insignificant.
- Although apatite is able to incorporate considerable quantities of heterovalent actinides isomorphically into its structure, its resistance to chemical and radiation impacts is not high compared with that of monazite.

Thus, there is a need to identify the best choice of additional host phases for which natural analogues were not found among accessory minerals of the Nizhnekansky massif. It is also important to continue to differentiate between minerals of different geochemical groups, in order to simplify the development of practical technological schemes. For example, the technology for producing phosphate ceramics is principally different from methods of obtaining crystalline matrices based an zirconium phases.

Among the stable minerals of zirconium, two forms of ZrO_2 - monoclinic (baddeleyite) and cubic (tazheranite), are acceptable for the immobilisation of actinides in granitic formations. Although ZrO_2 phases are not characteristic of granites, micro-inclusions of baddeleyite have often been discovered in crystals of accessory zircon separated from granite samples. The high isomorphic capacity of crystalline ZrO_2 with respect to heterovalent actinides makes this phase especially attractive for use in practical technologies.

The choice of titanium minerals, promising for use as actinide host phases which are geochemically compatible with granites of the Nizhnekansky massif, is somewhat more difficult to justify because, apart from rutile, these phases are not typical of granites. Nevertheless, there are no grounds for considering that zirkelite ($CaZrTi_2O_7$), sphene ($CaTiSiO_5$) and perovskite ($CaTiO_3$), should be significantly different from accessory minerals of the Nizhnekansky massif, in terms of geochemical behaviour. Although potentially all three titanium minerals could be used as host phases for actinides, zirkelite is particularly distinguishable in terms of (a) the level of research effort devoted to this mineral, (b) its high resistance to chemical perturbation and radiation, and (c) its isomorphic capacity.

Among phosphorus minerals, the best host phase for actinides is monazite ($CePO_4$), which is quite often found as a granitoid accessory mineral, although it has not been identified in granites of the Nizhnekansky massif.

6. The Choice of Host Phases for Iodine-129 Immobilisation

The problem of iodine-129 immobilisation is particularly complex, owing to the unusual geochemical behaviour of this element in nature.

Iodine is a widely-dispersed element and is relatively volatile. It concentrates mainly in residual salt solutions, in some species of algae, sponges and corals, in oil-bearing waters and in mineral formations of deserts. Iodine forms relatively few insoluble compounds and, therefore, is encountered infrequently as a mineral. The known iodates of Ca, Pb, Cu, are not chemically stable compounds. The more stable iodides of copper and silver also offer little promise for use as host phases.

Potentially, some chlorine-containing minerals (*e.g.* within the sodalite group $Na_8[AlSiO_4]_6Cl_2$) should be able to replace chlorine by iodine via isomorphic substitution. Of some interest is the structure of beryl ($Be_3Al_2[Si_6O_{18}]$) that can contain ions of Na, K, Cs (up to 7 % as oxides), as well as H_2O (up to 3 %) in the free channels formed by rings of silicon-oxygen tetrahedra.

Our attention was also attracted by the group of tourmaline minerals, especially those containing iodine [10]. Minerals united by the name tourmaline have a complex, multi-constituent stoichiometry, and may be represented by the general formula:

$$XY_3Z_6(BO_3)_3Si_6O_{18}(O, OH, F, Cl.....)_4$$

where: X = Na, Ca, K, Mg, Mn;
Y = Fe^{3+}, Fe^{2+}, Al, Ti, Mn, Li, Cr;
Z = Al, Fe^{3+}, Mg, V, Ti, Cr.

Tourmaline, with hardness 7-7.5 and density 2.9-3.3 g cm^{-3}, is a chemically stable mineral and does not decompose in acids. This mineral is typical of different kinds of granitic pegmatites. Some varieties of tourmaline are used as precious stones. During the destruction of host rocks, tourmaline accumulates in deposition zones, retaining its crystalline and optical properties. For example, in the island of Ceylon, precious, high-quality varieties of tourmaline are mined exclusively from deposition zones. An analysis of minerals from pegmatite veins revealed unexpectedly high contents of iodine in tourmalines (*Table 1* [10]).

The contents of iodine in these tourmalines – two orders of magnitude greater than its usual concentration in minerals, suggest the possibility of considering tourmaline as a concentrator of this element and, hence, a promising host phase for the immobilisation of iodine-129.

7. Conclusions

The investigations carried out by us lead to the following preliminary conclusions:

1. Separation of HLW into fractions allows maximum benefit from vitrification to be gained for the immobilisation of the caesium-strontium fraction. For such a waste form, geochemical compatibility between

TABLE 1. Iodine content in minerals of pegmatite veins ($\mu g\ g^{-1}$)

Mineral	Number of samples	Average content of iodine ($\mu g\ g^{-1}$)
Quartz	2	0.38
Microcline	1	0.36
Spodumene	8	0.38
Muscovite	5	0.40
Beryl	7	0.22
Tourmaline	15	17.36
Pollucite	1	0.36

glass blocks and host rock is not a problem owing to the (geologically) insignificant lifetime of this particular radioactive fraction.

2. The host phase(s) for the immobilisation of the HLW actinide fraction should be selected according to the geochemical characteristics of stable accessory minerals in the specific igneous rock. For example, for granitoid formations, 3 mineral groups are recommended, *viz.* the minerals of zirconium, titanium and phosphorus. Segregation into such groups will also simplify the development of a practical technology for synthesising ceramics based on the given host phases.
3. The following minerals are recommended as actinide host phases, being geochemically compatible with granitoids of the Nizhnekansky massif:

   - Zr - zircon ($ZrSiO_4$) + various forms of (ZrO_2);
   - Ti - zirkelite ($CaZrTi_2O_7$) + rutile (TiO_2);
   - P - monazite ($CePO_4$) [+ apatite $Ca_5(PO_4)_3(F,Cl,OH)$ possibly].

   During the actual synthesis of crystalline ceramics, combinations of mineral groups of zirconium and titanium are expected to be possible, *viz.*

   - zircon ($ZrSiO_4$) + sphene ($CaTiSiO_5$) + ZrO_2;
   - zirkelite ($CaZrTi_2O_7$) + rutile (TiO_2) + ZrO_2.

4. We consider that the tourmaline group represents the most promising host phase for the immobilisation of iodine-129 in granitoids.

Currently, experimental work is being performed on the synthesis of crystalline ceramics based on the above chosen host phases. At the V. G. Khlopin Radium Institute, a method is being developed for the synthesis of ceramics based on two co-existing host phases: zircon ($ZrSiO_4$) and ZrO_2 [6]. The synthesis of zirkelite ($CaZrTi_2O_7$) by the hot press method has been studied in detail and has even been carried out with the addition of curium [2]. Although monazite is quite easily synthesised in the form of powder by chemical methods, our experiments have so far failed to obtain a solid ceramic.

The synthesis of tourmaline has been achieved previously, although mainly under hydrothermal conditions, which is unacceptable for HLW immobilisation technologies. However, according to our estimates, there are no major obstacles for producing tourmaline via the hot press method. We recommend the use of sodium iodide in the starting blend for the subsequent incorporation of iodine in the tourmaline structure.

Acknowledgements

We are very grateful to Professor Vladimir Kirko (NIFTI, Krasnoyarsk) and Dr. Michael Stenhouse (QuantiSci, U.K.) for their support, and to Prof. German Kolonin (Institute of Geology, Geophysics and Mineralogy, Novosibirsk) for discussions and corrections to this paper.

References

1. GALKIN, B.YA, YESSIMANTOVSKIY, V.M., LAZAREV, L.N., *et al.* (1988) Extraction of caesium and strontium, REE and TRU from liquid volatile waste by means of an extractant based on a dicarboxylate of cobalt, in *Proc. International Solvent Extraction Conference ISEC-88*, Moscow.
2. WEBER, W.J., WALD, J.W., AND MATZKE, H. (1986) Effects of self-irradiation damage in Cm-doped $Gd_2Ti_2O_7$ and $CaZrTi_2O_7$, *Journal of Nuclear Materials* **138**, 196-209.
3. HEIMANN, R.B. AND VANDERGRAAF, T.T. (1988) Cubic zirconia as a candidate waste form for actinides: dissolution studies, *Journal of Material Science* **7**, 583-586.
4. BURAKOV, B.E. (1993) A study of high-uranium technogenous zircon $(Zr,U)SiO_4$ from Chernobyl "lavas" in connection with the problem of creating a crystalline matrix for high-level waste disposal, in *SAFE WASTE 93, Proc. International Conf.*, June 13-18, 1993, Avignon, (France), Vol. 2, pp. 19-28.
5. ANDERSON, E.B., BURAKOV, B.E., AND VASILIEV, V.G. (1993) The creation of a crystalline matrix for actinide waste in the Khlopin Radium Institute, in *SAFE WASTE 93, Proc. International Conf.*, June 13-18, 1993, Avignon, (France), Vol. 2, pp. 29-33.
6. BURAKOV, B.E., ANDERSON, E.B., PAZUKHIN, E.M., AND MIKHEEVA E.E. (1991) A method for obtaining crystalline zircon doped with isomorphous admixtures, Russian Patent Application No. 5020357/26, November 12, 1991 (in Russian).
7. EWING, R.C., LUTZE, W., AND WEBER, W.J. (1995) Zircon: a host-phase for the disposal of weapons plutonium, *Journal of Material Research* **10**, 243-246.
8. BURAKOV, B.E., ANDERSON, E.B., GALKIN, B.YA., STARCHENKO, V.A., AND VASILIEV, V.G. (1996) The crystalline host-phases for immobilization of weapons plu-

tonium and waste actinides, in E.R. Merz and C.E. Walter (eds.) *Disposal of Weapon Plutonium*, pp. 85- 89, Kluwer Academic Publishers, Dordrecht.
9. EWING, R.C., WEBER, W.J., AND LUTZE, W. (1996) Crystalline ceramics: waste forms for the disposal of weapons plutonium, in E.R. Merz and C.E. Walter (eds.) *Disposal of Weapon Plutonium*, pp. 65-83, Kluwer Academic Publishers, Dordrecht.
10. KRYLOVA, L.YA. (1978) Features of the geochemistry of iodine in rare-metal granite pegmatites, *Problems of Geochemistry and Typomorphism of Minerals, Leningrad*, Issue 2, pp. 36-41 (in Russian).

CAESIUM IMMOBILIZATION USING ZEOLITE-CONTAINING ROCKS AND HIGH TEMPERATURE TREATMENT

VALENTINA I. BOGDANOVA, B.A. FURSENKO, G.I. GALAI, I.A. BELITSKY, L.M. PREDEINA, V.S. PAVLYUCHENKO & I.V. DROBOT
Institute of Mineralogy and Petrography
Siberian Branch of the Russian Academy of Sciences
Novosibirsk, 630090, RUSSIA

Abstract.
A systematic study was undertaken of the chemical stability and structure of Cs-containing materials, prepared from Cs-saturated zeolite-containing rocks (ZCRs) by high temperature roasting, mainly in the absence of boron compounds.

Cs was introduced into the ZCRs via two procedures: (1) via an ion-exchange process between Cs in solution and the ZCRs; (2) by mixing ZCR powder with $CsNO_3$ solution. Cs-containing ZCRs prepared by the two methods were heated at various temperatures up to 1400 °C. The chemical stability of the resulting glass-crystalline material was determined by measuring the amount of Cs passing into solution at various pH levels ranging from 1 to 12. The experiments lasted from 1 to 28 or more days.

It was found that heating ZCR at 1000 °C results in the strong immobilisation of Cs, Al, Na, K, Mg regardless of the way in which Cs was introduced into the ZCR. The lowest quantities of leached cations Cs, Al, Si, Na, K, Ca, Mg were detected in neutral solutions. The amount of Cs leaching into solution at pH=5 from products roasted at tempreatures >1000 °C was $< 10^{-5}$ g cm^{-2} d−1 for the ZCR of all deposits studied. For the clay, the corresponding amount of Cs was about one order of magnitude higher. The crystal structure of the roasted material depends strongly on its Cs content: at a Cs content lower than 1 mg-equiv g^{-1}, the material comprised a predominantly amorphous phase; at higher Cs content, it was a mixture of amorphous and crystalline phases: ortho-rhombic ($CsAlSi_5O_{12}$) and/or cubic - pollucite ($CsAlSi_2O_6$). The quantity of crystalline phases was greater when the second procedure for introducing Cs into the rocks was used.

Thus, from the results obtained here, ZCRs are shown to be a good candidate for radioactive waste conditioning and final disposal. Such a wasteform may be used not only in radionuclide decontamination processes, but also as a flux additive resulting in the formation of chemically- and mechanically-resistant Cs-containing glass-crystalline materials.

M. J. Stenhouse and V. I. Kirko (eds.),
Defence Nuclear Waste Disposal in Russia: International Perspective, 69–84.

1. Introduction

According to [1,2], a number of countries have begun, or will soon start, operations towards the final disposal of both low-level (LLW) and intermediate-level (ILW) radioactive waste. By about 2010, a few countries plan to begin operations on the final disposal of high-level waste (HLW).

The currently accepted concept for radioactive waste disposal is deep disposal [3-5]. In addition, the presence of zeolitic tuff at a potential repository site has been recognised as an additional factor which is important for the safe isolation of radioactive waste [6].

Some scientists consider the geological formation itself to be the main barrier preventing the fast migration of radionuclides [4,5], whereas others emphasise the important role of the physico-chemical form in which the waste and its radionuclides are emplaced [3]. In reality, there are numerous inhomogeneities and fractures as well as the circulation of groundwater in rocks resulting from recent and contemporary tectonic activity, such that the migration of radionuclides may be unpredictable. Thus, for example, it was predicted [7] that contamination of groundwater and the Colombia river by radionuclides from the disposal of radioactive waste at Hanford would not occur for at least 175-180 years. Contrary to this prediction, however, traces of contamination were detected in the Colombia river even after 7-11 years.

Our previous results [8,9] also support the need for the reliable immobilisation of Cs and Sr prior to their disposal. It was shown experimentally that zeolite-containing rocks (ZCRs) safely retain absorbed Cs and Sr only in a neutral medium, around a pH of 5-7. Desorption of Cs and Sr from ZCRs increases rapidly with an increase in alkalinity and especially acidity. The concentration of Cs and Sr removed from a ZCR filter column under dynamic conditions at pH=4.5 reached 50 and 5 mg l^{-1}, respectively. It is worthwhile noting that the major component of activity of radioactive waste is attributable initially to Cs-137 and Sr-90 [1,10].

According to the generally accepted concept, the final disposal of only solid radioactive waste is allowable [11] in order to provide reliable immobilisation of radionuclides, including Cs. However, neither cement nor bitumen were believed by the authors of [12] to result in a stable material which was suitable for immobilisation of radionuclides, owing to their low chemical stability and mechanical strength and low thermal and radiation resistance. Bituminous materials may also be subject to bacterial degradation [2].

The immobilisation of Cs absorbed on clays by a high-temperature treatment was proposed in 1953 by Hatch [13]. At present, however, the only industrial-scale process which has been developed for radioactive waste immobilisation is the vitrification of radioactive waste into borosilicate glass [14]. The conversion of HLW into glass blocks is underway in a number of countries [15]. In 1981, the possibility of vitrifying ILW was reported [16] and this process was subsequently considered to be appropriate for both ILW and LLW [17-18].

Alternative materials such as ceramics, synthetic rocks *etc.* have been studied to a lesser extent. Such studies are mostly at the stage of preliminary investigations, for future development, of processes and materials which are more stable than borosilicate glass [19]. Another reason for seeking alternatives to borosilicate glass arises from the limited abundance of high-quality quartz sand [20] which is a key component of glass. Furthermore, the composition and properties of borosilicate glass are widely different from those of natural minerals.

A study concerning the utilisation of various natural materials for radioactive waste processing appears promising [21]. In this respect, natural sorbents are of particular interest. Such materials trap and confine radionuclides during the initial treatment stages, even before their conversion to ta vitrified form.

The aim of the present work, therefore, was to study the chemical stability (determined initially by the amount of Cs leaching into solutions at various pH) of glass-crystalline materials prepared by roasting Cs-saturated zeolite-containing rocks, and to determine the structure of these materials.

2. Experimental Study

2.1. INITIAL MATERIAL CHARACTERISATION

Zeolite-containing rocks of three main Siberian deposits: Kholinskoe (Buryatia region), Shivyrtuiskoe (Chita region) and Sakhaptinskoe (Krasnoyarsk region) were studied. The zeolite content in each rock was determined quantitatively by X-ray diffraction and the chemical composition was determined by X-ray fluorescence (analyst N.M. Glukhova, UIGGM, Novosibirsk).

The total cation-exchange capacity (CEC) was determined by measuring (a) the amount of absorbed ammonium ion (CEC_{NH_4}) and (b) the sum of exchanged cations ($CEC_{\Sigma cat}$; Σcat = Na, K, Ca, Mg). In both cases, special procedures were developed and officially approved as standard methods [22,23].

2.1.1. *Zeolite-containing rocks*

The main properties of ZCRs studied in this work were zeolite content and CEC (*Table 1*) and chemical composition (*Table 2*). The chemical composition of Obskaya clay (Novosibirsk region) is included in *Table 2* for comparison.

TABLE 1. Zeolite content and corresponding total cation-exchange capacity (CEC) in zeolite-containing rocks (ZCRs) studied.

Deposit	**Abbreviation**	**zeolite, weight %**	**CEC_{NH_4},**	**$CEC_{\Sigma cat}$,**
			mg-equiv. g^{-1}	
Kholinskoe (Buryatia)	ZCR-Kh	65	1.74	1.63
Shivyrtuiskoe (Chita region)	ZCR-Sh	73	1.85	1.85
Sakhaptinskoe (Krasnoyarsk region)	ZCR-Sakh	44–50	0.98	0.89

TABLE 2. Chemical composition of zeolite-containing rocks (ZCR) and Obskaya clay in wt. % of oxide determined by X-ray fluorescence analysis.

Sample	SiO_2	TiO_2	Al_2O_3	Fe_2O_3	MnO	MgO	CaO	Na_2O	K_2O	P_2O_5	*loss by roasting*	Σ
ZCR-Kh	67.46	0.12	12.44	1.11	0.10	0.42	2.33	0.66	4.39	0.03	10.85	99.88
ZCR-Sh	61.51	0.40	13.28	2.51	0.61	0.89	3.33	1.18	2.97	0.15	12.62	99.64
ZCR-Sakh	67.72	0.34	12.77	2.36	0.11	1.36	2.63	0.42	3.33	0.10	8.59	99.73
Obskaya Clay	60.07	1.53	23.29	1.56	0.09	0.61	0.48	0.17	0.74	0.05	11.33	99.93

2.2. PREPARATION OF CAESIUM-SATURATED ZCR

The Cs component was introduced into ZCR by two different procedures.

2.2.1. *Procedure 1*

Cs-containing ZCR was prepared by interacting ZCR with a solution of $CsNO_3$ over a sufficiently long time period (up to a few days), under both dynamic and static conditions, at room temperature. Experimental conditions were:

- solid:liquid ratio about 1:100;
- the weight of ZCR varied from 0.5 to 70 g;
- the grain fraction was always in the range 0.25-1.00 mm;
- Cs concentration in solution ranged from 0.1 to 0.5 M.

The total amount of Cs in solution was significantly higher (by about a factor of ten) than that corresponding to the total CEC of ZCR. The ZCR was washed in distilled water to remove non-exchanged Cs and subsequently dried in air at 100 °C – generally designated as Cs-ZCR[1].

2.2.2. *Procedure 2*

Cs-containing rock was prepared by mixing ZCR with a 1 M $CsNO_3$ solution. Experimental conditions were:

- rock:solution ratio varied from 2:1 to 1:3;
- the weight of ZCR was 3-10 g;
- the grain fraction was 0.25-1.00 mm (as in Procedure 1).

The $CsNO_3$ solution was added in such a way that liquid just covered the solid phase (if necessary liquid was added portion by portion). In addition to the original natural rocks, a 'treated' ZCR, prepared via Procedure 1, was used as starting material in Procedure 2.

The amount of Cs in each mixture prepared in this way did not exceed the maximum amount which could be absorbed by ZCR by 100% ion-exchange (about 2.5 mg-equiv g^{-1}). The mixture was dried in air at 100 °C (2–4 hours) and the resultant product was designated generally as Cs-ZCR*[2].

2.3. HIGH-TEMPERATURE TREATMENT OF CAESIUM-ZCR

The samples of Cs-containing ZCR prepared by the above procedures were subjected to high temperature treatment in the range 400 to 1400 °C. The typical exposure time was 30 min. after reaching a given temperature. A few experiments were carried out with the addition of boron compounds to Cs-ZCR at temperatures in the range 650-1000 °C. Powder of grain fraction 0.25-1.00 mm together with small columns were used for the high-temperature experiments. These columns, about 1 cm^3 in volume, were prepared from < 0.1 mm powder plus some water yielding a mass of about 1.5 g and surface area close to 5 cm^2. [3]

[1]For each sample, the abbreviated name of each deposit and its corresponding Cs content were also noted; for example, $Cs_{0.9}$-ZCR-Kh signified Cs-exchanged ZCR of the *Kh*olinskoe deposit containing 0.9 mg-equivalent of Cs per g of rock (mg-equiv g^{-1}).

[2]Additional identification for samples prepared using Procedure 2 included the name of the deposit and the amount of Cs added to the rock (in mg-equivalent g^{-1} rock). Thus, Cs-ZCR*-Kh-1 signifies 1 mg-equiv Cs added to each g of ZCR from the *Kh*olinskoe deposit. In those experiments where Cs was added to the Cs_x-ZCR (*i.e.* prepared by Procedure 1) the resulting product was designated as Cs_x-ZCR**. For example $Cs_{0.9}$-ZCR**-Kh-1 signifies that 1 mg-equiv Cs was added to 1 g of $Cs_{0.9}$-ZCR-Kh prepared by Procedure 1.

[3]Roasted materials, regardless of their structure, were named glass-crystallites (GC) and designated as Cs-GC or Cs-GC* according to the corresponding initial material (Cs-ZCR or Cs-ZCR*) followed by the roasting temperature. For example, Cs-GC*-Kh-1-1000 signifies glass-crystalline material prepared from Cs-ZCR*-Kh-1 at 1000 °C.

TABLE 3. Cation exchange capacity [CEC_{NH_4} and $CEC_{\Sigma cat+Cs}$] for Cs-containing zeolitic rocks prepared by Procedure 1 (Cs-ZCR) and Procedure 2 (Cs-ZCR*).

Sample	CEC_{NH_4}	$CEC_{\Sigma cat+Cs}$	difference, %
Cs-ZCR-Kh	1.37	1.24	-8
Cs-ZCR-Sh	1.40	1.53	9
Cs-ZCR-Sakh	1.01	1.01	0
Cs-ZCR*-Sh-1	1.23	2.20	79
Cs-ZCR*-Sakh-1	0.68	1.82	167
Cs-ZCR*-Sh-2	0.94	2.75	193
Cs-ZCR*-Sakh-2	0.57	2.30	309

The structure of glass-crystalline material was studied by X-ray powder diffraction (qualitative phase analysis). Chemical stability was determined by immersing 0.5 g powder (fraction 0.25-1.00 mm) or the whole column (see above) in 150 ml of solutions at pH values ranging from 1 to 12. Each experiment lasted 1 to 28 days (sometimes even more) and was carried out either at room temperature or in an oven at 50 °C. After specified time intervals, the cation content in solution was determined by chemical analysis. Cs, Na, K, Ca, Mg were determined by atomic absorption analysis, and Si and Al by spectrophotometry.

3. Results and Discussion

3.1. CHEMICAL STABILITY OF CAESIUM-SATURATED ZCR

The CEC_{NH_4} and $CEC_{\Sigma cat+Cs}$ values of the Cs-containing ZCR (Cs-ZCR) exchanged by Procedures 1 and 2 are presented in *Table 3*. The Cs-ZCR samples listed in *Table 3* were prepared under conditions which had been shown to provide the maximum possible saturation of ZCR by NH_4 [22]; the only difference was that the 1 M solution of NH_4^+ was replaced by 0.5 M solution of $CsNO_3$.

As shown in *Table 3*, the CEC_{NH_4} values for Cs-containing ZCRs are close to the $CEC_{\Sigma cat+Cs}$. For Cs-ZCR*, the $CEC_{\Sigma cat+Cs}$ is significantly higher than the CEC_{NH_4}, the difference being that in the former procedure,

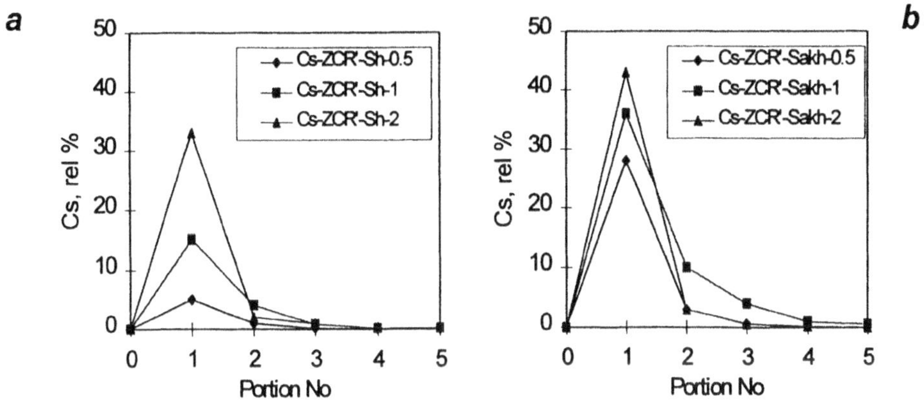

Figure 1. Amount of Cs which passed into sequential portions of distilled water washings from Cs-ZCR of the Shivyrtuiskoe (a) and Sakhaptinskoe (b) deposits, prepared by Procedure 2.

more $CsNO_3$ salt was added to the ZCR during preparation of the Cs-ZCR.

The fact that the CECs of Cs-ZCR determined by both methods are similar allows the assumption that equivalent ion exchange is occurring between NH_4 ions and Cs-ZCR cations, *i.e.* this result confirms that all Cs in Cs-ZCR was introduced by ion exchange.

In the case of Cs-ZCR*, it was difficult to judge the equivalency of the ion exchange by comparing CEC_{NH_4} and $CEC_{\Sigma cat+Cs}$ data since, unlike Cs-ZCR, Cs-ZCR* samples were not washed of excess $CsNO_3$ or cations Na, K, Ca, Mg, which could be replaced by Cs in ZCR, if equivalent ion exchange had occurred. Hence, the degree of Cs binding in Cs-ZCR* was determined by washing samples with five successive portions (100, 50, 50, 50, 50 ml) distilled water (at pH 5.7) (see *Figure 1*). The Cs which passed into solution during washing was considered to be non-bound Cs, whereas the Cs which did not pass into distilled water was taken arbitrarily as the bound component.

Figure 1 indicates that

1. Less than ~50% of Cs passed into solution. Thus, up to ~50% of Cs in Cs-ZCR* can be regarded as bound.
2. The less Cs added per g ZCR, the higher the quantity of bound Cs in Cs-ZCR*. The amount of bound Cs was larger in Cs-ZCR*-Sh than in Cs-ZCR*-Sakh for all values of added Cs (from 0.5 to 2.0 mg-equiv. g^{-1}). Such a difference is most pronounced for Cs-ZCR*-0.5.

From the data presented above, the conclusion can be drawn that the Cs in Cs-ZCR is bound - mainly as ion-exchanged form. In Cs-ZCR*, the Cs

TABLE 4. Amount of Cs and Al passed into solution at pH=1 in 1 day from Cs-containing glass-crystalline materials prepared by Procedure 1 and heated at 850 °C without boron (Cs-GC-850) and with boron additives (Cs-GC-B-850, grain fraction 0.25-1.00 mm).

Sample	Additive	Cs : ZCR to additive ratio	Amount in solution			
			Cs		Al	
			mg l^{-1}	rel. % *	mg l^{-1}	rel. % *
Cs-GC-Kh	–	1 : 0	177	100	7	100
Cs-GC-Sh	–	1 : 0	199	100	5	100
Cs-GC-B-Kh	B_2O_3	10 : 3	114	64	49	700
Cs-GC-B-Sh	B_2O_3	10 : 3	147	74	76	1520
Cs-GC-B-Kh	borax	10 : 3	213	120	86	1230
Cs-GC-B-Sh	borax	10 : 3	207	104	97	1940

*: Cs and Al concentration in solutions obtained with Cs-GC-850 (boron-free) was taken to be 100%.

exists both in bound and unbound form. It seems reasonable to assume that bound Cs in Cs-ZCR* is in ion-exchanged form as in the case of Cs-ZCR. The higher content of bound Cs in Cs-ZCR*-Sh compared with Cs-ZCR*-Sakh correlates well with the higher total CEC of the former and can be considered to provide good indirect evidence of the above assumption.

Hence, interaction of the Cs solution with an inorganic absorber having an ion-exchange function (ZCR in this case) results initially in the binding of some Cs by ion-exchange. This process would result, in turn, in a lower release of Cs to the gaseous phase during subsequent high temperature treatment.

3.2. CHEMICAL STABILITY OF MATERIALS AFTER HIGH-TEMPERATURE TREATMENT

3.2.1. *Experiments with boron additives*

Essentially no immobilisation of Cs was obtained by heating Cs-ZCR at 450–850 °C either with or without boron compounds. However, in the case where B_2O_3 was added to Cs-ZCR prior to heating to 850 °C, the amount of Cs leaching into solution from Cs-GC-B-850 was slightly lower compared with that of boron-free Cs-GC-850. At the same time, the presence of boron increased the mobility of aluminium significantly (see *Table 4*).

According to the X-ray diffraction patterns, both Cs-GC-B-850 and Cs-

GC-B-1000 were amorphous. The X-ray pattern of Cs-GC-850 (in the absence of boron) consisted of a halo corresponding to the amorphous phase, together with a few small peaks corresponding to (crystalline) clinoptilolite and opal-cristobalite.

No boron compounds were used in the subsequent experiments.

3.2.2. *Experiments without boron additives*

Heating Cs-ZCR, Cs-ZCR* and Cs-ZCR** having a Cs content up to 2 mg-equiv g^{-1} at 900 °C and especially at 1000 °C resulted in the strong immobilisation of Cs which was most pronounced in a neutral medium (see *Figure 2*).

Figure 2 indicates that:

- The smallest quantity of Cs leached was detected in solutions of pH in the range 5–7.
- Cs mobility in the acid medium is essentially higher for Cs-GC*-Clay ([**b**] curves) than for the Cs-ZCR*-Sh ([**a**] curves) of the same Cs content. This difference decreases as pH increases above 3.
- Cs differential leaching curves have a maximum at the first day.
- The amount of Cs passed into acid solution from Cs-GC*-2.5 ([**c**] curves) is essentially greater than that passed from Cs-GC*-1 ([**a**] curves).
- The amount of Cs passed into solution from pollucite ($CsAlSi_2O_6$, containing 33 weight% Cs or 2.5 mg-equiv g^{-1}; [**d**] curves) was slightly lower than that passed from Cs-GC*-2.5 ([**c**] curves). This result occurred despite the fact that the latter was shown by X-rays to consist of pollucite and contains nearly the same amount of Cs as pollucite. The result is apparently due to the relatively short heating time which was not sufficient to achieve complete crystallisation of Cs-GC*-2.5 material into pollucite.

Comparison of the differential leaching curves obtained for $Cs_{0.9}$-GC-1000 and Cs-GC*-1-1000 both of which contained nearly the same amount of Cs, shows that there is a tendency for a lesser amount of Cs leaching from Cs-GC* than from Cs-GC, especially in alkaline medium. This observation may be attributed to better Cs binding in crystalline phases which were shown to dominate in Cs-GC*.

The results of Cs leaching from the columns prepared according to procedure 2 by mixing $CsNO_3$ with ZCR and Cs-ZCR and roasted at 1000 °C, are presented in *Table 5*. The results indicate that stable Cs-containing matrixes were obtained in both cases. The amounts of Cs which passed into solution from Cs-GC' were slightly smaller than those from Cs-GC**. This result is probably associated with the higher degree of crystallisation detected in Cs-GC* compared with essentially amorphous Cs-GC**. An in-

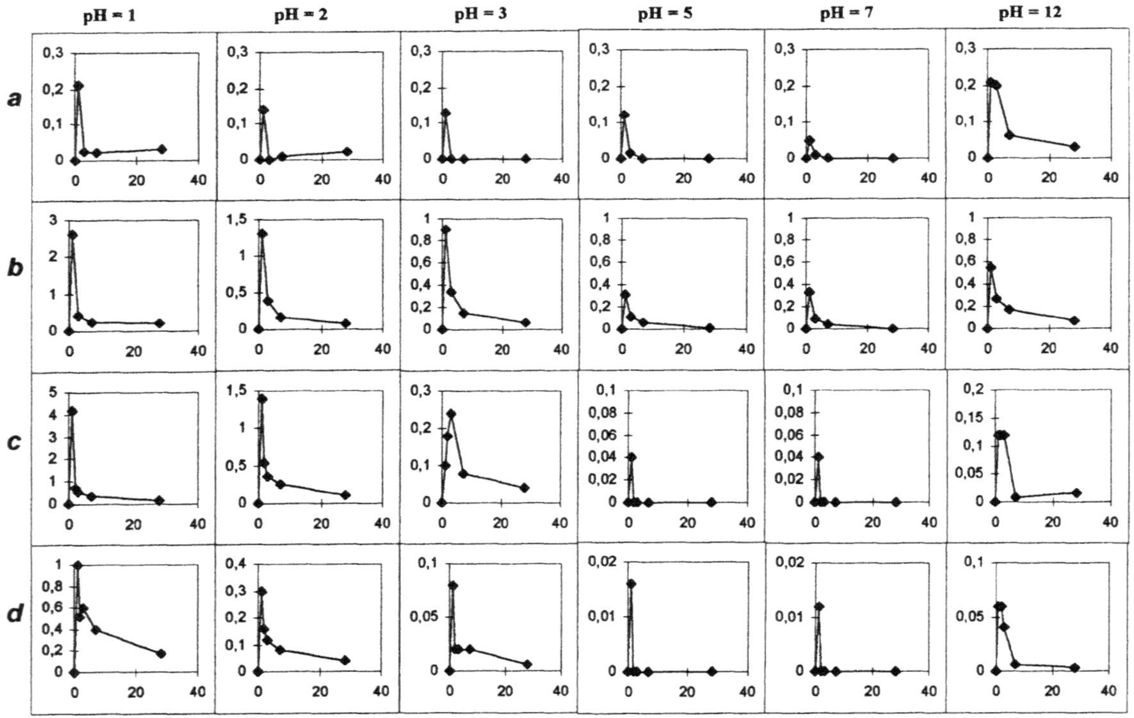

Figure 2. Differential curves of amount of Cs leaching into solutions at various pH values and at 17 °C

Cs-containing glass-crystalline (Cs-GC) materials (fraction 0.25-1.0) prepared at 1000 °C (quantity of Cs passed into solution during 1 day expressed as percent of initial Cs content in Cs-GC). X-axis: days, Y-axis: Cs, %.

[a] = Cs-GC*-Sh-1, [b] = Cs-GC*-Clay-1, [c] = Cs-GC*-Sh-2.5, [d] = pollucite. Note the different Y-axis scales at different pH.

TABLE 5. Amount of Cs passed into solution at pH=2 at 20 °C during first day from columns of Cs-GC*-1000 and Cs-GC**-1000 prepared using various initial samples.

Initial sample		Preparation procedure		Material obtained	
ZCR or Cs-ZCR	*Amount of Cs added, mg-equiv./g*	*Total Cs in a sample, mg in column*	*Sample designation*	Material *designation*	*Cs amount in solution, g cm^{-2} d^{-1} × 10^4*
ZCR-Kh	2	312	Cs-ZCR*-Kh-2	Cs-GC*-Kh-2	1.4
$Cs_{0.8}$-ZCR-Kh	1	321	$Cs_{0.8}$-ZCR**-Kh-1	$Cs_{0.8}$-GC**-Kh-1	3.6
ZCR-Sh	2	291	Cs-ZCR*-Sh-2	Cs-GC*-Sh-2	1.3
$Cs_{1.0}$-ZCR-Sh	1	274	$Cs_{1.0}$-ZCR**-Sh-1	$Cs_{1.0}$-GC**-Sh-1	4.2
ZCR-Sakh	2	308	Cs-ZCR*-Sakh-2	Cs-GC*-Sakh-2	2.7
$Cs_{0.3}$-ZCR-Sakh	1.5	318	$Cs_{0.3}$-ZCR**-Sakh-1.5	$Cs_{0.3}$-GC**-Sakh-1.5	2.6

crease in leaching temperature up to 50 °C results in an increased amount of Cs leaching into solution from all types of material (*cf.* differential leaching curves in *Figures 2* and *3*).

The total amount of Cs which passed into solution at pH=1 from $Cs_{0.9}$-GC-1000 at 25 °C (grain fraction 0.25–1.0 mm) after 63 days was found to be less than 3% of the total Cs per gram of GC. At 50 °C, the amount reached 12%.

In neutral solutions, the rate of Cs leaching from both powder (grain size fraction 0.25–1.0 mm) and ZCR columns was slow. Thus, Cs was shown to pass into solution at pH=5 at room temperature from Cs-GC-1000 and Cs-GC*-1000 at a rate $< 10^{-5}$ g cm^{-2} d^{-1}. The corresponding value for Cs-GC-Clay-1000 was about one order of magnitude higher.

Similarly, heating Cs-ZCR and Cs-ZCR* to 1000 °C resulted in the immobilisation of Al. Silicon mobility did not change significantly on heating to 1000 °C. In contrast to the results for Cs, no significant difference was observed in the amounts of Si and Al which passed into solutions at various pH values (1 to 12) from Cs-GC and Cs-GC* prepared using both clay and zeolite. The differential leaching curves of Al and Si also exhibited a maximum over the first day, which did not exceed 0.5% of their initial amount.

Heating to 1000 °C also resulted in the immobilisation of Na, K, and Mg. The extent of Ca immobilisation was incomplete; in the case of Cs-

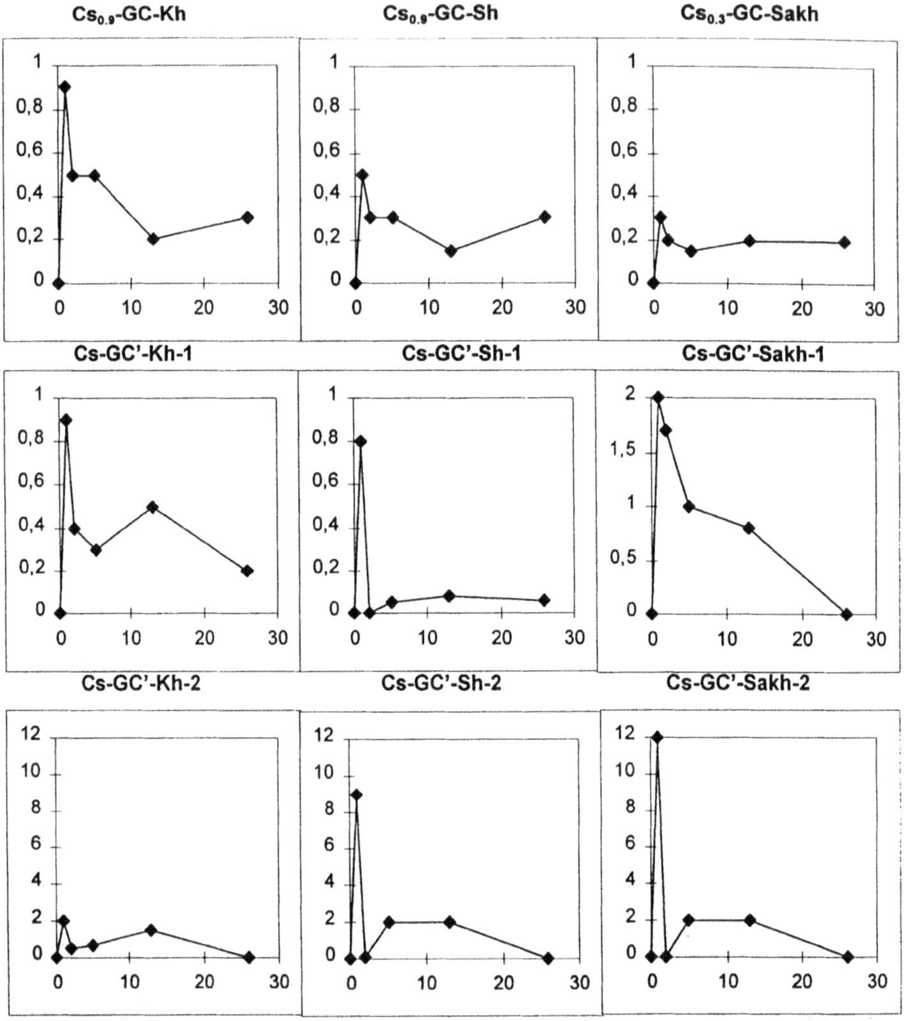

Figure 3. Differential curves of Cs leaching into solution at pH=1 at 50 °C from Cs-containing glass-crystalline (Cs-GC) materials (columns of 1.5g in mass) prepared at 1000 °C (expressed as an amount of Cs passed into solution during 1 day in percents of initial Cs content in Cs-GC); X-axis = days, Y-axis = Cs %.

GC-Sh-1000 and Cs-GC*-Sh-1000, about 50% of Ca passed into solution at pH=1, probably due to the presence of Ca compounds (*e.g.* $CaCO_3$) in ZCR-Sh, which decomposed during heating to 1000 °C, yielding CaO which is relatively soluble in acid media.

Heating Cs-ZCR and Cs-ZCR* to temperatures above 1000 °C resulted in the further immobilisation of all cations. The amounts of Cs, Si, Al, Na,

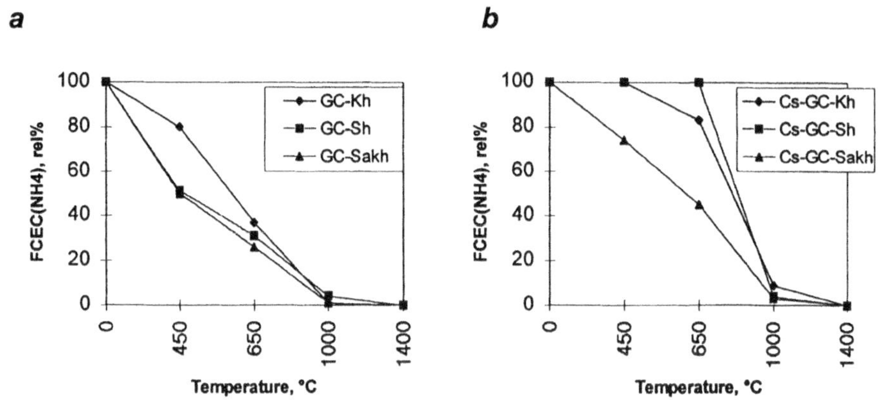

Figure 4. Full cation exchange capacity (CEC_{NH_4}) of (a) Cs-free and (b) Cs-containing glass-crystalline materials versus roasting temperature.

$FCECNH_4$ is expressed as a percent of $FCECNH_4$ measured on the corresponding ZCR or Cs-ZCR prior to heating. Kh – Kholinskoe deposit, Sh – Shivyrtuiskoe, Sakh – Sakhaptinskoe.

K, Ca, and Mg which passed into solutions at pH values in the range 1 to 12 (and even concentrated HCl) were below detection limits. The combined mobility of Cs, Na, K, Ca, Mg may be estimated by the values determined via CEC_{NH_4} [22] (*Figure 4*).

Figure 4 shows that heating Cs-ZCR at 1000 °C results in a drastic decrease in CEC_{NH_4}. The decrease in CEC_{NH_4} was slightly smaller for Cs-ZCR than for the original non-exchanged ZCR at the same roasting temperature. This result may be due to the greater thermostability of Cs-exchanged ZCR.

3.2.3. *Structure of roasted materials*

The X-ray diffraction patterns of roasted materials indicate that they consist either of amorphous or crystalline phases: orthorhombic ($CsAlSi_5O_{12}$) and/or cubic ($CsAlSi_2O_6$) – pollucite. These phases may coexist in various proportions. At present, it can only be stated that the total amount of Cs in a sample is one of the factors controlling the phase composition of material. At a Cs content < 1 mg-equiv g^{-1}, an amorphous phase was usually formed, whereas at higher Cs content, crystalline phases were formed, with or without an amorphous phase. Crystalline phase formation is favoured by following Procedure 2 to incorporate Cs into ZCR. It is quite possible that in this case, more favourable conditions were established for critical crystal nuclei formation owing to a less homogeneous distribution of Cs (compared with that obtained via Procedure 1, entirely by ion-exchange).

4. Conclusions

In summarising our experimental results, we conclude that:

1. Heating ZCR up to 1000 °C (regardless of the way in which Cs was incorporated) results in the strong immobilisation of Cs, Al, Na, K, and Mg. The immobilisation of Ca at this temperature was not complete in some cases (ZCR of Shivyrtuiskoe deposit). Heating up to 1000 °C has a minor effect on the Si mobility. The amounts of cations Cs, Al, Si, Na, K, Ca, and Mg passing into solution occurred from roasted material in neutral media.
2. The amount of Cs which passed into solution at pH=5 from materials roasted at temperatures > 1000 °C for the ZCR of all deposits studied was $< 10^{-5}$ g cm^{-2} d^{-1}. For the clay, the corresponding amount was about one order of magnitude higher.
3. The chemical stability of materials prepared by Procedure 2 using ion-exchanged Cs-ZCR (prepared by Procedure 1) was almost the same as that of materials prepared from the original ZCR.
4. The crystal structure of roasted material depends on its Cs content: at a Cs content < 1 mg-equiv g^{-1}, the material was predominantly amorphous; at higher Cs contents, a mixture of amorphous and crystalline phases was produced. The crystalline component was greater when Procedure 2 was used for Cs incorporation in the rocks.

The experimental procedure described above for Cs-ZCR* preparation (*i.e.* Procedure 1) followed by high-temperature treatment can be considered as a model of a one-stage treatment for processing Cs-containing radioactive waste [14]. The preparation of Cs-ZCR** followed by roasting can be considered as a model for the utilisation of ZCR which was used initially as ion-exchangers (*e.g.* in filtration columns) and now is used as a flux additive. The latter approach to ZCR utilisation appears to be more worthwhile because the CEC of ZCR cannot always be fully realised due to the relatively slow kinetics of ion-exchange.

High-temperature (above 1000 °C) roasting of Cs-ZCR yields matrices in which Cs is confined reliably regardless of the initial procedure for incorporating Cs in the rock. The amount of Cs passing into solution from roasted glass-crystalline materials containing up to 2 mg-equiv g^{-1} Cs was $< 10^{-5}$ g cm^{-2} d^{-1} in neutral medium (pH=5–7).

The above findings support the use of ZCR to process Cs-containing waste, both as decontaminating agent and as a flux additive. In the latter case, the ion-exchange capacity and mechanical properties of the rock are of less importance. In addition to the original ZCR, one can utilise the 'spent' Cs-ZCR, previously used as an ion exchanger. The large world resources of ZCR - more than $3{\cdot}10^9$ Mg, is an additional justification for using this material for the treatment and conditioning of Cs-containing waste.

References

1. LIN, K.H. (1991) An overview of radioactive waste management technology development, Current status and trends, *J. Environ. Sci. Health*, **A26**, (3), 373-393.
2. IAEA (1995) Handling of radioactive waste, IAEA Publication, Vienna, 42pp. (in Russian).
3. NIKIFOROV, A.S. (1990) Technical policy of radioactive wastes handling in the USSR, *Teploenergetika*, 1990 (8), 37-41 (in Russian).
4. KEDROVSKII, O.L., SHISHIC, I.YU., GUPALO, T.A. *et al.* (1991) Establishment of isolation requirements for high-level radioactive wastes in geological formations, *Atomnaya energiya*, **70**, (5), 294-297 (in Russian).
5. KEDROVSKII, O.L., RYBALCHENKO, A.I., PIMENOV, N.K. *et al.* (1991) Deep sited disposal of liquid radioactive wastes in porous geological formations, *Atomnaya energiya* **70**, (5), 298-300 (in Russian).
6. DUBLYANSKY, YU, LAPIN, B.N., REUTSKI, V. *et al.* (1996) Geological history of Yucca Mountain (Nevada) and the problem of the high-level nuclear waste repository, Abstracts of International Workshop Meeting *Defence Nuclear Waste Disposal in Russia*, 24-26 June, 1996, Krasnoyarsk, pp. 194-195.
7. ENERGIYA (1992) U.S.A. Program on nuclear military plant waste removal *Energiya: ekonomika-technika-ekologiya*, 1992 (2), 47-49 (in Russian).
8. BELITSKY, I.A., BOGDANOVA, V.I., AND FURSENKO, B.A. (1992) Ecological and chemical-technical problems of possible utilisation of zeolite-containing rocks for the recovery of territories subjected to radioactive pollution, in *Prirodnye Zeolity Rossii*, Abstracts of Republican Conference, November 25-27, 1991, Novosibirsk, 49-54 (in Russian).
9. BOGDANOVA, V.I., BELITSKY, I.A., FURSENKO, B.A. *et al.* (1996) Structure and chemical resistance of thermally treated zeolite-containing rocks and the dependence on the method of introducing Cs before heating, *Proceedings 13th Conference on Experimental Mineralogy*, Sept 12-15, 1995, Chernogolovka, in press (in Russian).
10. STEFANOVSKY, S.V. (1993) Immobilisation of sulphate-containing radioactive waste in glass, *Physika i khimiya obrabotki materialov* 1993 (2), 63-64 (in Russian).
11. SOBOLEV, I.A., KORENKOV, I.L., KHOMCHIK, L.M. *et al.* (1989) *Environmental Protection During Radioactive Waste Treatment*, Energoatomizdat, Moscow, 154 pp. (in Russian).
12. DIMITRIEV, S.A., STEFANOVSKY, S.V., KNYAZEV, I.A. *et al.* (1993) Plasma-chemical processing of radioactive waste, *Physika i khimiya obrabotki materialov* 1993 (4), 65-74 (in Russian).
13. HATCH, L.P. (1953) Ultimate disposal of radioactive wastes, *Am. Scientist* **41**, 410.
14. BAER, V. (1989) Industrial processes of high-level waste solution vitrification, *IAEA Bulletin* **31** (4), 47-51.
15. ZHU, G.L. AND CHAN, CH.Y. (1989) Radioactive waste treatment: international review (contemporary tendency and development), *IAEA Bulletin* **31** (4), 5-14.
16. ZAKHARENKO, V.N., KULICHENKO, V.V., AND KRYLOVA, N.V. (1981) On the possibility and conditions of intermediate-level waste vitrification, in *Issledovaniya v oblasti pererabotki obluchennogo topliva i obezvrezhivaniya radioactivnych otchodov*, Prague, ChSK AE, Volume 2, pp. 96-103 (in Russian).
17. STEPANOVSKY, S.V., SOBOLEV, I.A., LIFANOV, F.A. *et al.* (1988) Microstructure and chemical resistance of glass containing intermediate-level waste, *Radiokhimiya* **30**, (6), 820-824 (in Russian).

18. MOLNAR, L., LUKAVSKY, J., VANCL, V. *et al.* (1989) Present state of nuclear reactor waste treatment in Czechoslovakia, in *Management of Low and Intermediate Level Radioactive Waste*, Vienna, Volume **1**, pp. 87-95.
19. LAVEROV, N.P., KANCEL', A.V., LISICIN, A.K. *et al.* (1991) Basic problems of radioecology in connection with radioactive waste disposal, *Atomnaya energiya* **71** (6), 523-534. (in Russian).
20. BABAEV, M.K. AND GANBAR, D.M. (1989) The prospects of utilisation of mordenite raw material for preparing glass-crystalline materials, *Doklady AN Azerbaidzhan SSR* **45**, (10), 36-39 (in Russian).
21. KONUKHIN, V.P. AND KOMLEV, V.N. (1995) *Nuclear Technologies and the Ecosphere*, Apatity, 335p (in Russian).
22. BOGDANOVA, V.I., BELITSKY, I.A., PREDEINA, L.M. *et al.* (1993) Cation-exchange capacity determination of zeolite-containing rock by absorbed ammonium, *Institute of Mineralogy and Petrography Instruction No. 24*, Novosibirsk, 18 pp.
23. BOGDANOVA V.I. BELITSKY, I.A., PREDEINA, L.M. *et al.* (1993) Cation-exchange capacity determination of zeolite-containing rock by the sum of exchanged cations removed, *Institute of Mineralogy and Petrography Instruction No. 25*, Novosibirsk, 16 pp.

USE OF NATURAL MATERIALS FROM NORTHERN RUSSIA FOR THE ISOLATION OF RADIOACTIVE WASTES AND SPENT NUCLEAR FUEL

V. N. KOMLEV
Mining Institute of the Kola Scientific Centre
Russian Academy of Sciences
24 Fersman str., 184200 Apatity
Murmansk Region, RUSSIA

Abstract.
The application of natural materials to the isolation of radioactive waste and spent nuclear fuel is being assessed, together with possible isolation technologies. The operational requirements for such materials are identified and a proposal for an inter-regional ecological and technological project is discussed.

1. Statement of the Problem

The situation concerning the management of radioactive waste and spent nuclear fuel in Russia requires a relatively straightforward, economical, and technologically feasible solution which guarantees the long-term isolation of radioactive material from the biosphere. Materials 'created' by nature largely fulfil such requirements.

Technologies for radioactive waste and spent fuel isolation can be modified to incorporate relatively inexpensive natural materials of the European North of Russia, thereby performing several functions, *viz.* by:

- providing a barrier at the base of the repository,
- filling the voids between individual packages of conditioned or unconditioned radioactive waste or spent fuel, as well as the voids between waste packages and the walls of the repository;
- fulfilling a similar purpose when emplacing highly radioactive and hazardous waste packages outside the normal repository zone;
- completely filling the voids and shafts of underground excavations at repository closure;

M. J. Stenhouse and V. I. Kirko (eds.),
Defence Nuclear Waste Disposal in Russia: International Perspective, 85–98.

- completely filling surface trench-type repositories by forming a 'green lawn';
- forming a water isolation or filtration barrier within the host rock for a deep repository (or soils in the case of a near-surface repository),
- fulfilling a similar purpose for the cavities created by underground nuclear explosions;
- incorporating radionuclides in rock-like matrices or making containers for radioactive waste materials based on such matrices;
- containing specific radioactive wastes, *e.g.* the emergency discharges of surface or sub-surface nuclear plants (steam-gas mixture, water, reactor core melt) with the help of geotechnical barriers.

Naturally, making use of all of these, or similar applications requires the development and technical specification of a number of different types of material. Preferably, such projects should be carried out in the context of establishing a regional radioactive waste and spent fuel repository for the north-west region of the Russian Federation, taking into consideration a possible site location (Novaya Zemlya or Kola peninsula), the conditions for its development, as well as taking into account the construction of an underground nuclear power plant in the town of Apatity.

The use of similar local natural materials would allow, for example, the possibility of siting, for safety reasons, a deep repository close to the surface disposal site of the All-Russian Research and Design Institute for Integrated Power Technology (VNJPJEhT), or allow a combined repository for spent fuel, the reactor-compartments of nuclear submarines and solid radioactive waste on a coastal site of the Kola peninsula (Murmansk, *Polyarnaya Pravda*, V. A. Perovsky, May 29, 1996), thereby maintaining the obvious advantages of simplicity and economy.

An evaluation of the availability and characteristics of local natural materials for constructing protective engineered barriers for the isolation of radioactive wastes and spent nuclear fuel, is envisaged in the project, agreed in 1996, involving the VNJPJEhT and the Mining Institute of the Kola Scientific Centre (KSCMI) addressing, as one of the tasks, the preparation of a technical and economic basis for managing the disposal of radioactive waste and spent nuclear fuel in the region.

The list of possible applications described above is probably incomplete. Completion of all of those mentioned is quite difficult, however, considering the substantial needs of the NW region and Russia as a whole, as well as the fact that requirements are not always standard.

Practical examples of the application of natural materials are well-known throughout the world. According to disposal concepts in Germany, Sweden and Finland, once waste packages have been emplaced in underground radioactive waste repositories, mine workings and tunnels are back-filled with various natural materials [1-4]. In France, trench-type repositories are covered with gravel, sand and clay [3]. The U.S.A. uses natural materials to isolate radioactive wastes in pits or trenches, including the disposal of reactor compartments of nuclear submarines in pits at the Hanford site.

The void spaces of reactor-compartments (reactors) of Soviet nuclear submarines containing the undischarged core, were filled with either an organic caustic agent based on furfurole, or with steel and concrete, prior to flooding. According to one conclusion of A.V. Yablokov's commission, sea water destroys the majority of metallic containers within 10 years, and structural concrete containers within 30 years. Specialists, in developing plans to isolate the sunken nuclear submarine *Komsomolets*, decided against filling up individual submarine parts with concrete; however, a method of filling compartments with radionuclide sorbents, including natural ones, as part of a polymerising gel may be chosen.

Similar processes are adopted in Germany for the surface disposal of a wide range of toxic industrial wastes. German specialists use natural materials as an isolation barrier for the base and sides of disposal sites, creating 'walls of soil' [5]. The trench repositories for solid radioactive waste and the ponds/lakes which serve as catchment basins for liquid radioactive waste at industrial sites of some nuclear centres of Russia, are filled with clay and soil. Almost half of the research funding of the International Stripa Project focussed on an investigation of natural isolating materials [2].

Solidification of radioactive waste in rock-like matrices is illustrated by the SYNROC technology of Australia, solidification of plutonium-liquid wastes in tanks in Japan [6], solidification of liquid radioactive waste injected into the disposal horizons, into a rock-like mass under disposal conditions at the Integrated Mining and Chemical Company (MCC) in Krasnoyarsk [7], the binding of liquid radioactive waste without containers in the former German Democratic Republic directly at the discharge-point in mines of the repository Morsleben [8], and by other wasteforms developed in the U.S.A., the former U.S.S.R. and Czechoslovakia [2].

In addition, we can expect that in the near future, the development of technologies for fabricating spent fuel canisters in preparation for a repository [9], will generate the need for long-lived (*i.e.* of the order of hundreds of thousand of years), durable and inexpensive containers for the disposal of plutonium and highly-concentrated uranium from nuclear military stockpiles, including gamma-active containers, in meeting the strict requirements of the principle of non-proliferation of nuclear weapons, as well as 'the standard of spent fuel' [10,11]. This need may well be satisfied by the application of processes for making rock-like matrices [12] and/or glass-ceramics [13].

Geotechnical systems have been established in Sweden, France, U.S.A.,

U.S.S.R. and other countries cited in [14], developed or designed on the basis of the application of rocks and minerals to the containment of emergency discharges from nuclear stations. In the tragic Chernobyl accident, in addition to adding lead, caoutchouc, trisodium phosphate, and polymerisation liquids, boron carbide, dolomite, sand and clay were used to fill up the destroyed reactor (*Energiya*, 1996, **N4**). In 1996, as part of the initial work in the project to construct an underground nuclear station at MCC, Krasnoyarsk, via an agreement between the KSCMI and VNJPJEhT in Moscow, construction arrangements and volume-planning decisions are being optimised in order to make recommendations concerning the use of geotechnical barriers for localising water-transported radionuclides, to evaluate the methods of localising core-melt using natural materials, and to make recommendations concerning the construction of corresponding "traps".

For various applications such as those described above or others of a similar nature, a number of operating requirements must be imposed on the selection of natural materials:

- stability to radiation;
- ability to contain radionuclides, including sorption during the purification of draining water or escaping gases;
- water exclusion;
- prevention of gas discharge or escape;
- cooling or heat-dissipation;
- protection in the context of nuclear safety,
- ability for mechanised, automatic transportation to the 'hot' radioactive zone ;
- solidification of liquids (chemically) without using heat;
- sintering or alloying into ceramics or glass;
- fire-proof;
- compatibility with construction-materials of artificial origin and with the geosphere;
- corrosion stability.

The water-isolation characteristics and the ability of materials to retain radionuclides as they migrate, are the most important properties, even if water eventually reaches the radioactive waste. Protective engineering barriers based on natural materials should adhere to the principle:

Water can get into the disposal area (repository or nuclear test cavity) but radionuclides must not leave it.

In general, in the region under consideration, vermiculite and alkaline nepheline-containing rocks enriched with rare-earth minerals meet the design requirements best of all. Vermiculite and rare-earth minerals of the local rocks are efficient with respect to the sorption of radionuclides from percolating water and from escaping gases. Gamma-irradiation and heating do not reduce, but generally improve, the sorption characteristics of minerals in the relevant radiation and temperature ranges (up to 10 MGy and 80 °C). Minerals effectively retain radionuclides by delocalising influences. After gamma-irradiation, nepheline is able to bind additional water, thereby displaying water isolation properties.

The disaggregated state of the materials allows their easy transportation, for example, in a stream of liquid or gas, and allows gases to escape as required. Furthermore, the presence of bound water in vermiculite provides for the cooling of radioactive waste and spent nuclear fuel under special circumstances attributed to the endothermic effects of dehydration. The major minerals of prospective alkaline rocks (except nepheline) have sufficient radiation stability and the strong mechanical stability of natural materials as a whole. The presence in rocks of elements able to absorb thermal neutrons can achieve nuclear safety with respect to isolated radioactive waste and spent fuel. Vermiculite and nepheline barely interact with acids. Nepheline-containing natural materials offer the potential for making low-temperature glass ceramics having a strong tendency to isolate radionuclides.

Most of the physical and chemical characteristics of minerals described above have been characterised in detail [2,15]. We focus here on some technological aspects of obtaining the characteristics required to meet the required function.

2. Filling to Required Capacity with Dispersed Solid Mixtures or Slurries of Natural Materials

Practical examples exist where a dry minerals mixture has been fed into an isolation zone. A process for preparing and transporting pulp or the necessary solutions is carried out routinely for cement-containing mixtures and the floating maintenance base *Lepse* [16]. In general, there is no reason why, in principle, such schemes for feeding water-thin dispersion-mineral pulps cannot be achieved. We note in passing that natural aluminosilicate sorbents can be considered as analogues of traditional cements. Such materials are suitable for solidification as alkaline- and alkaline/alkaline-earth aluminosilicate astringents (geocements) under normal humidity conditions, and under conditions of drying or autoclaving [17,18]. Using either nepheline concentrates directly or the products obtained from processing them via either an acid technique or a method of sintering with limestone, technologists have obtained binding materials which can harden water-containing mixtures to a compressive strength 18-30 MPa after hardening for 30 days [19].

In general, the typical feature of either the acid- or alkali-processing of nepheline (possibly also vermiculite) is the production of materials exhibiting astringent and sorbent properties.

The presence of aluminosilicates in contact with concrete structures or concrete containers will favour the development of alkaline conditions, advantageous in the long-term. Such conditions are experienced widely in general building practice [20] and are part of the near-field design concept, *e.g.* in the proposed repository at Sellafield (U.K.) achieved in part by the emplacement of concrete radioactive waste containers.

Similar technologies for preparing and transporting by pipe dispersed materials in dry condition as well as pulp, have been tested in mining and underground buildings [21]; for example, the industrial application of methods of filling excavated spaces in mines using solidifying and slurries of alkaline nepheline-containing and vermiculite-containing materials. Extensive experience indicates that such a system is effective for annual feed volumes up to 20,000–25,000 m^3 and horizontal feed distances up to 400–500 m [19].

Useful experience can be derived in the field of pipe transportation of acid pulps via the geotechnical application involving the underground leaching of uranium ores [22].

Other methods of transporting mixtures and filling a required volume also exist, including routine operations in mining, concentrating mineral ores and construction. These operations typically generate multiple products, including ones containing nepheline and vermiculite.

3. High-Temperature and Low-Temperature Rock-like Materials

R.V. Bogdanov (Radiochemistry Department of the Chemistry Research Institute, St. Petersburg State University), working with specialists of the M.V. Lomonosov Porcelain Facilities (St. Petersburg), has successfully developed an experimental-production facility for fabricating ceramic containers from natural materials (including nepheline-containing materials) from the NW region of the Russian Federation, which are suitable for the long-term (*i.e.* equivalent to geological timescales) isolation of spent nuclear fuel fragments.

A.V. Kossinskaya has tested a stone-casting process for the fabrication of full-size containers for radioactive waste, including the use of raw materials from the region under consideration. Corresponding industrial-scale production facilities are available at the Kondopoga stone-casting works (Karelia), and at the Pervouralsk plant for repairing mining equipment [12]. According to data obtained by E.K. Nazimova (Institute of Chemistry, KSC), the experimental- and industrial-scale stone-casting based on a nepheline-containing charge, has been achieved at a number of Research Institutes as well as at the Borsk glass, Saransk electric-bulb and Kondopoga stone-casting facilities. Nepheline glass and nepheline ceramics, suitable as various technical materials, were produced at some facilities in Russia as far back as the 1930's.

By interacting nepheline with nitric acid (aluminium nitrate) under conditions of heating at a rate of 13 °C per minute, the process of nepheline glass formation is complete at a temperature of 530–648 °C [23].

Radioactive waste containers for a repository for, *e.g.* plutonium-containing materials, in accordance with conditions which meet the non-proliferation principle in the best possible way, could be fabricated using 'spent' (saturated with radionuclides) nepheline or vermiculite sorbent thereby fulfilling a double function. Prototypes for making ceramic or stone-cast containers can be found in the manufature of containers for radioactive waste and spent nuclear fuel in Germany.

The use of vermiculite and nepheline may allow the incorporation of radioactive waste during conditioning without the need for high-temperature processing, analogous, for example, to the production of heat-resistant, bakeproof and relatively durable ceramic-vermiculite plates produced at the Chelyabinsk Metallurgical Works [241. Acid modification of minerals may fulfil a positive function in this context.

4. Technologies for the Decontamination and Recycling of Products

Based on the application of vermiculite and nepheline, it is possible to anticipate a number of dual-purpose processes suitable for radiochemical and/or petrochemical production, *viz.* technologies for the purification, condensation or solidification of acid solutions (including nitric acid) of liquid radioactive waste of various concentrations, of oil products and other organic compounds, as well as techniques for fixing dust-generating surfaces, for reducing the permeability of the bases of slurry ponds and sedimentation tanks, and for remediating sites contaminated with radionuclides or oil products.

Several applications exist involving analogous, though inactive materials and processes, *e.g.* silicon-containing products produced from the above-mentioned minerals have been successfully tested under experimental conditions as well as being exploited commercially in mining companies and at plants manufacturing oil products in the Kola peninsula (V.J. Zakharov, D.S. Podozersky, A.P. Zossin, T.J. Priymak, S.P. Mesyats *et al.*).

Apart from silica being produced from the interaction of nepheline with any of the principal mineral acids, other byproducts of practical importance are alumino-phosphate bands (interaction with phosphoric acid) and various coagulants for the purification of drinking water and sewage (interaction with sulphuric and hydrochloric acids) [23].

As an example, we cite a process for fixing dust-generating surfaces of slurry ponds, based on the use of products from the acid processing of nepheline. During a snow-free period, under real conditions of alternating wetting and drying cycles, with a wind speed of 30 m s^{-1}, the protective covering preserves stability for loadings up to 5-10 kg cm^{-2}, retaining dust, but allowing atmospheric precipitation to drain away [25].

In the uranium-extraction industry of the former U.S.S.R., isolation of dust-generating slurry ponds was achieved by covering them with layers of inert recycled rocks of total thickness 3 m [2,26]. Crushed rocks containing the minerals discussed above or derivatives of them, could serve the basis for such layers, or one component of them at least, and form a cover for the dust-generating surfaces of open areas for containing liquid radioactive waste.

There is a known adverse effect associated with the spontaneous condensation and solidification of liquid radioactive waste with polymerising silicon acid in the full-scale repository at the MCC, Krasnoyarsk [27]. Such a process could, under certain circumstances, be beneficial, however. The production of a gel from the reaction of silicon acid with nepheline and vermiculite is not difficult and has been obtained under industrial conditions [23]. Based on the substantial experience involving the blocking of pores owing to rock silicatisation, such an effect could be used to advantage for the purposes of radionuclide fixation and water isolation [28].

In the collector-horizons of liquid radioactive waste repositories at MCC and SCC, interactions of radioactive waste with rocks of the collector-horizon result in the retention of radionuclides by the solid phase and, hence, in the self-isolation of wastes within the given horizon [29,30]. Similarly, wastes draining through underground layers at the Mayak site also interact with mineral components [31]. However, at the Mayak site, interactions occurring in the base of the Staroye Boloto reservoir cause less pollution of the groundwater than at lake Karachay although the former contains more activity [32]. In the latter case, therefore, the deliberate placement of nepheline- or vermiculite-containing rocks, or derivatives of them, in contact with liquid radioactive waste can make interactions even more effective, thereby increasing the natural protection offered by the binding properties of rocks containing radioactive waste.

It has been established conclusively that the production of gel-formed products by the acid decomposition of nepheline (by the contact of nepheline with the acid solutions) is an irreversible process which severely disrupts the ability to distinguish and separate individual components, mainly because of the poor filtration (water-yield) of pulps produced. Specialists have tried to remove the gel or make it water-repellent using various types of physical and chemical influences (chemical stimulation of coagulation, ultrasound, change of temperature over a wide range, pressure, *etc.*). All these influences yield precipitates with a water content no less than 72%, however. Although this situation is unfavourable for normal nepheline-processing technologies [23], the unusual, constant ability to bind water could be a benefit in certain circumstances.

Processes for purifying solutions and coagulating precipitates using ferrocyanide- or oxihydrate-sorbing precipitating agents, are used widely in radiochemical industries in Russia and abroad, as well as in other industries [32-35]. On the other hand, the corresponding potential of products from the acidification (complete or incomplete) of aluminosilicates discussed above is increasing [23,34].

5. Conclusions and Recommendations

From the previous discussion, applications of most of the possible types of natural materials (or their derivatives) considered here in the context of radioactive and hazardous nuclear materials, either have been tested throughout various branches of industry, in stages or as a whole, or such tests are under development and are not expected to present any difficulties. Many applications involving, for example, nepheline and its derivatives, have been identified already [36,37]. Thus, a familiarity exists with inactive 'nepheline'-materials, for which technologies are well established and have been well tested. This background offers promise for their successful application to the isolation of radionuclides under similar production conditions.

Vermiculite and alkaline nepheline-containing rocks are mined in sufficient quantities in the Murmansk region. In addition, saponite, a waste product of the Pomorsk Mining Company responsible for developing a large diamond deposit in the Arkhangelsk region, can probably serve as a potential isolation material in the NW region of Russia. Saponite is a mineral of the montmorillonite group, and has bentonite-like properties.

Modification of the final materials is possible in order to accommodate the mineral-based raw materials local to other regions of Russia; in particular, the raw materials mined near the large nuclear centres of Mayak (Chelyabinsk), MCC (Krasnoyarsk) and SCC (Tomsk). Such raw materials include vermiculite of the Potaninskoye deposit, nepheline of the Kusnetskoye Alatau, of the Vishnyovye and Ilmenskiye Mountains. Similarly, in the far eastern region of Russia, the potential for radioactive waste and spent nuclear fuel isolation is offered by vermiculite of the Koksharovskoye and Tatyanovskoye deposits in the Primorsky Krai. The nepheline (Zhdanovskoye) and vermiculite (Volyn) deposits are located in the Ukraine and also in the majority of countries which have well-developed nuclear technologies.

The use of nepheline and vermiculite can allow the unification of industries associated with radionuclide-isolation, involving the same type of application. This uniformity in processing can be applied to wastes arising in the north-west and far eastern regions of Russia; at all three abovementioned nuclear centres of Russia, and at the Murmansk, Chelyabinsk, Ekaterinburg, Novosibirsk, Khabarovsk groups of special companies *Radon*. The application could possible be extended to the radioactive waste repositories in the Ukraine.

In carrying out experiments, it is expedient to consider what wastes can be isolated with the recommended natural materials, *viz.*:

- liquid radioactive waste from radiochemistry plants in natural or artificial storage facilities, including the underground collector-formations;
- unprocessed spent fuel, including damaged fuel onboard floating maintenance bases and in onshore storage facilities of the Navy and nuclear ice-breaker fleet;
- fission materials from dismantled nuclear military loads requiring long-term storage,
- damaged reactors from nuclear submarines with undischarged cores,
- solid radioactive waste dumps outside civilian storage facilities, for which liquefaction followed by waste-transfer by pipe is difficult;
- other nuclear and hazardous radioactive materials.

Raw materials, methods and equipment exist for such experiments, including the large-scale ones. The results of the present work can be used as a basis for developing plans for the management of radioactive waste and nuclear materials.

It is necessary to increase efforts to bring to the attention of the majority of relevant scientists, the technological aspect of radioactive waste and spent nuclear fuel isolation testing. V.G. Khizhnyak is correct in stating: "A repository will be required in any case" [38, p.29]. We note that not just one repository is required and not only in the vicinity of Krasnoyarsk. This is a key point – we need technological decisions to be made which are agreed upon, scientifically-based, discussed widely, and subsequently conveyed to the public.

An inter-regional, possibly international, project "Natural materials and technologies of radioactive waste and spent nuclear fuel isolation" conceived on the Russian side by representatives of the Russian Academy of Sciences, Minatom, and ecologists and geologists of the territory, could help to consolidate the search for a solution.

In the case of international collaboration, German firms, such as DBE (Deutsche Geselischaft zum Ban und Betrieb von Endlagern for Abfallstoffe mbfl) would be appropriate as foreign partners for several reasons. Firstly, Germany is one of the few countries which have managed radioactive wastes effectively and over a long time. Secondly, the country has practical experience of various international ecological and technological projects in all "radiation significant" regions of Russia, as well as in Kazakhstan and the Ukraine [39]. The form of Germany's participation in the present inter-regional, international project could be considered, too. If necessary, the project could be discussed by a group of IAEA experts devoted to international projects in the sphere of radioactive waste management.

Various sources of funding are possible, *viz.* the federal programme "Handling of radioactive wastes and spent nuclear materials, their utilisation and repository for 1996-2005" ratified by resolution of the Government of the Russian Federation N 1030 from October 23, 1995; regional and All-Russian ecological funds; Russian conversion funds and programmes; programmes of the European Union and NATO aimed at supporting the regions of Russia and All CIS-countries; programmes of the International Scientific and Technical Centre (by analogy, for example, with the project of the Institute of Geology of Ore Deposits, Petrography, Mineralogy and Geochemistry of the Russian Academy of Sciences, British Nuclear Fuels Ltd. in the U.K. and the U.S. Geological Service, entitled "Handling of radioactive waste and the repository at Mayak, Poisk, 1996, N20); and funds from individual countries, in particular U.S.A., Japan and Norway.

We suggest the planning and realisation of an inter-regional project "Natural materials and technologies of radioactive waste and spent nuclear fuel isolation". Such a project is currently supported by the Co-Director of the NATO Workshop V.J.Kirko (Krasnoyarsk 1996), A.M. Adam, the deputy head of the administration of the Tomsk region, and the administration of the SCC.

ACKNOWLEDGEMENT

The authors would like to thank Professors N.N. Melnikov and V.P. Konukhin for their support and opinions which allowed this work to be carried out.

References

1. W. Bechthold. Direct Disposal Demonstration Experiment. - Kernforsehungszentrum Karlsruhe GmbH, 1990. - 11p.

2. Конухин В.П., Комлев В.Н. Ядерные технологии и экосфера. - Апатиты, 1995, - 340 с.

3. Мельников Н.Н., Конухин В.П., Комлев В.Н. Подземное захоронение радиоактивных отходов . - Апатиты, 1994. - 214 с.

4. Pierau B. Untertagedeponie Konrad - Konzept fur die Verfullung und Abdichtung der Schachte // Vortrage des Seminars "Geotechnische Fragestellungen bei Untertagedeponien und Transportvorgangen": Veroffentlichungen des Jnstitutes fur Grundban, Bodenmechanik, Felsmechanik und Verkehrswasserbau der RWTH Aachen. - 1990. - H. 18. - S. 182-203.

5. Abdichtung von Deponien und Altlasten // EF-Verlag fur Energie - und Umwelttechnik GmbH. - Berlin, 1992. - 639 S.

6. РЖ Chem. abstracts. - 1971. - V. 74. - N 12. - Реферат 59901.

7. Отверждение среднеактивных отходов на месте хранения / Г.Б. Борисов, А.С. Поляков, В.В.Куличенко и др. // Атомн. энергия. - 1994. - Т.77. - Вып. 6. - С. 449-451.

8. Bericht des Bundesministers fur Umwelt, Naturschutz und Reaktorsicherheit an den Ausschuss fur Umwelt, Naturschutz und Reaktorsicherheit des Deutschen Bundestages fur die Sitzung am 4. Dezember 1991 zum Endlager fur radioaktive Abfalle Morsleben (ERAM).

9. Кривохатский А.С., Рогозин Ю.М., Савоненков В.Г. Радиохимические проблемы создания подземных хранилищ радиоактивных отходов и подземных атомных станций // Использование подземного пространства страны для повышения безопасности ядерной энергетики: Матер. Межд. конф., часть 3, Апатиты, октябрь 1992 г. - Апатиты, 1995. - С. 27-33.

10. Плутоний в России / Г.В.Воробьев, А.М.Дмитриев, А.С. Дьяков и др.- М/: СоЭС, 1994. - 144 с.

11. Макхиджани А., Макхиджани Э., Ядерные материалы сквозь тусклое стекло. - Вашингтон: IEER PRESS, 1995. - 132 с.

12. Косинская А.В. Камнелитые материалы для объектов атомной энергетики // Технология и организация производства. - 1992. - N2. - С. 43-44.

13. Изучение возможности включения высокорадиоактивных отходов в керамические матрицы на основе естественных горных пород / Р.В.Богданов, Р.А.Кузнецов, А.С.Сергеев и др.// Радиохимия. - 1994. - т.36. - Вып.5. - С.470-479.

14. Комлев В.Н., Конухин В.П. Радионуклиды и минералы - сосуществование в природных и техногенных условиях. - Апатиты, 1992. - 146с.

15. Комлев В.Н., Конухин В.П. Комплексное экспериментальное исследование материалов защитных геотехнических барьеров для хранилищ радиоактивных отходов // Использование подземного пространства страны для

повышения безопасности ядерной энергетики: Матер. Межд. конф., часть 3, Апатиты, октябрь 1992г. - Апатиты, 1995. - С. 97-108.

16. Судаков В.Б., Костыря Г.З. Основные положения технологии создания крупногабаритных упаковок для тепловыделяющих радиоактивных отходов // Там же. - С.75-81.

17. Иммобилизация РАО отверждением в устойчивые минералоподобные геоцементные камни / Э.Е. Коновалов, П.В.Кривенко, Л.В.Лавриненко и др. // Тез. докл. Первой Российской конф. по радиохимии, 17-19 мая 1994г., Дубна. - М, 1994. - С.105.

18. Исследование закономерностей сорбции радионуклидов цезия-137 и стронция-90 природными алюмосиликатными сорбентами / Н.Г.Богданович, В.Д.Шумская, Э.Е.Коновалов и др.// Там же. - С.119.

19. Закладочные материалы из отходов производства / И.И.Бессонов, А.А.Леонтьев, В.П.Конохов и др. - Апатиты, 1988. - 71с.

20. Розенталь Н.К. Бетоны для строительства подземных атомных станций и хранилищ радиоактивных отходов // Использование подземного пространства страны для повышения безопасности ядерной энергетики: Матер. Межд. конф.,часть2, Апатиты, октябрь 1992г. - Апатиты, 1995. - С.264-266.

21. Мостков В.М. Строительство подземных сооружений большого сечения. - М.: Госгортехиздат, 1963. - 308с.

22. Мосинец В.Н. Геотехнические методы добычи цветных и редких металлов // Цветная металлургия. - 1992. - N2. - С.24-30.

23. Химико-технологические основы и разработка новых направлений комплексной переработки и использования щелочных алюмосиликатов. Часть 1 / В.И.Захаров, В.Т.Калинников, В.А.Матвеев, Д.В.Майоров. - Апатиты, 1995. - 180с.

24. Технологические особенности производства безобжиговых керамовермикулитовых изделий/ П.П.Гедеонов, В.С.Спирина, М.И.Флерова, Л.Д.Кабыш // Технология и свойства строительных материалов и бетонов на основе отходов промышленного производства и попутных продуктов. - Челябинск, 1977. - С.27-32.

25. Разработка технологии закрепления пылящих поверхностей хвостохранилищ новыми вяжущими веществами: Отчет о НИР / Институт химии и технологии редких элементов и минерального сырья КНЦ РАН; Руководитель В.И.Захаров. - Апатиты, 1982. - 93с.

26. Котенко Е.А., Мосинец В.Н. Свойства горных пород и охрана окружающей среды при добыче и переработке урановых руд // Горный вестник. - 1995. - N2. - С.49-56.

27. Некоторые аспекты проблемы обращения с отходами из емкостей-хранилищ радиохимического производства горно-химического комбината / Ю.А.Ревенко, Ю.П.Сорокин, К.Г.Кудинов и др. // Тез. докл. Первой Российской конф. по радиохимии, 17-19 мая 1994г., Дубна. - М., 1994. - С.112.

28. Тампонаж обводненных горных пород: Справочное пособие / Э.Я.Кипко, Ю.А.Полозов, О.Ю.Лушникова и др. - М.: Недра, 1989. - 318с.

29. Поглощение плутония и тория породами пласта-коллектора при подземном захоронении жидких высокоактивных отходов/В.Г.Балахонов, В.А. Белов, Ю.В.Буров и др. //Тез. докл. Первой Российской конф. по радиохимии, 17-19 мая 1994 г., Дубна. - М, 1994. -С. 120.

30. Носухин А.В., Ревенко Ю.А. Результаты эксплуатации, обоснование безопасности захоронения жидких радиоактивных отходов и перспективы дальнейшего использования полигона "Северный" Красноярского горно-химического комбината // Сборник докладов II Межд. радиоэкологической конференции "После холодной войны: разоружение, конверсия и безопасность" (Красноярск, 12-16 сентября 1994 г.).- Красноярск, 1995. - С. 60-66.

31. Плохих Г.П. Почему тревожится общественность // Там же. - С. 112-117.

32. Кохрэн Т.Б., Норрис Р.С., Бухарин О.А. Создание русской бомбы. - Вествью Пресс, 1995. - 145 с.

33. Печенюк С.И. Сорбционно-гидролитическое осаждение платиновых металлов на поверхности неорганических сорбентов.- Л.:Наука, 1991. - 248 с.

34. Адсорбционно- активные материалы для промышленной экологии / А.П.Зосин, Т.И.Приймак, Л.Б.Кошкина, Т.Ф.Мартынова. - Апатиты, 1991. - 114 с.

35. Преображенская Л.Б. Политика Великобритании в области обращения с радиоактивными отходами // Инф. бюлл. Центра общ. инф. по атомн. энергии за 1991 г. - М., ЦНИИатоминформ. - 1992. - С.169-175.

36. Хибинские апатиты. - Ленинград: Ленхимсектор, 1931. - 319с.

37. Хибинские апатиты и нефелины. - Ленинград: Госхимтехиздат, 1932. - 309с.

38. Сборник докладов II Межд. радиоэкологической конференции "После холодной войны: разоружение, конверсия и безопасность" (Красноярск,12-16 сентября 1994г.). - Красноярск, 1995. - 298с.

39. Х.Энгельман, Б.Мартенс. Доклад о германо-российской кооперации в области обращения с радиоактивными отходами на Международном семинаре МАГАТЭ "Сотрудничество в вопросах обращения с ядерными отходами в России" (Вена, 15-17 мая 1995г.) и на совещании Контактной экспертной группы МАГАТЭ по международным проектам в области обращения с РАО (Москва,14 марта1996г.).

GEOLOGICAL ASPECTS OF THE DEEP DISPOSAL OF RADIOACTIVE WASTE

T J MCEWEN
QuantiSci Limited
47 Burton Street
Melton Mowbray, LE13 1AF
U.K.

Abstract.
Various environments have been selected throughout the world for the potential deep disposal of long-lived radioactive waste. The selection of these environments has been carried out using a variety of methods, some of them more logical and defensible than others. The raison d'être for their selection also varies from country to country.

Important lessons have been learnt from the site selection programmes, the site characterisation activities and the accompanying performance assessments that have been carried out concerning the suitability of geological environments for the disposal of long-lived waste. These lessons are the subject of this paper.

1. Introduction

The main points of interest regarding the geological aspects of suitable disposal environments that need to be considered in the disposal of radioactive waste are as follows:

- the role of the geological barrier;
- the definition of potentially suitable disposal environments;
- the requirements of a disposal environment for long-lived waste;
- the properties and behaviour of the geological barrier in relation to:
  - site selection,
  - site characterisation,
  - engineered barriers,
  - the waste form.

Several of these points of interest are discussed below.

M. J. Stenhouse and V. I. Kirko (eds.),
Defence Nuclear Waste Disposal in Russia: International Perspective, 99–120.

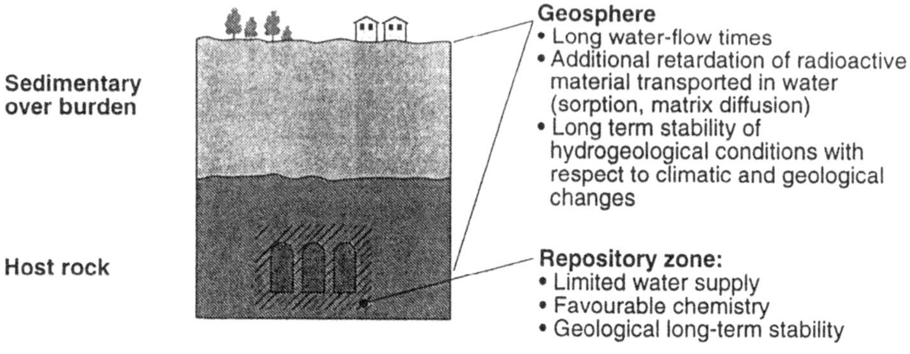

Figure 1. Schematic illustration of the geosphere or geological barrier (based on the Nagra concept).

1.1. THE GEOLOGICAL BARRIER

The geological barrier (also variously known as the far-field or geosphere) can be defined as all the rock mass surrounding the repository near-field as far as the surface (*Figure 1*). The rock mass surrounding a deep repository in any type of rock will be affected, out to a certain distance, by the presence of the repository. This affected zone, which will vary with rock type, repository design, heat loading from the repository, *etc.* is likely, at least in part, to be included within the near-field. Thus the near-field is usually considered to include the excavation damage zone (EDZ) and the zone of altered chemistry.

The first questions that need to be asked when considering the geological barrier are:

- What is the role of the geological barrier and what can, or should, be expected of such a barrier?
- What are the desirable characteristics of the geological barrier?
- Where are we most likely to find such a barrier?
- How necessary is a good geological barrier and could long-lived waste be disposed of almost anywhere at depth?

Some of the requirements of a geological barrier are intuitively obvious, such as the fact that the rock mass isolates the waste from man's environment and provides adequate shielding from radiation emitted from the waste. Beyond these lie more significant aspects of the barrier, in particular, for deep disposal:

- deep disposal places the waste out of reach of major disruptive events;

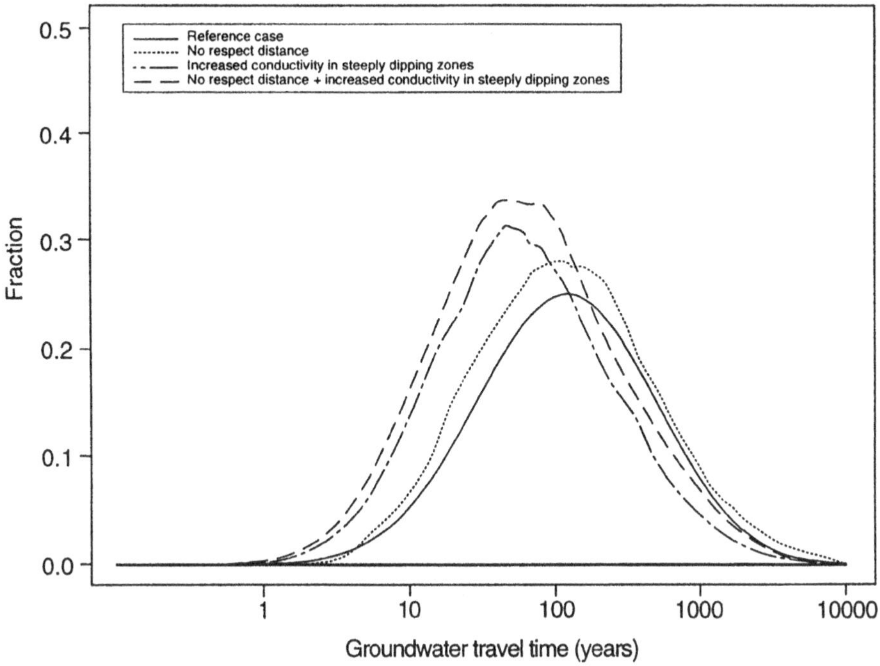

Figure 2. Dependence of groundwater travel times on the distance between repository and fracture zones and on the ratio between hydraulic conductivity of the fracture zones and the surrounding rock mass (k_f/k_r) (from [1]).

- deep disposal can provide an environment where the rates of natural processes affecting the waste are slow.

SKB's Safety Assesment Review [1, 2] came to the possibly rather provocative conclusion that *"SKB 91 has shown that the safety-related requirements on a site where a final repository is to be built are such that they they are probably met by most sites SKB has investigated in Sweden"* [1, 2]. SKB also considered that the: *"primary function of the rock is to provide stable mechanical and chemical conditions over long periods of time, so that the long-term performance of the engineered barriers is not jeopardised"* [1, 2]. These conclusions are specifically related to SKB's KBS-3 disposal concept [1, 3] in which the waste canister and the engineered components of the near-field are assumed to prevent absolutely, or at least very greatly minimise, for considerable periods of time the release of radionuclides from any part of the near-field or engineered barrier system (EBS). Performance assessment calculations had shown that in some situations, the return times for radionuclides from a repository at 500 m depth were extremely short and the retardation offered by the geological barrier was minimal [1] (*Figure 2*). The delay offered by the geological barrier was, therefore, assumed to be

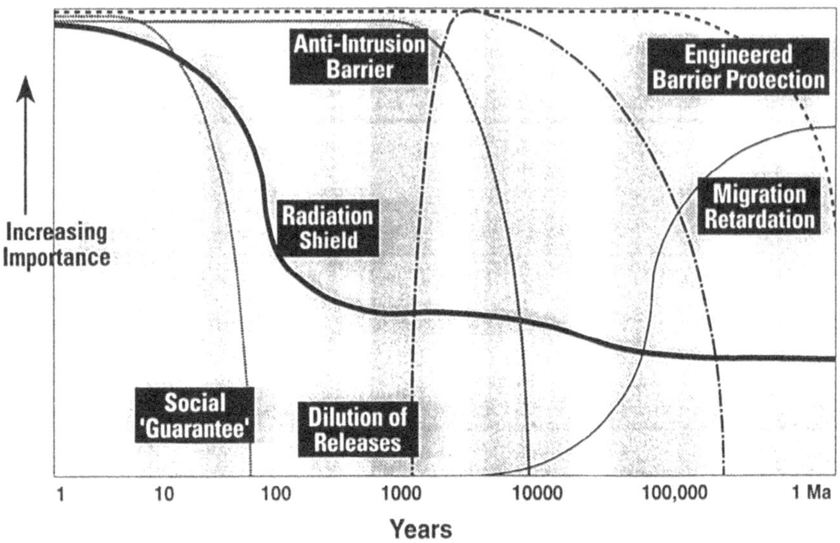

Figure 3. The relative importance of the various functions of the geological barrier against time. The protection of the engineered barriers, by maintaining stable or only slowly evolving chemical and mechanical conditions at depth, can be seen to be its most significant function over long time periods.

minimal in many situations and could not be relied on. Performance assessments by Nagra [4] and TVO [5], assuming the same types of EBSs, have also arrived at similar conclusions for the disposal of HLW in crystalline rock.

In other types of geological environment, such as in argillaceous rocks and evaporites, radionuclide transport from a repository at depth can be far more restricted, with much longer return times for radionuclides to the biosphere, *e.g.* [6, 7, 8, 9, 10]. Nevertheless, even in these environments there are likely to be some fast pathways for radionuclide transport which could effectively bypass, to some extent, the geological barrier, *e.g.* [11, 12].

AECL's recent PA [13], based on the concept of disposal within the crystalline rocks of the Canadian Shield, assumes extremely long advective groundwater flow times through the geological barrier (in excess of 7 Ma). This appears to be due to the assumptions made regarding the properties of the granite found at the Canadian URL site, which is effectively unfractured at depth beneath a large low-angle fracture zone (*Figure 3*). It is unclear whether any other parts of the Canadian Shield possess these unusual properties, which are very different from other apparently similar rocks in Sweden and Finland (see review in [15]). Groundwater travel times from repositories within different rock types, therefore, vary by considerable

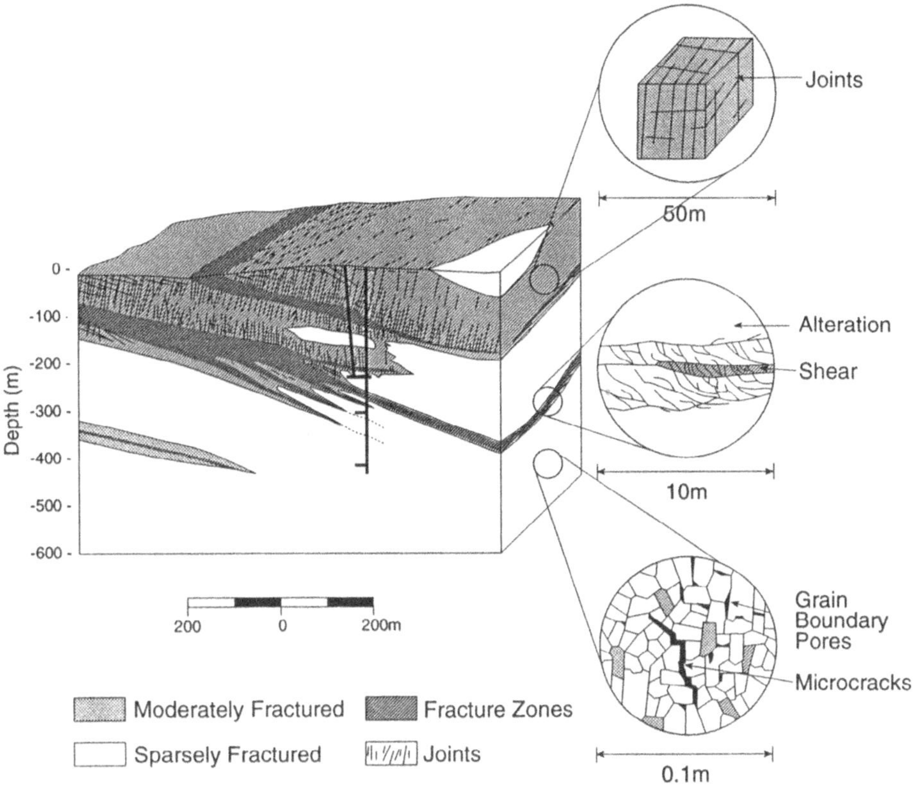

Figure 4. The three fracture domains showing the conceptual model of the Canadian Shield used for modelling groundwater flow (from 13, 14]). Well-fractured granite, with predominately high-angle fractures, lies above a major low-angle fracture zone, with considerably less fracturing below this zone.

amounts; but it still seems possible to develop an acceptably safe repository concept in rocks where such travel times are relatively short.

The purpose of the geological barrier is, therefore, six-fold and is not confined to delaying or preventing the arrival of radionuclides at the surface (*Figure 4*):

- acting as a *social guarantee* (*i.e.* institutional control);
- acting as a radiation shield
- acting as an anti-intrusion barrier (*i.e.* limiting inadvertent human intrusion)
- acting to dilute releases from the repository
- acting to retard the migration of radionuclides
- acting to protect the engineered barriers (*Figure 5*).

It can be seen from *Figure 5* that one of the most significant attributes of

Canada (AECL 94)

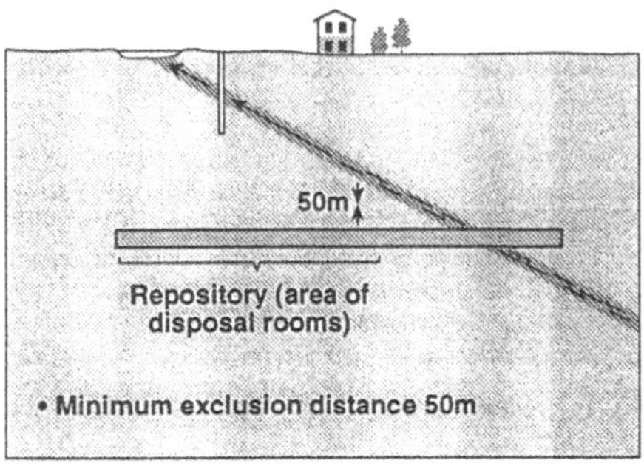

Sparsely fractured host rock modelled as porous medium around repository. Very low hydraulic gradients below major fault LD1 mean that transport from repository to fault is by difusion over minimum distance of ~50 meters. Gives very long transit times.

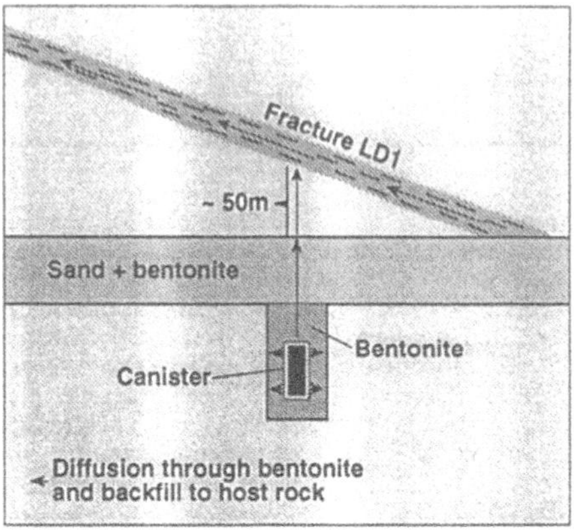

Probabilistic failure of canisters over 500 to 10,000 years. Canisters "disappear" (i.e. offer no transport resistance) after failure. Many release pathways are calculated from different repository sectors - most important are those closest to the LD1 fracture, which provides advective transport to overburden and biosphere.

Figure 5. Gross conceptual model choice for the geological barrier and the EBS (see Figure 4) [13].

the geological barrier is the protection of the EBS by limiting the changes in the chemical and mechanical environments that can take place at depth within the rock. This is in contrast to the retardation of long-lived radonuclides, which is also of importance, but which is commonly believed to be the only long-term requirement of the barrier.

1.2. THE DEFINITION OF POTENTIALLY SUITABLE DISPOSAL ENVIRONMENTS

Chapman *et al.* [16] set out a programme of work for defining geological environments that were potentially suitable for the deep disposal of long-lived waste. This approach differed from that previously followed, particularly within the U.K., as it concentrated on the definition of hydrogeological and geological environments and on their potential containment characteristics, rather than on any intrinsic properties of the rocks themselves. Chapman *et al.* [*op. cit.*] changed the emphasis from the intrinsic property of the host rock to consider instead environments that were characterised by the following general features:

- predictable groundwater flow paths, preferable long and resulting in mixing with older, deeper waters or leading to discharge at sea;
- very slow local and regional groundwater movements in an area with low regional hydraulic gradients;
- ease of construction to allow for economic repository design
- conformity with the many accepted restrictions regarding seismicity, depth, *etc.*

Using these characteristics as a guide, it was possible to define environments that might fulfil these requirements and to map out their occurrence in the U.K. (*Figure 6*). Thus, the application of fundamental hydrogeological principles made it possible to define what may be the most geologically suitable environments for the deep disposal of long-lived waste in a typical northern European setting. The rationale behind the choice of these environments is presented in Chapman *et al.* [1986][1]. Modifications to this rationale have taken place due to the development of ideas on groundwater flow, the effects of climatic change and, in particular, groundwater geochemistry. Apart from these changes of emphasis, the concept behind the selection of these environments in the U.K. is believed still to be valid and is transferable to most parts of the world (areas with large unsaturated zones, *e.g.* Yucca Mountain, being an obvious exception).

Other countries have carried out site selection programmes for deep disposal sites [17], and several of these programmes are described in sufficient detail to allow a comparison to be made with the approach followed in the U.K. In some cases, notably Sweden, Finland and Germany, only one type of disposal environment, was considered, whereas in the U.K., France, U.S.A. and Switzerland, several different environments, or rock types, have been considered.

In some countries, notably Sweden and Finland, only crystalline rocks have ever been considered for radioactive waste disposal and both countries have followed similar paths in their search for disposal sites. The essential elements of the first few stages of the site selection process in Finland are summed up in *Figure 7*, where suitable environments for disposal were determined by the expected size of the repository for spent fuel and by the location of regional and more local-scale fracture zones, which define the area for investigation. SKB followed a similar approach in their selection of sites for R&D, as described in KBS-3 [3] and are currently following a similar approach in their site selection programme [18]. In both countries, no account was initially taken of geochemical factors in the selection of environments and subsequently sites for investigation. However, in both

[1]Salt was not included as a potential disposal medium as there are no suitable salt deposits in the U.K.

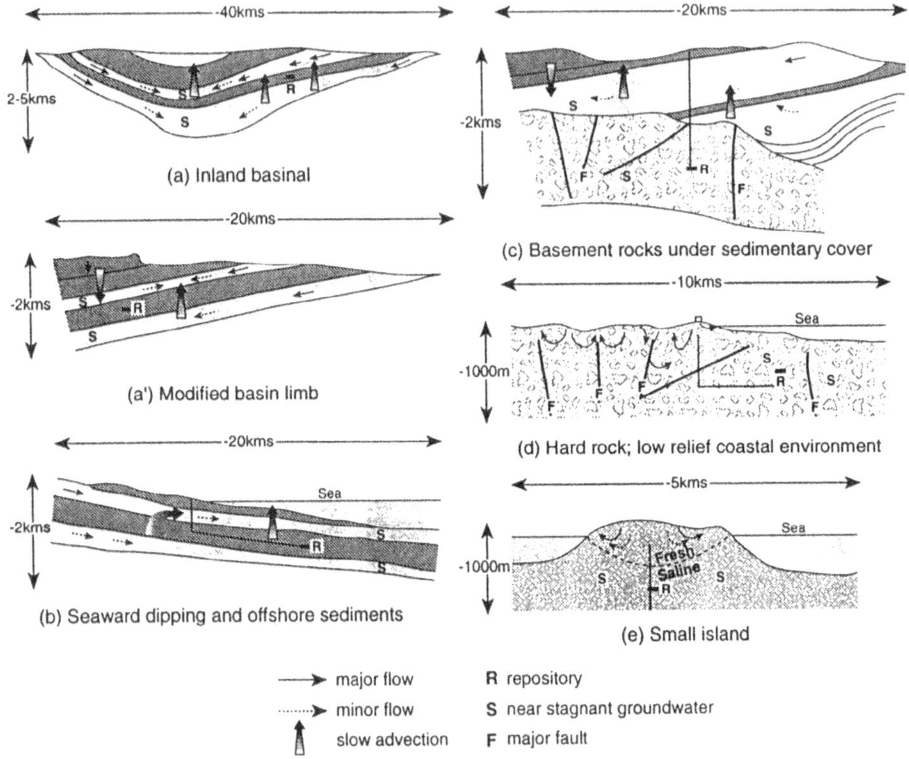

Figure 6. Potentially suitable environments for the deep disposal of long-lived radioactive waste (from [16]).

countries, the regulatory authorities, following independent reviews of the waste disposal programmes, asked SKB and TVO to evaluate basic rock masses for disposal purposes.

AECL have recently produced their Environmental Impact Statement (EIS) for Environment Canada [19], together with the accompanying reports, *e.g.* 13, 14], and the EIS contains an outline of the proposed site selection programme for Canada, which follows essentially the same approach as that used in Finland and Sweden.

1.2.1. *Revision of definition of suitable environments*

Since 1986, there have been some revisions to the thinking on what constitutes a suitable environment for the deep disposal of long-lived wastes. The differences in thinking are most apparent in two areas:

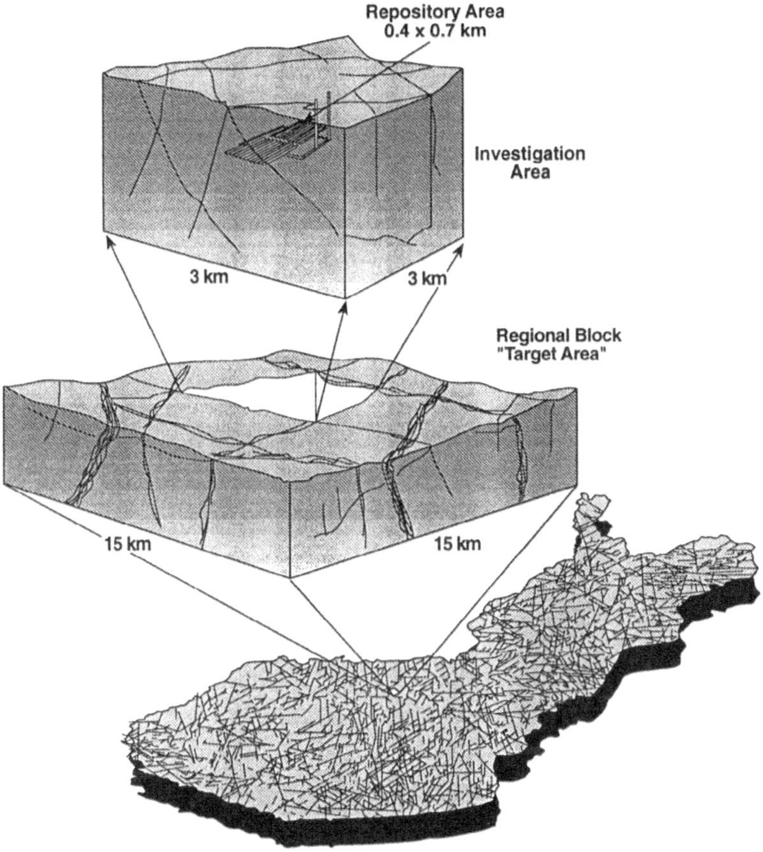

Figure 7. Structures within the Precambrian basement of Finland defined mainly from aerial surveys, together with a limited amount of field work, showing the classification of lineaments and the selected area for future investigation (from [5]).

- the importance of the geochemical environment in controlling the mobilisation and transport of radionuclides and
- the significance of future climate and geologically-mediated changes on the performance of a deep repository

Recent performance assessments, such as Kristallin-1 [4], TVO-92 [5], SKB-91 [1] and AECL 94 [13] all emphasise the role of the geological barrier in preserving suitable geochemical and geotechnical conditions at depth in the vicinity of the repository and Nirex's preliminary safety case for Sellafield rests on the assumption of chemical conditioning of the near-field by cement [7, 20]. All these disposal concepts rely on the assumptions that groundwater flow in the vicinity of the repository is of limited extent and that chemical conditions remain reducing for long periods.

The chemical control that the geological environment can offer can be described in terms of its control of pH and Eh. The capacity for Eh buffering and the rate of this buffering is of particular significance in determining how the disposal system is likely to react to changes in the external environment. For example, influx of more oxygenated groundwater to depth, which could take place during glacial conditions, due to seismic pumping or after erosion of the overlying rocks has taken place, could have an important influence on the mobilisation of actinides. The capacity of the rock mass to buffer such effects could play an important role in determining its capacity to act as a barrier to radionuclide transport. The significance of geochemical attributes in selecting environments for waste disposal is discussed in greater detail in [21].

Chapman *et al.* [16] selected type environments at the coast and beneath the sea for two main reasons:

- The constant hydraulic head imposed by the sea reduces groundwater flow rates in sub-seabed formations. Hydraulic heads generated inland can decrease relatively rapidly offshore.
- Releases from a repository to the sea will be subject to large dilutions, compared with similar releases to onshore locations.

The definition of these type environments did not discuss and incorporate the important changes that occur to sea levels and ice loadings during glacial periods. For the majority of the next million years, the sea level is likely to be substantially lower than its present value (current estimates suggest that the mean sea level over the next million years is expected to lie between -30 and -50 m [22]) and the possible benefits of a coastal or sub-sea location need to take this factor into account. In addition, coastal locations will be subject to movements of the seawater/freshwater saline interface, due to sea level changes. Other hydrochemical interfaces, which may be present at depth due to the presence of different groundwater types, may also be subject to movement due to climatic and long-term geological changes. Such interfaces are likely to move at considerably lower rates than the seawater/freshwater interface and the hydrochemical system at depth may be out of equilibrium and, therefore, normally in a transient state. A coastal environment may, therefore, be superior, with respect to long-term safety, only when the sea level is close to the current level. Coastal environments may, however, be associated with complexities that are not displayed by inland environments (*e.g.* future increases in sea level which could flood coastal sites, thereby introducing more saline groundwater into the repository environment).

Nirex investigated the offshore environment as part of their site selection programme for a deep repository (but did not consider the option of a long undersea tunnel) [6] and carried out modelling and initial PA exercises. Based purely on the results of Nirex's PA and neglecting the difficulties associated with this type of disposal strategy, environments in which releases always took place to the marine environment exhibited high levels of long-term safety.[2]

In terms of long-term radiological safety, it is preferable if releases from a repository take place to the marine environment during all future climate states, because of the massive dilution potential of the sea and because the normal water well drinking scenario does not have be considered. 140 m is normally taken as the maximum eustatic decrease in sea level at times of glacial maximum (see *Figure 8*) and there are additional complications that may need to be considered, for example the influence of any vertical crustal movement. The water depth of 140 m is, of course, an absolute minimum to guarantee releases to a marine environment. A more credible water depth in this instance would be one greater than approximately 180 m. Such water depths may be accessible within a reasonable distance of the present coastlines of several countries.

1.3. REQUIREMENTS FOR A HLW DISPOSAL ENVIRONMENT

Table 1 (and cont.) outlines the desirable and/or essential attributes of a disposal environment and/or host rock for the deep disposal of HLW, in a similar manner to that used by Nagra [23]. All the attributes listed in *Table 1* (and cont.) are desirable, but none of them can be specified in terms of absolute values or conditions. In fact it would be counterproductive to try and include any of these desirable, or even essential, attributes in any regulations regarding site selection, *etc.* They are best left as *guidelines*; see, for example [23].

In defining suitable type environments and sites, account needs to be taken of these desirable and essential attributes, but a balance needs to be struck between the various attributes, since no one environment is likely to possess only desirable attributes. For example, a basic or even an ultrabasic rock mass may be thought to exhibit exemplary geochemical conditions at depth, but it may also lie in a much more structurally complex environment than, say, a post-tectonic granite. The granite, on the other hand, is far less likely to possess these geochemical advantages, but its structural setting is likely to be simpler, its hydrogeological regime may well also be simpler and easier to investigate and it is likely that construction at depth within the granite would be easier that in the basic rock.

[2] Levels of long-term safety were estimated to be of the order of 10^6 times greater for the situation in which releases always took place to the marine environment, compared with the situation where releases took place onshore.

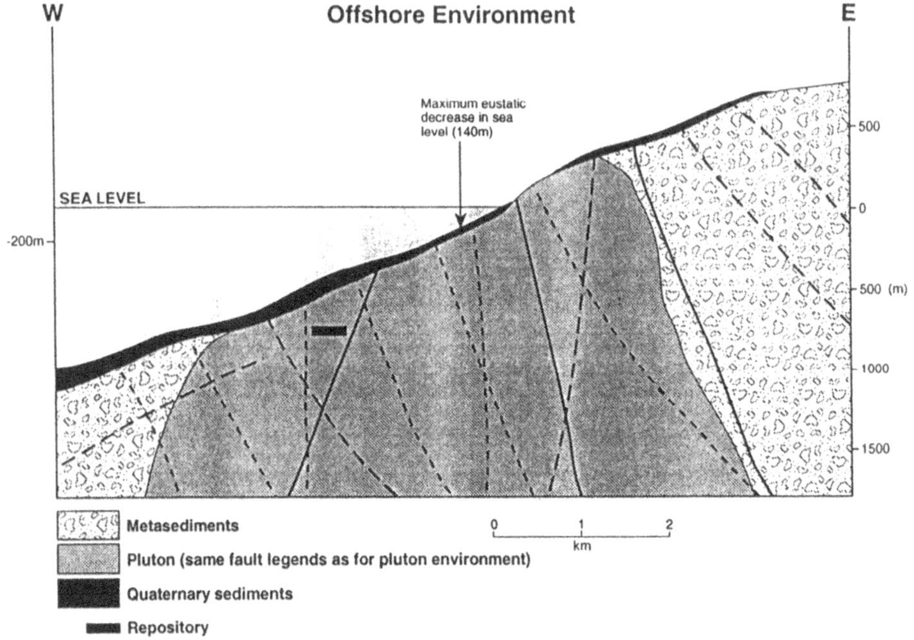

Figure 8. Cross section of an offshore disposal environment showing the maximum eustatic decrease in sea level of -140 m and a possible location for a repository offshore. Releases from such a repository would always take place to the marine environment.

Table 2 (and cont.) outlines the undesirable attributes of environments that should be avoided wherever possible. The over-riding requirement of a geological environment is that it should protect the EBS for long periods of time, be as simple as possible, be amenable to analysis and investigation and suffer from the smallest amount of future changes. All the undesirable attributes included in *Table 2* (and cont.) can be considered to cause problems in these areas. The disposal environments selected should try to minimise the influence of these undesirable attributes and maximise the desirable attributes.

There are, however, very significant differences between the tectonic environments of some countries, such as Japan, and that of northern Europe, which was implicitly assumed in Chapman *et al.* (*op. cit.*). Tectonic rates and seismic and volcanic activity have very much greater levels of intensity in some countries than in the majority of Europe and, therefore, assume far more significance in radioactive waste disposal. Chapman *et al.* (*op. cit.*) only discussed such aspects in passing and concentrated on defining potentially suitable hydrogeological environments. Because of the relative simplicity and less dynamic nature of the geological environment in much of Europe (except for parts of southern Europe, such as Greece and Italy), a situation which has persisted in northern Europe for more than 10 Ma, several different types of disposal environment could be defined relatively easily.

It is, therefore, more difficult to define suitable type environments and sites in tectonically more active countries than was the case for the U.K. Using the types of constraints that have been applied in Europe by Nirex, TVO, SKB, Andra, Nagra and GSF in the selection of environments and sites for the deep disposal of long-lived waste and those that are likely to be applied in Canada, large parts of Russia and Japan, for example, would automatically be excluded from consideration, but this does not imply that suitable disposal environments cannot be found in such countries. Examples of proposed disposal sites and environments that are undergoing, or could undergo, significant geological changes during the time over which a performance assessment might be expected to be applied are:

1. Yucca Mountain, for example, is in an area of active faulting and in its immediate neighbourhood there are several volcanic cinder cones which have erupted within the last few tens of thousands of years.
2. Nagra's proposed site for LLW and shorter-lived ILW at Wellenberg is in an area of nappe movement and hydrostatic heads well out of equilibrium, due either to the presence of natural gas or/and to the rate of deformation of the host marls. Rapid uplift and erosion rates are also expected, due to marked topography and continuing Alpine uplift. In addition, the proposed area for the deep disposal of HLW in the crystalline basement of northern Switzerland in very limited by structural considerations.
3. Disposal sites in Sweden, Finland, Canada and the U.K. will all be repeatedly glaciated during the period considered in their respective PAs. Very thick ice will be present (possibly up to 3 km) over repositories in Sweden and Finland, possibly once every 10^5 a. Very considerable depression of the land surface, by more than 800 m in northern Sweden and Finland, occurred during the last glacial maximum. Displacement is expected on fault and fractures, permafrost could develop to depths of several hundred metres and glacially-derived groundwater could be forced to considerable depths.
4. One of ANDRA's proposed URL sites, in the Gard in southern France, lies in the Rhone Valley, which suffered very considerable erosion, followed by extremely rapid sedimentary infilling as part of the *Messinian Crisis*, 5 Ma, due to the flooding of the Mediterranean Sea. A similar type of event is believed, by ANDRA, to be possible,

sometime within the next few Ma and ANDRA are studying its consequences for disposal safety.

These examples demonstrate that potential disposal sites can be selected in areas where considerable future climatic and geological changes are considered likely.

2. Summary

Important lessons have been learnt from the site selection programmes, the site characterisation activities and the accompanying performance assessments that have been carried out concerning the suitability of geological environments for the disposal of long-lived waste. This paper has discussed and briefly summarised what are believed to be the most important attributes of geological environments that need to be taken into account when considering the safe disposal of such long-lived wastes. The *requirements of the geological barrier* have been discussed with reference to the EBS and to long-term geological processes. The preferred properties of the geological disposal environment and the host rock are presented in two groups as being essential/desirable for long-term safety and in relation to the technical feasibility of repository development. These requirements can be used

References

1. SKB (1992a) SKB 91 – Final disposal of spent nuclear fuel. Importance of the bedrock for safety, *SKB Technical Report* 92-20, Swedish Nuclear Fuel and Waste Management Company, Stockholm, Sweden.
2. SKB (1992b) Treatment and final disposal of nuclear waste. The Äspö Hard Rock Laboratory, *Background Report to the R&D Programme 92*, Swedish Nuclear Fuel and Waste Management Company, Stockholm, Sweden.
3. SKB (1983) KBS-3, Final Storage of Spent Nuclear Fuel, Volumes I–IV, Swedish Nuclear Fuel and Waste Management Company, Stockholm, Sweden.
4. NAGRA (1994) Kristallin-1 Safety Assessment Report, *Nagra Technical Report* NTB 93-22E, Nagra, Wettingen, Switzerland.
5. TVO (1992) Final disposal of spent fuel in the Finnish bedrock, Technical plans and safety assessment. Report YJT-92-31E, Nuclear Waste Commission of Finnish Power Companies, Helsinki, Finland.
6. NIREX (1989) Deep Repository Project. Preliminary Environmental and Radiological Assessment and Preliminary Safety Report (PERA/PSR), *U.K. Nirex Report* No. 71. U.K. Nirex Ltd, Harwell, U.K.
7. NIREX (1995) Post closure performance assessment. Nirex-95: A preliminary analysis of the groundwater pathway for a deep repository at Sellafield; 3 volumes, *U.K. Nirex Science Report* S/95/012. U.K. Nirex Ltd, Harwell, U.K.
8. BILLINGTON, D.E., LEVER, D.A., AND WISBEY, S.J. (1989) Comparative assessment of concepts and areas for deep disposal (Phase 2): Summary, *U.K. Nirex Safety Studies Series Report* NSS/A301, U.K. Nirex Ltd, Harwell, U.K.
9. BONNE, A. (1991) Geoscientific investigations in the Belgian R&D programme concerning the disposal of radioactive waste in clay, in P.A. Witherspoon (ed.) *Geological Problems in Radioactive Waste Isolation – A World-Wide Review, Lawrence Berkeley*

Laboratory Report LBL-29703. Lawrence Berkeley Laboratory, Berkeley, California, U.S.A.

10. CHAPMAN, N.A. AND MCEWEN, T.J. (1991) Geological and hydrogeological aspects of the deep disposal of nuclear waste in Britain, in R A Downing and W B Wilkinson (eds.) *Applied Groundwater Hydrology*, Oxford Science Publications, pp177- 198.
11. PAGIS (1984) Performance Assessment of Geological Isolation Systems - Phase 1 report for the Salt Option, *CEC R&D Programme on Radioactive Waste Management and Storage*. Commission of the European Communities, Luxembourg.
12. NAGRA (1988) Sediment study - Intermediate Report 1988: disposal options for long-lived radioactive waste in Swiss sedimentary formations, Executive Summary, *Nagra Technical Report* 88-25E, Nagra, Wettingen, Switzerland.
13. AECL (1994b) The disposal of Canada's nuclear fuel waste: The geosphere model for post-closure assessment, *AECL Report* AECL-10709, COG-93-9, AECL, Pinawa, Canada.
14. DAVISON C.C. *et al.* (1994) The disposal of Canada's nuclear fuel waste: The geosphere model for post-closure assessment, AECL-10709, COG-93-9, AECL, Pinawa, Canada.
15. NAGRA (1995) Disposal programme for HLW. *Nagra Bulletin*, No. 25, March 1995, Nagra, Wettingen, Switzerland.
16. CHAPMAN, N.A., MCEWEN, T.J., AND BEALE, H. (1986) Geological environments for deep disposal of intermediate level wastes in the U.K., in *Proc. of IAEA Symposium on Siting, Design and Construction of Underground Repositories for Radioactive Wastes*, Hannover, March 1986. IAEA, Vienna, Austria, pp 311-328.
17. MCEWEN, T.J. (1995) Review of site selection programmes for the deep disposal of radioactive waste. *QuantiSci Report* to MMC (Japan), IG4432.
18. SKB (1995) Treatment and final disposal of nuclear waste, RD&D Programme 95, Swedish Nuclear Fuel and Waste Management Company, Stockholm, Sweden.
19. AECL (1994a) Environmental Impact Statement (EIS) on the concept for disposal of Canada's nuclear fuel waste, *AECL Report* AECL-10711, COG-93-1, AECL, Pinawa, Canada.
20. ATKINSON, A., WILLIAMS, S.J., AND WISBEY, S.J. (1992) NSARP Reference Document. The near field, *U.K. Nirex Safety Studies Series Report* NSS/G117, U.K. Nirex Ltd., Harwell, U.K.
21. SAVAGE, D. AND STENHOUSE, M.J. (1995) Geochemical data requirements for performance assessment, *QuantiSci (formerly Intera Information Technologies) Report* IG2855-TR1 Report to HMIP (in press).
22. GOODESS, C.M., PALUKITOV, J.P., AND DAVIES, T.D. (1991) Studies of climatic effects and impacts relevant to deep underground disposal of radioactive waste, *U.K. Nirex Safety Studies Series Report* NSS/R267, U.K. Nirex Ltd., Harwell, U.K.
23. THURY, M., GAUTSCHI, A., MAZUREK, M., MÜLLER, W., NAEF, H., PEARSON, F.J., VOBORNY, O., VOMVORIS, S., AND WILSON, W. (1994) Geology and hydrogeology of the crystalline basement of northern Switzerland. *Nagra Technical Report* NTB 93-01E. Nagra, Wettingen, Switzerland.
24. MCCOMBIE, C., MCKINLEY, I., AND ZUIDEMA, P. (1991) Sufficient validation: the value of robustness in performance assessment and system design, in *Proceedings of Geoval-90*, Stockholm, May 1990, NEA/OECD, pp. 598-610.

TABLE 1. Requirements for HLW disposal environments (continued over).

Requirements for HLW disposal environment	
1. Essential/desirable for long-term safety:	**Preferred properties of the environment and/or the host rock**
(a) Physical isolation of the engineered barriers	(i) Good mechanical stability of the host rock. (ii) Host rock volume sufficient for repository plus surrounding buffer zone (*i.e.* separation from fracture zones, sufficient thickness of sedimentary host formation). (iii) Low rates of erosion and uplift
(b) Low fluxes of groundwater through repository.	(i) Low permeability of the repository host rock. (ii) Low hydraulic gradients at repository depth.
(c) Favourable thermal conditions	(i) None or limited hydrothermal activity; limited geothermal fluxes; not excessive temperature gradient. (ii) Suitable thermal properties of host rock (*i.e.* sufficiently high thermal diffusivity).
(d) Favourable hydrochemistry	Reducing conditions; pH neutral to alkaline, good redox and pH buffering capacities and rates.
(e) Adequate dilution of releases from repository	Possibility of dilution by presence of near-surface aquifers (*e.g.* Nagra's disposal concept in northern Switzerland) or release to sea (expected for substantial periods of the next 1 Ma, *e.g.* release to sea where current sea depth is >140 m)

TABLE 1 (cont.) Requirements for HLW disposal environments (continued over).

Requirements for HLW disposal environment	
1. Essential/desirable for long-term safety (cont.):	**Preferred properties of the environment and/or the host rock**
(f) Long-term stability:	
i) Tectonic	Low rate of deformation; no (or limited) active faults; any known/expected deformation occurring on existing faults; no expected local volcanic activity; low rate of uplift/subsidence, low rate of erosion.
ii) Hydrogeological	Limited change in hydrogeological properties (especially of the more transmissive parts of the rock mass) and conditions at depth; no young groundwater at repository depth; no, or only limited, hydrothermal activity expected.
iii) Chemical	Eh, pH, buffering, retardation*etc.* conditions remain favourable
(g) Radionuclide retardation in the far-field	Ensure delayed release of radionuclides by sorption along flow paths; retardation (and decay) possible due to long flow paths and/or low transport rates.
(h) Low risk of human intrusion	Lack of natural resources of all types (*e.g.* mineral, hydrocarbon, geothermal), including at present non-economic resources.

TABLE 1 (cont.) Requirements for HLW disposal environments.

Requirements for HLW disposal environment	
2. Technical feasibility of repository development	
(a) Explorability (*cf.* Nagra [15]	Site characterisation not involving excessive complications (*i.e.* practicable), not requiring development of new, esoteric investigation techniques.
(b) Development of PA case	(i) Possibility of development of robust geological, hydrogeological and transport conceptual models (*i.e.* geological and hydrogeological environment must not be excessively complex). (ii) Robust environment should allow justifiable development of simple PA conceptual models, result in lower uncertainties, *etc.*
(c) Economic feasibility	(i) Suitable geotechnical properties of host rock; excessive support not required; disposal possible within reasonable depth constraints. (ii) Repository design, construction and operation without excessive expense and difficulty. (iii) Characterisation programme not excessively expensive.
(d) Safety during construction and operation	Proven excavation techniques, proven disposal engineering techniques
(e) International consensus	Better if planned disposal concept similar to those in other countries, or at least not involving contentious assumptions.

TABLE 2. Undesirable factors and attributes of environments that need to be avoided, where possible, when selecting suitable disposal environments and sites (continued over).

Undesirable factors and attributes of environments	
Factor or attribute	**Undesirable properties of the environment and/or host rock**
(1) Geological environment	(i) Complex structure: results in difficulties in definition of structure, expensive and time-consuming site characterisation programme (requiring excessive numbers of boreholes in potential volume of repository - could cause problems with regulators); problems in upscaling; increased spatial inhomogeneities; large uncertainties in geological structure inevitable; difficulty in defining robust geological models (possibly irreconcilable problems in this area, *i.e.* multiple conceptual models necessary); geological uncertainties almost certainly transmitted to uncertainties in hydrogeological and transport models and hence to increased uncertainties in PA results. (ii) Volume (surface area and depth) limitations resulting in reduced potential disposal volumes and limitations on degrees of freedom. (iii) Thick and continuous Recent and Quaternary sediments (tend to possess complex spatial heterogeneity), make site characterisation much more difficult, *e.g.* seismic reflection surveys, mask underlying rocks of interest. (iv) Offshore investigations, especially those in deeper water, more difficult, expensive and time-consuming (an important consideration for coastal, small island and offshore environments). (v) Large size of environment requiring investigation (at various levels of detail): directly related to complexity of geological environment.

TABLE 2 (cont.) Undesirable factors and attributes of environments that need to be avoided, where possible, when selecting suitable disposal environments and sites (continued over).

Undesirable factors and attributes of environments	
Factor or attribute	**Undesirable properties of the environment and/or host rock**
(2) Hydrogeological environment	(i) Complex hydrogeological environment- resulting in difficult, expensive and time-consuming investigation (*e.g.* controls on groundwater flow difficult to determine, boundary conditions difficult to resolve). (ii) Large size of environment requiring investigation (*i.e.* large catchment area, large area required for groundwater flow modelling). (iii) Difficulties in measurement of hydraulic properties (*e.g.* soft rocks, unstable boreholes, potential two-phase flow (*e.g.* due to gas phase and/or anomalous pressures due to tectonism causing partially saturated conditions). (iv) Hydrogeological environments a long way from equilibrium (especially those due to tectonic activity and gas evolution): makes modelling difficult. (v) Directions of groundwater flow and fluxes sensitive to external forcing factors (*e.g.* changes in *in-situ* stress, changes in hydraulic properties of fault zones). (vi) Uncertainties in controlling factors on flows at depth, *e.g.* significance of control of geological structures on flow uncertain.

TABLE 2 (cont.) Undesirable factors and attributes of environments that need to be avoided, where possible, when selecting suitable disposal environments and sites (continued over).

Undesirable factors and attributes of environments	
Factor or attribute	**Undesirable properties of the environment and/or host rock**
(3) Chemical factors	(i) Chemical environment sensitive to external changes (*e.g.* seismic events, lowering of sea level, uplift, deformation, *etc.*). (ii) Chemical environment sensitive to changes during construction and operation (*e.g.* production of acidic solutions by oxidation, H production by low temperature weathering of ultrabasics). (iii) Oxidising and acidic conditions at depth. (iv) Low pH and Eh buffering capacities and rates. (v) Chemically-sensitive minerals present at depth.
(4) Tectonics and seismicity	(i) Rapid rates of uplift (and possibly also subsidence, if it is rapid, since this implies a dynamic geological environment with the possibility of increased uncertainties in future behaviour). (ii) Presence of active faulting. (iii) Significant deformation rates. (iv) Significant seismic activity (vibrational damage, fault movement, elongation, reactivation, seismic pumping, *etc.*)

TABLE 2 (cont.) Undesirable factors and attributes of environments that need to be avoided, where possible, when selecting suitable disposal environments and sites.

Undesirable factors and attributes of environments	
Factor or attribute	**Undesirable properties of the environment and/or host rock**
(5) Volcanic and related activities	(i) Presence of Quaternary volcanos and their associated effects. (ii) Geothermal activity, now or during the Quaternary. (iii) Volcano-tectonic faulting and minor intrusions.
(6) Rock mechanics (constructability)	(i) Difficult construction conditions: poor stability conditions of underground structures (resulting in excessive support requirements; limitations on disposal volumes; limitations on vault dimensions [important for TRU]); underground structures sensitive to seismic damage; backfilling and sealing problems; increased cost and time to construct repository. (ii) Lack of safety during pre-closure phase. (iii) Retrievability problems (if ever required by regulations). (iv) Complex monitoring problems both pre- and post-closure (if ever required by regulations).
(7) Future climate states	(i) Thick glaciers (>few hundred metres), resulting in the potential for glacially-driven oxygenated groundwater to depth. (ii) Permafrost possibly problematic if well-developed over large area for substantial periods. (iii) Areas sensitive to future climate changes (*e.g.* significantly increased erosion rates, fluvial downcutting, increased hydraulic gradients at depth, *etc.*)

GEOLOGICAL FORMATIONS IN BULGARIA SUITABLE FOR A DEEP HLW REPOSITORY

D. EVSTATIEV, D. KOZHOUKHAROV
Geological Institute of the Bulgarian Academy of Sciences, Acad. G. Bonchev Str., Bl. 24, Sofia 1113, BULGARIA

Abstract.
The total volume of vitrified high-level waste (HLW) and of intermediate-level wastes containing significant amounts of long-lived alpha emitters, generated over the lifetime of *Kozloduy* Nuclear Power Plant (NPP), is estimated to be about 2,100 m^3. Selection of a suitable formation to host a deep HLW repository in Bulgaria presents serious difficulties owing to the small area of the country, its complex geology, and the expected negative reaction of the local population. After a preliminary categorization of the Bulgarian territory, more than 40 areas were identified initially as potential disposal sites. This number was reduced to 20 using a screening method. A large portion of the selected regions comprises thick (up to 1,000 m) Lower Cretaceous marls, some at a distance of 60–80 km to the south of Kozloduy NPP. There are no active faults in the vicinity of the chosen areas and the seismicity is of degree VII on the MSK scale. Several regions in south-east Bulgaria containing magmatic and metamorphic rocks, have been suggested as prospective sites for deep HLW disposal: the Byala Reka structure of Precambrian highly metamorphic gneisses, the Avren and Zalti Chal areas in meta-ultrabasites, the Sakar Palæozoic granite pluton, and two areas in volcanogenic-sedimentary rocks. The possibility of subsea geological disposal has also been discussed. At present, a programme is being developed for more detailed investigations of the suitable geological formations.

1. Introduction

The Bulgarian Nuclear Power Plant (NPP) of *Kozloduy* is equipped with 6 nuclear reactors of the VVER type (produced in Russia) and has an output of 3760 MW of electric energy, equivalent to 35–40% of the total output for the country. The construction of a second NPP at *Belene* has ceased

M. J. Stenhouse and V. I. Kirko (eds.),
Defence Nuclear Waste Disposal in Russia: International Perspective, 121–134.

because of public protests. The volume of vitrified high-level waste (HLW) associated with the lifetime of Kozloduy NPP is estimated to be about 310 m^3. In addition, about 1800 m^3 of low- and intermediate-level wastes (L/ILW) containing a significant amount of long-lived alpha emitters will be generated during reprocessing, which also have to be disposed of in a deep geological formation.

The lack of a solution to the problem of the processing and disposal of radioactive waste is one reason for the negative public opinion and for opposition by ecological organizations to nuclear power generation. In 1991, the Council of Ministers of the Republic of Bulgaria imposed on the Bulgarian Academy of Sciences (BAS) the task of developing a concept for the construction of a national radioactive waste repository with the intention of a speedy solution to the problem. A large team of scientists from the Earth Sciences Division of BAS was established to study foreign experience and to make a thorough analysis of the natural environment in Bulgaria. As a result, areas and geological complexes potentially suitable for the construction of radioactive waste repositories were identified for further investigation [1].

The following strategy for the selection of a geological formation for radioactive waste disposal was adopted:

- to take maximum advantage of the experience of countries which have a well-developed nuclear power generation programme. The delay in solving the radioactive waste management problem in Bulgaria has one positive aspect against a background of numerous negative effects, *viz.* the possibility to consider in detail the rich foreign experience and adapt it to the specific conditions of the country;
- to make maximum possible use of the scientific information available (Bulgaria, being a small country, has been well studied with respect to the earth sciences);
- to adopt the deep (geological) disposal concept for solidified HLW, with application of the multibarrier principle and to use the same approach for both HLW and long-lived radioactive wastes;
- to apply a recognised screening method to the choice of a suitable formation;
- to follow quality assurance procedures and make information available to the general public;
- to take into consideration a proposal to dispose of HLW and L/ILW at the same site.

A two-stage process was developed for selecting sites and geological formations suitable for HLW disposal. Firstly, a preliminary division of the country into regions was performed, individual regions being identified where suitable geological formations were expected to be found. Subsequent screening of these regions was carried out to select 20 sites offering considerably more favourable conditions.

2. Methodology

A team of specialists in the different earth sciences was formed within the Bulgarian Academy of Sciences, in order to make a preliminary categorization of the national territory and to identify regions with natural environments potentially favourable for HLW repository construction.

The geological assessment was based on published and unpublished data using geological and tectonic maps. The new geological map of Bulgaria at a scale of 1:100,000 based on lithostratigraphic principles, provides a summary of all available data concerning the geology of the country and displays the distribution, composition and structure of rock complexes. Geological maps at a scale of 1:25,000 were also used to provide more detailed information.

This process allowed the selection of a limited number of rock complexes with suitable properties: clays, marls, granites, gneisses, volcanic rocks and serpentines. The data from tectonic and seismic investigations allowed the delineation of regions with mild tectonic history from the Neogene up to the present day, having few active faults and foci of earthquakes of relatively small destructive power. Details of the thermal field and thermal regime in Bulgaria were also considered.

The hydrogeological conditions were analysed using maps of different scale showing the hydrogeological basins and their hydraulic parameters. From this analysis, zones having favourable hydrogeological settings were defined.

The seismic characteristics of the Balkan Peninsula and Bulgaria are presented in maps which designate strong earthquakes and the predicted degree of seismicity according to the MSK scale, for the past 1,000 and 10,000 years.

The emphasis of the geotechnical analysis was on the structure and texture of rock massifs, on their physical and mechanical properties, and on the presence of exo-geodynamic processes. A map of the geological hazard in Bulgaria (scale 1:500,000) was also used in the geotechnical assessment [2].

Geographical analysis of the country took into consideration details of relief, socio-economic factors and demography. Hydrological factors (orography, hydrographic network, *etc.*) as well as climatic factors, were also summarized.

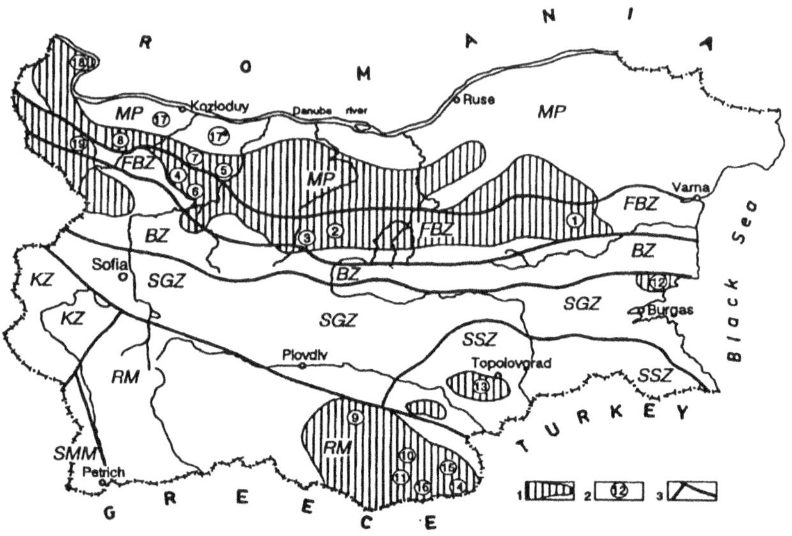

Figure 1. Map of the potential areas for further investigations concerning a national radioactive waste repository in Bulgaria; after [3].

1 - prospective areas for further investigation; 2 - a suitable site and its number; 3 - boundaries between the morphotectonic units according to E. Boncev (MP - Moezian Plate; FBZ - Fore-Balkan zone; BZ - Balkan zone; SGZ - Sredna Gora zone; SSZ - Sakar-Strandzha zone; KZ - Kraishte zone; SMM - Serbian-Macedonian massif; RM - Rhodope massif.

As a result of the analysis of earth sciences data, a map categorizing the Bulgarian territory with regard to the siting of a HLW repository, was developed at the scale 1:500,000 [3].

The next step in the effort to resolve the problem of site selection was an attempt at the preliminary selection of a limited number of sites, including abandoned mines, which appeared promising for additional investigation. About 40 areas were considered initially, this number being reduced to 20 by means of a screening procedure (*Figure 1*). These 20 sites were subjected to additional analysis taking into account all available data concerning geological, geomorphological, geographical, hydrological, climatic, hydrogeological and geotechnical factors. The conclusions from the investigation of each site were presented in a special report.

The site selection procedure performed was based on the criteria accepted in [4], adapted to Bulgarian conditions [5]. Criteria were divided into four groups according to:

- rock-linked criteria (sorption capacity, thermal conductivity and solubility);
- formation-linked criteria (thickness, minimum depth, surface area, homogeneity and uniformity, permeability and geotechnical properties;
- formation environment criteria (hydraulic gradient, seismicity and tectonics);
- supplementary factors: accessibility of the site and its potential use for economic purposes, the sensitivity of the zone to climatic and hydrological changes, and socio-geographic and demographic conditions.

The criteria for each site were evaluated on the basis of expert conclusions using direct and indirect parameters, or by analogy with similar cases. A comparison of the sites was made by means of an expert-rating selection model [5].

The site selection for HLW repository in Bulgaria is a problem which is under development and a methodology has been developed with the financial support of NATO [6], for site selection in marl terrains, which are potentially suitable for further investigation.

3. Geology of Bulgaria

Bulgaria's territory has a complex and varied geology, the age of its rocks and formations ranging from the Precambrian to the Quaternary. Metamorphic and magmatic rocks are distributed mainly in South Bulgaria - the Rhodopes massif, the Sredna Gora and the Sakar-Strandzha zone, and the sediment complexes are in the region of the Balkanides, the Kraishte and in the Moezian platform (*Figure 1*). The volcanogenic-sediment and volcanic complexes are well developed in the Sredna Gora (Upper Cretaceous age) and in the periphery of the Rhodopes massif, situated in the Paleogenic and Neogene depressions (Eocene-Pliocene age).

The Bulgarian territory falls entirely within the Alpine-Himalayan orogeny [7]. Deformations have been related to the great Precambrian stage of development of the highly metamorphic complexes as well as the Caledonian-Varistcian deformations,the early and late Alpine deformations and deformations related to the neo-tectonic stage.

The superposition of various tectonic movements in the course of time has led to the formation of folded structures varying in type and amplitude as well as terrains of block or block/thrust structures. The associated rock complexes are strongly deformed and possess hydrogeological and geotechnical conditions which are difficult to model. Such complexity applies especially to the Balkanides, the Kraishte, the Sakar-Strandzha zone and parts of the Rhodopes massif (*Figure 1*). Fault zones exist in these morpho-tectonic units, along which neo-tectonic movements have occurred of amplitude up to 3,400 m, accompanied by the crushing of rocks and by the development of exogenic geological processes.

4. Suitable Geological Formations

4.1. LOWER CRETACEOUS MARLS

A vast depression was formed in the Fore-Balkan in North Bulgaria during the Lower Cretaceous where sediments composed mainly of marl were deposited [8]. These sediments have been extensively investigated both at surface and at depth in the course of exploration for oil and gas, and their tectonic conditions and lithostratigraphic structure have been studied in detail [9]. The sediments have been subjected to additional analysis for the purposes of the problem considered in three reports which are supplements to the BAS Concept for a national radioactive waste repository [10,11,12].

The information presented below is obtained from these reports and the results from the analysis show that several areas situated in North-West and North-East Bulgaria are most promising for future, more detailed investigations (*Figure 1*). The regions consist of comparatively pure clayey marls of great thickness - up to 1,000 m and more. The marls in Central North Bulgaria comprise thick stacks of limestones and sandstones. Some of these rocks are water-bearing and for this reason are excluded from further consideration.

4.1.1. *Prospective Regions in North-West Bulgaria*

Several regions between the Iskar and Ogosta rivers, located 60-80 km to the south of the Kozloduy NPP, fall into the category of prospective regions. The locations exist in thc northern strip of the West Fore-Balkan, which is characterized by relatively calm neo-tectonics. Intensive tectonic activity had ceased after the Pyrenean phases, when the Fore-Balkan was formed. There are no active faults in the vicinity of the sites and the seismicity is of degree VII (MSK scale) for a period of 10,000 years.

Gas deposits have been discovered in the region and for this reason, more than 150 deep boreholes have been drilled, allowing lithostratigraphic and geophysical investigations as well as a study of the geological structure both at the surface and at depths up to 3,500 m.

The Sumer (Aptian-Albian) and Mramoren (Barremian-Aptian) Formations, comprising extremely thick clayey marls, are of greatest interest from the point of view of the construction of a deep repository (*Figure 2*). The available hydrogeological data prove that the marls are dry and possess a low water permeability.

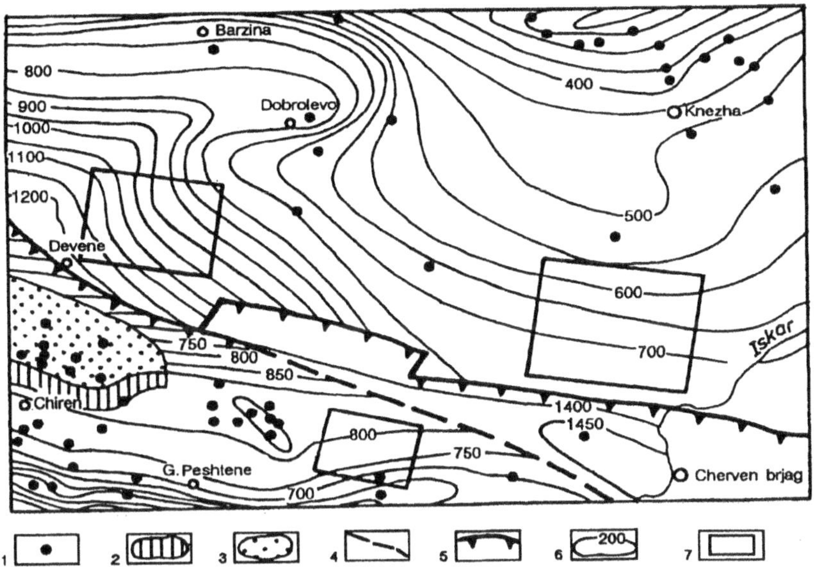

Figure 2. Map of equal thickness of the Mramoren Formation (West Fore-Balkan); after [10].

1 - borehole; 2 - natural outcrop of the Mramoren Formation; 3 - a zone of pre-Neogene washing; 4 - fault; 5 - boundary between the Balkan and the Fore-Balkan morphotectonic zones; 6 - isopachs; 7 - prospective sites for further investigation.

4.1.2. *Prospective Sites in North-East Bulgaria*

The Gorna Orijahovica Formation is the most interesting of all the Lower Cretaceous sediments in North-East Bulgaria with respect to its suitability for the deep disposal of HLW [9]. The formation consists of clayey marls where terrigenous intercalations are almost absent. The selected area near the village of Zlatar, Shoumen district, is situated in a shallow anticline and is characterized by calm neotectonics and lack of active faults. The marl thickness at this location exceeds 600 m. The Ticha Formation, which is a thick buffer padding also containing thick marl layers, occurs under the Gorna Orijahovica Formation. The seismicity of the region is of degree VIII (MSK scale) according to the seismic map for a 1,000 year period.

The BAS Concept accepts marl terrains as prospective host formations for both a deep HLW disposal site and a surface-type repository for L/ILW. With respect to difficulties relating to the attitude of the population concerning a radioactive waste repository, it would appear advantageous to build repositories for both types of wastes at the same place.

Two types of repositories are proposed for consideration for the deep disposal of HLW: mine and borehole type. The comparatively small volumes of this type of waste allows the possibility for the construction of the second type of repository, which involves less expense and possesses certain technological and ecological advantages. Moreover, considerable experience exists in Bulgaria in the drilling of deep boreholes in marl rocks including, relating to the exploration for oil and gas deposits.

4.2. NEOGENE CLAYS

Neogene clays cover vast areas in North-West Bulgaria and are located from about 10 km to 100 km from Kozloduy NPP (*Figure 1*). These clays are described in a special report on the BAS Concept for a national repository [13].

The Neogene sediments from the Brusartsi, Smirnenski, Florentin, Krivodol and Deleina Formations exhibit considerable lithologic variability - clays, sands, limestones and sandstones. In some areas the clays have great thickness - up to 100 m, but the available geological information does not permit the countouring of individual areas which are potentially suitable for further investigation. A negative characteristic of the Neogene clays is the existence of water-bearing sand lenses which are used as a source of potable water.

4.3. MAGMATIC AND METAMORPHIC COMPLEXES

Several sites in South-East Bulgaria consisting of magmatic and highly metamorphic rocks, are proposed in the BAS Concept. After a comprehensive analysis of natural conditions in this part of the country, the following sites were selected (Nos. 9, 10, 11, 13, 14, 15 and 16 in *Figure 1*):

- the Byala Reka formation, containing Precambrian highly metamorphic gneisses;
- the Avren and Zalti Chal sites, consisting of meta-ultrabasites;
- the Sakar site, built of Palæozoic granites; and
- two sites in the East Rhodopes Paleogene depression, built of volcanic and volcanogenic-sediment rocks.

Preliminary explorations have been performed on these sites and the results are presented in papers and in the respective reports of the Concept [14,15,16,17,18]. In addition, the Precambrian granite-gneiss formations discovered in the core of the Madan-Davidkovo structure are also an interesting site for investigation.

The Sakar site in south-east Bulgaria (No 13 in *Figure 1*) fits entirely within the mountain region of the same name (*Figure 3*). The region com-

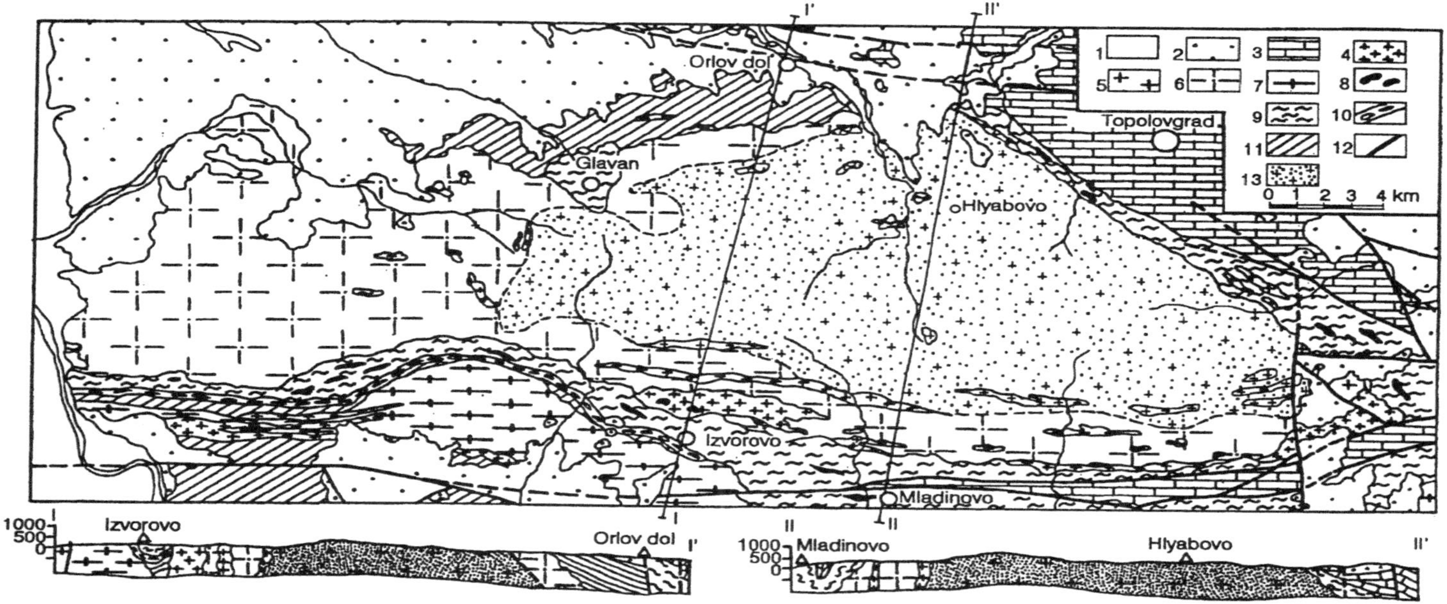

Figure 3. Geological map of the Sakar pluton; after [14].

1 - Quaternary; 2 - Pliocene sediments; 3 - Upper Palæozoic and Triassic metasediments; 4 - leucocratic (Alexandrovo) granites; Sakar Pluton: 5 - medium-grained granites; 6 - porphyritic granites; Precambrian: 7 - Lessovo gneiss-granites; 8 - meta-ultrabasites; 9 - Zalti Chal variegated Formation; 10 - Konstantinivo metaconglomerate Formation; 11 - Panovo porphyro-blastic Formation; 12 - faults; 13 - a suitable area for further investigations.

prises Precambrian high-crystalline metamorphites, Palæozoic metasediments and granites, Triassic metasediments and Pliocene formations [19,20]. Tectonically, the region falls within the Sakar unit, a component of the Sakar-Strandzha tectonic zone. About 60% of the site consists of Palæozoic granites, known as the Sakar pluton. These granites have been extensively studied at the surface but there are no data relating to conditions at depth. There are no active faults ifrom the neo-tectonic stage - faults of this type are observed along the periphery of the mountain outside the limits of the pluton. The seismicity is of degree VII (MSK scale) for a period of 1,000 years.

4.3.1. *Characteristics of the Sakar Pluton*

The Sakar pluton intrudes entirely within the Precambrian high-crystalline ortho- and para-metamorphites of the Pre-Rhodopian supergroup [14,15], assumed to be of Archæozoic - Lower Proterozoic age (*Figure 3*). The pluton intersects and has partially metamorphized (by contact) the porphyritic blastic magmatites of the Panovo Formation, the various gneisses, gneiss-schists and schists, amphibolites, graphite- and iron-bearing quartzites, *etc.*, of the Zalti Chal variegated Formation. Numerous xenoliths of the rocks listed above occur in the peripheral parts of the pluton. The Sakar pluton intersects and has altered by contact metamorphism the Upper Carboniferous sediments at its western end.

There are three basic varieties of granitic rock of the Sakar batholith (*Figure 3*), *viz.*

- uniformly-grained;
- porphyritic and porphyroid; and
- aplitoid-pegmatoid (Alexandrovo).

The porphyritic granites are found in the western parts of the pluton. The porphyries consist of potassium feldspar and phenocrysts are characterized by a great variety of sizes and composition. The length of these crystals typically varies from several centimeters to 7–8 cm, sometimes reaching 12–15 cm. The porphyritic granites consist of both biotite and two-mica varieties.

The uniformly-grained granites are medium to coarse crystalline. They make up the region between the villages of Glavan, Balgarska Polyana, Hlyabovo, Sakartsi and the neighbourhood of the Kayraka and Karaburun peaks. The mica content consists mainly of biotite, although two-mica varieties also occur.

The mineral composition of the two facial varieties is the same, *viz.* quartz, plagioclase, microcline, biotite and, less frequently, muscovite an the accessory minerals of apatite, zircon, garnet, sphene and clinozoisite.

Biotite has often been replaced by muscovite and some microclines exhibit later crystallization and active erosion of plagioclases.

The Alexandrovo leucocratic aplitoid-pegmatitic granites are distributed more widely than shown in the map (*Figure 3*). These granites occur in the form of elongated bodies and veins mainly in the east-west direction in the southern parts of the pluton, where they intersect rocks of the Panovo and Zalti Chal Formations and the two facial varieties of the Sakar batholith.

The granites are fine- and medium-grained and, less frequently, coarse-grained. Three varieties of granite are distinguishable: medium-grained muscovite, coarse-grained pegmatoid and fine-grained aplitoid. Although the individual varieties are interconvertible, the medium-grained granites are dominant, while the aplitoid granites mainly occupy the peripheral parts of the rock formations or contribute to some of the veins. In terms of mineral composition, the latter granites differ from the other two varieties of the Sakar batholith only in the biotite content, which is quite rare or completely absent from the Alexandrovo granites.

Different opinions exist concerning the age of the Alexandrovo and Sakar granites, covering the range from the Archæzoic to the Jurassic. On the basis of the direct geological relationships between the Alexandrovo and Sakar granites, which intersect and alter (contact metamorphism) rocks of the Palæozoic Klokotnitsa Formation, in conjunction with the presence of granite fragments in the Lower Triassic metasediments (Sakar type), there is justification for assuming that their age is Upper Palæozoic.

Three categories of terrains are distinguished in the Sakar pluton according to its internal structure. The site, consisting of medium-grained granites, indicated by a special raster in the geological map (*Figure 3*), is considered to be the most suitable for further investigation and possible construction of a radioactive waste repository.

5. Conclusions

The following conclusions can be drawn on the basis of the studies and investigations carried out:

It is accepted in Bulgaria, as in the other countries, that a HLW repository should be located deep under the earth's surface. It is recommended that future investigations continue in two alternative regions:

- north-west Bulgaria, which is nearer to Kozloduy NPP but is more densely populated and of greater economic importance, and
- south-east Bulgaria, which is not so densely populated and of smaller economic importance, but at a greater distance from Kozloduy NPP.

Several prospective sites located in north-west Bulgaria consist of thick formations (more than 1,000 m), well studied clayey marls, some of them having been found at the surface, *e.g.* the Sumer Formation. These sediments are practically dry, have not been affected by tectonic movements during the last geological ages, and fall into regions with seismicity lower than degree VIII (MSK scale) according to the 10,000-year seismic map. The formations are suitable for the construction of a repository of both mine and borehole type. In the zones where marls have been discovered at the surface, a repository for both HLW and LLW radioactive wastes can be constructed, which is advantageous from a socio-economic and ecological perspective.

The prospective sites in south-east Bulgaria are the Sakar granite pluton and some massifs in the East Rhodopes consisting of gneisses, volcanic rocks and serpentinites. These massifs are extremely thick and have a vast surface distribution but, owing to the lack of deep boreholes, the hydrogeological conditions at depth are unknown.

The number of the selected sites should be reduced to 3-4, since the detailed investigations preceding repository design are expensive and of long duration. In this regard, additional investigations are necessary in order to collect the required site-specific information, which is indispensable for the screening and selection of the most suitable areas.

The national legislation and strategy (national programme) for radioactive waste management should be developed in parallel with investigative and technical activities. Long-term work should also be carried out with the public and social organisations to explain the real and imaginary hazards related to radioactive wastes. An independent (autonomous) specialist organisation devoted to the management of the whole range of problems associated with radioactive waste should be established, in the same way as in western countries which have well-developed nuclear power generation programmes.

Acknowledgements

The authors express their special gratitude to Prof. T. Nikolov, Corr. Members of the Bulgarian Academy of Sciences, Prof. B. Monov, Prof. E. Kozhoukharova, Prof. N. Popov, Prof. N. Ruskova and to all colleagues who have made a substantial contribution to the characterisation of prospective areas towards the selection of a site for the national radioactive waste repository.

References

1. MILANOV, M., EVSTATIEV, D., AND KOZHOUKHAROV, D. (1993) Concept for National Repository for Radioactive Wastes in the Republic of Bulgaria (Summary), Bulgarian Acad. of Sci., Sofia, 40 pp.
2. Map of the Geological Hazards in Bulgaria. (1994), I. Iliev. Solit, VTS, Troyan.
3. KOSTADINOV, V. AND KOZHOUKHAROV, D. (1992) Categorization of terrains in the Bulgarian territory, Concept of Bulgarian Acad. of Sci. for a national radioactive waste repository.
4. COMMISSION OF THE EUROPEAN COMMUNITIES (1979) European catalogue of geological formations having favourable characteristics for the disposal of solidified high-level and/or long-lived radioactive wastes, Introduction and Summary, Volume compiled by BRGM, France, 79 pp.
5. EVSTATIEV, D. AND VACHEV, B. (1993) Site selection procedure for high level radioactive waste disposal in Bulgaria, IAEA regional project for Technical Assistance PEP/9/010, Recommendations on management of radioactive VVER reactor wastes, Meeting in Sofia.
6. EVSTATIEV, D., ANGELOVA, R., KARASTANEV, D., MCEWEN, T.J. AND STENHOUSE, M.J. (1996) Site selection methodology for deep disposal of high level radioactive wastes in marl terrains, Report to NATO under Linkage Grant No. OUTR.LG 941382.
7. BONCEV, E. (1971) *Problemi na balgarskata geotektonika (Problems of Bulgarian Geotectonics)*, Technika, Sofia, 203 pp. (in Bulgarian).
8. NIKOLOV, T. AND KHRISCHEV, K. (1965) Osnovi na stratigrafiata I litologiata na dolnata kreda v Predbalkana (Principles of the stratigraphy and lithology of the Lower Cretaceous in the Fore-Balkan), *Geology of Bulgaria, Series: Stratigraphy and Tectonics* **6**, 77-175.
9. NIKOLOV, T., RUSKOVA, N. AND KHRISCHEV, K. (1991) Osnovi litostratigrafii nizhnemelovogo otdela v Bolgarii (Principles of the Lower Cretaceous lithostratigraphy in Bulgaria), *Geologica Balc.* **21**, 6, 3-47 (in Bulgarian).
10. MONOV, B. (1992) Lithostratigraphy of the Lower Cretaceous in North-West Bulgaria (the Lom depression and the West Fore-Balkan), Concept of Bulgarian Acad. of Sci. for a national radioactive waste repository.
11. MONOV B. (1993) Prospective sites for investigations concerning the construction of a surface-type repository for low- and medium activity radioactive wastes in the Lower Cretaceous marls in North-West Bulgaria, Concept of Bulgarian Acad. of Sci. for a national radioactive waste repository.
12. NIKOLOV, T. AND RUSKOVA, N. (1992) Geological structure and a complex geological assessment of the Lower Cretaceous in the Fore-Balkan between the rivers of Ossam and Louda Camchia with suggestions for the location of a radioactive waste repository, Concept of Bulgarian Acad. of Sci. for a national radioactive waste repository.
13. POPOV, N. (1992) Geological characteristics of terrains in the clay formations in Vidin, Montana and Pleven regions, prospective for National Radioactive Waste repository investigation, Concept of Bulgarian Acad. of Sci. for a national radioactive waste repository.
14. KOZHOUKHAROV, D. (1984) Lithostratigraphy of Precambrian metamorphics from Rhodope supergroup in the Central Rhodopes, *Geologica Balc.* **14** (1), 43-88.
15. KOZHOUKHAROV, D. (1987) Lithostratigraphy and structure of the Precambrian metamorphics from the core of the Byala-Reka dome, East Rhodope Mountains, *Geologica Balc.* **17** (2), 15-38.
16. KOZHOUKHAROV, D. (1992) Composition, structure and complex geological assessment of the Byala-Reka structure in the East Rhodopes, Concept of Bulgarian Acad. of Sci. for a national radioactive waste repository.

17. Kozhoukharova, E. and Kozhoukharov, D. (1992) Petrographic characteristics and complex geological assessment of the ultrabasite bodies in the region of the Zalti cal village, Ivailovgrad district, Concept of Bulgarian Acad. of Sci. for a national radioactive waste repository.
18. Kozhoukharova, E., Kozhoukharov, D., Pristavova, S., and Ichev, M. (1992) Geological structure and complex geological assessment of the 'Sakar' site, Concept of Bulgarian Acad. of Sci. for a national radioactive waste repository.
19. Kozhoukharov, D., Savov, S., Catalov, G., Kozhoukharova, E., Goranov, A., and Bojanov, I. (1994) Geological map of Bulgaria at scale 1:100,000, sheet Topolovgrad.
20. Kozhoukharov, D., Bojanov, I., Kozhoukharova, E., and Goranov, A. (1994) Geological map of Bulgaria at scale 1:100,000, sheet Svilengrad.

COMPLEX GEOLOGICAL INVESTIGATIONS TO SELECT A SITE FOR HIGH-LEVEL WASTE DISPOSAL IN THE KRASNOYARSK REGION (RUSSIA)

E.B. ANDERSON, V.G. SAVONENKOV, S.I. SHABALEV & YU.M. ROGOZIN
V. G. Khlopin Radium Institute
St. Petersburg, RUSSIA

E.F. LYUBTSEVA
St. Petersburg State University
St. Petersburg, RUSSIA

YU.A. REVENKO, YU.I. SABAEV & A.V. NOSUKHIN
Chemical Mining Company, Krasnoyarsk-26
Krasnoyarsk, RUSSIA

V.L. MILOVIDOV
VNIPI Promtechnologii
Moscow, RUSSIA

N.V. LUKINA
GIN RAN, Moscow, RUSSIA

A.P. LOPATIN
State Centre "Priroda", Krasnoyarsk branch
Krasnoyarsk, RUSSIA

V.M. DATSENKO
KNIIGiMS
Krasnoyarsk, RUSSIA

AND

V.A. KRYZHANOVSKI
"Krasnoyarskgeolsyomka"
Krasnoyarsk, RUSSIA

M. J. Stenhouse and V. I. Kirko (eds.),
Defence Nuclear Waste Disposal in Russia: International Perspective, 135–141.

1. Introduction

The disposal of high-level radioactive waste (HLW) in deep geological formations is now considered by the international community as the ultimate stage in the optimum solution to the general problem of radioactive waste management, although other, more hypothetical, possibilities, (*e.g.* transmutation, space disposal) have been considered.

It is no exaggeration to state that the creation of an underground HLW repository in the Krasnoyarsk region is the greatest of the key environmental and economic problems governing the very existence of the large nuclear company, the Mining & Chemical Company (MCC), which generates considerable quantities of radioactive wastes, including HLW. On the other hand, radioactive waste disposal is not merely a problem for the Krasnoyarsk region, but is one of the key problems determining the future development of the Russian nuclear power industry in general, particularly if the nuclear fuel cycle includes the reprocessing of spent nuclear fuel which is planned for the RT-2 plant currently under construction in the Krasnoyarsk region.

It is no accident, therefore, that the Russian Federal programme on radioactive waste management includes a special topic devoted to characterisation work in the Krasnoyarsk region with a view to selecting a site for the construction of a repository.

The results of a number of international scientific research programmes have demonstrated that in the case of long-term storage of HLW and intermediate-level wastes (ILW) on the earth's surface, the risk to the population and to the biosphere in general is too great and, consequently, unacceptable. Thus, the concept of an underground radioactive waste repository has wide support among the leading nuclear countries, including the U.K., France, U.S.A., Sweden, Finland and Germany. In these countries, projects to construct repositories for the geological disposal of radioactive waste are either underway or are planned.

2. Methodology for Site Investigations

The RPA V. G. Khlopin Radium Institute (VGKRI), together with leading specialists of the Atomic Power Ministry, the Russian Academy of Sciences, and institutes and organisations of the Krasnoyarsk region, have been carrying out complex scientific research works for several years. In this context, the methodology for the selection of a site is based on criteria and concepts for the safe isolation of radioactive waste in geological formations, developed by the IAEA [1].

These criteria and basic requirements embrace all aspects of the construction and operation of underground disposal systems, and describe:

- site configuration,
- geology,
- hydrogeology,
- radionuclide migration characteristics,
- tectonics and seismicity,
- man-made and natural features,
- resource potential, and
- land surface considerations.

In parallel with geological criteria, the physical, geographical, socio-economic and ecological features of the region under consideration are taken into account.

The work is carried out in consecutive stages, each successive stage aimed at providing ever-increasing confidence in demonstrating the safety of a proposed site. This sequential approach also incorporates successive changes in the scale and level of detail of the investigations.

3. Geological Investigations in the Krasnoyarsk Region

Initially, large structural elements of the continental crust were analysed within a radius of 500 km from the MCC, *viz.*

- the ancient Siberian platform to the east;
- the young West-Siberian plate to the west;
- the Altai-Sayan orogenic area to the south;
- the Yenisei ridge to the north and the location of the MCC site.

The most promising area for field work was determined to be the Yenisei ridge, a large outcrop of the ancient Archæan-Proterozoic basement of the Siberian platform.

The next stage of work limited the area of search, but broadened the range of methods used and increased the level of detail of the investigations. The most promising geological formation for waste disposal was chosen from the metamorphic and igneous rocks of the ridge. Accordingly, the following studies were carried out:

- analysis of complex geophysical data, and
- interpretation of aerial and space photographs and of the topographic characteristics of the area.

Via a combination of features, the most uniform and monolithic areas were selected from within the southern part of the Yenisei ridge. Initially, within an area of 9,000 km^2, about 20 'plots' within the oldest gneiss bedding and granitic rock masses were selected for further consideration. The reliability of the supporting information was checked by comparing the results obtained by different organisations using different methods.

During the first stage of the site selection process, several formations were considered as possible candidate host rocks for the isolation of radioactive waste:

- the oldest metamorphic rocks of the Kanian series;
- the granitic rock mass of the Taraksky massif;
- the granitic rock mass of the Belogorsky massif; and
- certain other geological formations.

For various reasons, however, all of the above options were rejected, and the granite of the Nizhnekansky massif was selected as the most promising in which to construct an underground disposal facility. In reality, all groups of specialists involved in the work reached this conclusion essentially independently, using different methods and approaches.

The name of the rock mass is determined from its geographical location: within the low flow of the Kan river which discharges into the Yenisei river, on its right bank, 90 km downriver from Krasnoyarsk city. With respect to structural aspects, the massif intersects the Baikal-Yenisei deep fracture zone. The host rocks comprise Archæan crystalline schists and gneisses. The massif outcrops at the Earth's surface, with individual areas overlain by Jurassic and Devonian deposits.

On the whole, the Nizhnekansky massif is a large granitic pluton of 'ethmolith' (bowl-like) shape, with a steep crack-like feeding channel in the axial part. The granitic rock, about 15–30 km wide, is elongated in a north-west direction, coinciding with an extension of the Archæan host rocks; outcrops more than 60 km long can be observed. The exposed area extends to more than 1,500 km^2 which, together with the buried portion, lead to an overall area of 3,500 km^2. Thus, the Nizhnekansky massif is one of the largest granitic masses of Middle Siberia, with its north-western extremity located 3 km from the MCC.

The Nizhnekansky massif was formed over several discrete stages:

- a first (early) phase represented by quartz diorites and granodiorites;
- a second (main) phase represented by biotite granites (often having porphyritic aspect) and adamellites;
- a final stage, during which vein deposits of aplites, pegmatites, diabases, diorite-porphyres, lamprophyres, *etc.* appeared, their most probable age being 850±60 Ma.

The north-western part of the Nizhnekansky massif, proposed for detailed investigation, is a forested low-relief area divided by valleys formed by tributaries of the right-bank of the Yenisei (the Big and the Small Tels) and of the left-bank of the Kan (the Big and the Small Itats). A reconnaissance study, carried out in 1993, allowed the topographic and photographic elements for carrying out surface testing of the granitic rock to be specified, as well as carrying out reconnaissance geophysical surveys using the radio-electromagnetic profiling and audio-magnetotelluric sounding methods.

The most recent tectonic disturbances of the investigated part of the massif are reflected in the river network, thereby indicating the most monolithic blocks of rocks in the terrain. Such a type of regional geomorphological criterion may be used to establish the possibility of locating the construction site on the massif's watershed areas.

The field work yielded the necessary information for mapping the massif's tectonic disturbance on a 1:100,000 scale and subsequently, on a larger scale, in order to characterise the location and configuration of the most homogeneous areas of granite. Analysis of the data obtained showed that a series of active faults had developed within the massif studied, predominantly of submeridional extension, with a relatively pronounced step of 4-5 km. These faults limit the dimensions of 'uniform' rock blocks to a maximum of 20 km in length. The most uniform portions of these blocks have dimensions 3 km by 4 km. In principle, such dimensions are sufficient to permit the construction of any type of burial facility (borehole or shaft facilities).

At present, several sites in the Nizhnekansky massif are being considered for further detailed investigations. According to preliminary data, the thickness of granites in this part of the massif is at least 4-5 km, the depth of the rock fracture zone reaching 300-350 m.

4. Geochemical Studies

The surface and ground waters of the region are 'fresh', with a total dissolved solids content in the range 0.13–0.44 g l^{-1} of Na-HCO_3 type, and having a low degree of mineralisation. The latter points to infiltration as the predominant source of groundwater. Based on hydrogeochemical investigations carried out in 1995, no anomalies were discovered within the boundaries of the massif to indicate the presence of open active fractures through which water could penetrate from deep horizons.

For a subsurface disposal system intended for long-lived radioactive waste, one of the main components for isolating radionuclides is provided by the host rock itself. Thus, investigation of the barrier properties of the granite of the Nizhnekansky massif is important.

Radio-geochemical studies carried out at VGKRI indicated that for typical rock samples, the following values for the distribution ratio, R_d (experimental measurement of the distribution coefficient, K_d), for certain radionuclides were obtained:

Radioisotope	R_d $cm^3\ g^{-1}$
plutonium-238	2000–5000
americium-241	500–900
caesium-137	60–100
strontium-90	5–150
neptunium-237	3–6

Thus, the sorption on the rock (granite) appears to be adequate for plutonium, americium, caesium and strontium, but rather low for neptunium. It was also found that the chemical 'form' (speciation) of plutonium and americium in solution has great significance on the sorption behaviour of these elements.

In addition, the available experimental data lead to the conclusion that for the granites, the formation of weathering zones and migration channels in fracture zones is accompanied by the appearance of secondary sorbent minerals that act as a barrier to natural migration. A similar retardation function will be fulfilled by the argillaceous rocks (Jurassic deposits) which overlie the granite in some areas of the Nizhnekansky massif.

5. Summary

The investigations carried out so far in the Krasnoyarsk region have allowed the following conclusions to be made in favour of the suitability of the Nizhnekansky massif as host rock for a repository:

- The granitic rock of the Nizhnekansky massif was chosen after analysis and comparison of several regional structures, geological formations and massifs, using the IAEA basic criteria, and consideration of alternative options and safety requirements.
- Granites are universally recognised as a favourable medium for radioactive waste isolation. These rocks continue to be the focus of various studies including site-specific investigations, within the scope of national and international scientific programmes.
- The area of the earth's crust which includes the Nizhnekansky massif is stable tectonically and is characterised by slow uplift, currently ≤ 1 mm a^{-1}.
- According to expert evaluation, the territory which includes the north-western portion of the Nizhnekansky massif, exhibits low seismic activity, with the maximum predicted earthquake in the region under investigation not exceeding magnitude 5 (Richter scale).

- The dimensions of the comparatively uniform granite blocks selected are large enough to allow all surface and underground structures associated with the disposal facility to be located in one site.
- Economic and industrial activities within the designated areas of the massif are relatively limited, since the areas are sparsely populated
- With regard to potential sites, a location in the vicinity of the MCC (within 10 to 30 km) would appear to be essential for solving transportation and communicational problems associated with the characterisation, construction and operation of the underground disposal facility.

Further development of the complex investigations will involve:

- detailed geological and geophysical surveys (audio-magnetotelluric, radio-electromagnetic, seismic sounding) of one of the massif's rock masses;
- continuation of geochemical and hydrogeological investigations;
- acquisition of preliminary data for computer modeling of migration processes and of the complete waste isolation system within the granite host rock;
- continuation and broadening of investigations of the barrier properties of the granite and its weathering zones as well as of the overlying Jurassic deposits;
- preparation of the justification for the siting and drilling of a structural borehole with suporting multi-purpose scientific investigations;
- the development of a concept and associated criteria for the site-specific (Nizhnekansky granite) safe disposal of HLW and long-lived wastes generated by the MCC.

In order to provide a scientific justification for the suitability of areas within the Nizhnekansky massif for designing and constructing an underground repository in accordance with safety principles and technical criteria of Russia and of international authoritative bodies (*e.g.* IAEA), it will be necessary to carry out extended, labour-intensive and expensive investigations, and to participate in international programmes.

References

1. IAEA (1983) *Criteria for Underground Disposal of Solid Radioactive Wastes*, International Atomic Energy Agency Safety Series Report No. **60**, Vienna.

ON THE QUESTION OF PREDICTING THE SAFETY OF THE NORTHERN 'POLYGON' OF THE KRASNOYARSK MINING-CHEMICAL COMPANY BASED ON A GEOLOGICAL AND GEOPHYSICAL MODEL

F.B. BAKSHT, A.A. DERGACHEV, N.P. DOROFEEV,
S.V. COMISSAROV, L.A. SEDYH, B.P. CHESNOKOV &
M.A. CHURILIN
Krasnoyarsk Scientific Research Institute of Geology and Mineral Resources, Krasnoyarsk Geological Committee, Geophysical Section of Russian Academy of Sciences
Krasnoyarsk, RUSSIA

Abstract.
This paper raises an important public issue concerning the need for a detailed survey of the geology around the *Northern Polygon* with the purpose of defining the potential hazard to, and possible disruption of, the present stability of the zone 'created' for liquid radioactive waste disposal, as well as other rock masses within the territory of the MCC.

1. Introduction

Land disposal of liquid radioactive waste from the Krasnoyarsk Mining-Chemical Company (MCC) has been carried out for more than 30 years at the *Northern 'Polygon'*, located on the right bank of the Yenesei river, 50 km from Krasnoyarsk. The boundaries of this polygon are located 2.5 km and 4 km from the rivers Yenesei and Large Tel, respectively. Owing to limited geological investigations having been conducted in this region, knowledge of the geological structure is extremely poor. Geological and geophysical investigations of the territory came to a finish at the end of the 1950's and, since this time, have not been renewed.

Liquid radioactive waste was pumped into 'collector-layers', occurring at depths of about 150 m and 450 m. These layers consist of weakly-cemented, sand-gravel Jurassic deposits, separated bv clay horizons. Jurassic deposits

M. J. Stenhouse and V. I. Kirko (eds.),
Defence Nuclear Waste Disposal in Russia: International Perspective, 143–146.

were formed in the Telsk erosion-tectonic depression in the rocks overlying the Precambrian metamorphic complex (gneisses, granites *etc.*). To the west, and possibly to the north, a depression, approximately 10 km × 15 km, is bounded by remnants of downward sloping regional fractures. The depth of this depression extends to 600 m in places of maximum displacement.

According to a number of researchers (Nosuhin A.V. and Rybalchenko A.I., personal communications), fracture zones and pre-Jurassic weathering of crust, can perform a barrier role in certain circumstances, hindering the penetration of liquid radioactive waste from the weakly-cemented rocks of the disposal zone into underlying metamorphic rocks and the river systems.

According to a map of the region, the region under consideration lies in the area adjacent to the Altai-Sayan and Baikal seismic zones, within the Trans-Asia seismic zone. The northern front of the Altai-Sayan mountains, distinguished by the highest modern geodynamic activity (up to 12 mm per year in the area between Omsk and Baikal), passes through the same zone. The general (average) seismicity of the Krasnoyarsk region lies in the range 5-6.

The geodynamic features of the region, combined with its deep geological structure, provide input for an interpretation of its modern tectonic structure.

The *Northern Polygon* is located in the centre of a downthrust block of crystalline rock of the ancient basement, strongly jointed and underlying the Telsk aquifer. This block is itself located in the centre of a complex geological system, resulting from the effects of centripetal-centrifugal stress owing to compression-stretching. Interpretation of satellite photographs and various geological maps established the presence of a number of closed zones containing intensive tectonic deformations, as reflected in the relief. The most recent zones characterise the occurrence of a local morphosculpture-depression (topographic feature), the origin of which is well-established as seismo-tectonic.

As an indirect illustration of this geologically complex structure, unfavourable from the point of view of tectonic activity, numerous local heterogeneities exist in Jurassic and Quaternary rocks, as revealed by geological and geophysical survey methods. Such heterogeneities are accompanied by various geomorphological and geophysical anomalies, most of an uncharacterised geological nature. The most recent anomalies were investigated by seismic prospecting (platform 27), geophysical prospecting by electric methods, and magnetic prospecting (in the location of the Polygon). Presumably, the anomalies can be attributed to the heterogeneity of the basement, by the unevenness of its surface owing to weathering and tectonic alteration of loose rocks, containing the collector-layers.

Given the lack of studies of the geological environment of the region of interest, which performs the major role of isolating liquid radioactive waste, there is an obvious need for an extensive geological investigation using modern, sophisticated techniques. We propose that such research should focus on three areas:

- an objective, reliable and accurate assessment of the geological environment;
- an evaluation of the risk of potential ecological danger from the artificially-created deposits containing radionuclides under current 'static' conditions;
- an evaluation of the probability of a technological accident occurring, caused by natural events such as strong, or even weak, earthquakes.

It should be noted that in the Krasnoyarsk region over the last 150 years, at least two appreciable earthquakes have occurred of magnitude 4 and 6. According to our calculations, such events can cause the movement of rocks which are acting as a barrier, by up to 40 mm. In addition, it is impossible to exclude the occurrence of potentially catastrophic situations owing to the cumulative effect of small deformations in non-uniform Jurassic rooks, causing the movement of sediments even under weak natural or man-made earthquakes.

We consider that the geological and geophysical investigations carried out in the area belonging to the MCC is rather limited and certainly does not meet the typical requirements of current regulatory bodies. Currently for the region, only an out-of-date (1963) geological map of scale 1:200,000 exists, although, owing to the need for an improved geophysical survey, a gravimetric survey at an intermediate scale has now been started. Other geological and geophysical survey techniques such as investigation of geodynamic conditions, micro-seismic zoning, *etc.*, have not been carried out, with the exception of a series of non-systematic engineering-geological surveys. Such a situation exists owing to the lack of agreement concerning the specific responsibilities of the many (12) organisations concerned with a study of the geological environment of the region.

A programme is now underway involving more complex geological and geophysical surveys of the MCC territory, coordinated by the Krasnoyarsk Geological Committee (*Krasnoyarskgeolcom*). This programme comprises four tasks:

1. Surveyg of the structure of the deep crust of the region using relatively complex, regional geological and geophysical investigations with modern techniques, within a framework of detailed seismic zoning.
2. Volumetric prospecting of the geological structure of the territory at depth (up to 12 km), to provide the input necessary to evaluate

natural and man-made seismicity. For this purpose, a geological-geophysical survey at an intermediate scale combined with detailed micro-seismic zoning, is underway.

3. Detailed volumetric prospecting of the *Northern Polygon* region and boreholes using surface, subsurface and borehole geophysical methods to identify rock structure, and to determine the distribution pattern of the liquid radioactive waste, as a means of solving hydrogeological and subsurface excavation problems.
4. Monitoring of the geological environment around the *Northern Polygon* and the subsurface excavations of MCC, in order to study natural and man-made seismicity in the region; also monitoring of the movement of liquid radioactive waste distributed in collector-layers. This task provides for the re-activation and modernisation of the seismograph for the *Northern Polygon*, establishing a programme for its development and upgrade.

The final selection of monitoring sites and the parameters to be measured, will be established based on the information received once the first tasks have been completed. At present, only preliminary conclusions can be made based on mathematical modelling using poor experimental data generated using relatively primitive techniques (in particular, borehole surveys).

Completion of the proposed programme coordinated by the Krasnoyarsk Geological Commitee will allow the seismic and radioactivity hazard of the rock mass to be evaluated with high precision, a necessary condition for the prediction of potentially extreme situations in the area and for the prevention of relatively severe stress problems among the population.

SITE CHARACTERISATION METHODS AND DATA INTERPRETATION – THE DRIGG SITE CHARACTERISATION

P M BOOTH

Environmental Assessments Partnership
BNFL, Risley, Warrington
Cheshire, WA3 6AS
United Kingdom

Abstract.

Further site characterisation work is being undertaken to underpin the impact assessment of BNFL's Drigg low level radioactive waste (LLW) disposal site. This site has been in operation for about forty years and is located approximately seven kilometres south of the Sellafield reprocessing plant in West Cumbria. The site accepts the majority of LLW produced in the U.K. which may be generated by hospitals and general industry as well as the nuclear industry itself. Historically, waste was tumble tipped into trenches which were then covered with an interim cap incorporating a plastic membrane. Waste is now placed in a concrete vault within steel Isofreight containers. This practice will continue through to the end of operations which is anticipated to be in the middle of the next century. Subsequently, a final closure cap will be emplaced over the disposal area.

The five year Drigg Site Characterisation programme was, therefore, set up in order to improve and build upon our current understanding of the geology and hydrogeology of the site. The first one and a half years are devoted entirely to data acquisition with the greatest activity surrounding the drilling and geophysical aspects of the programme. Four years will be devoted to collecting a high quality time series data set of water levels, stream flows and meteorological variables and to produce a comprehensive interpretation of the hydrogeology of the site.

The programme to date has been very successful and the acquired information has enabled us to start refining our geological model of the site which will then be used to underpin our eventual conceptual hydrogeological model.

M. J. Stenhouse and V. I. Kirko (eds.),
Defence Nuclear Waste Disposal in Russia: International Perspective, 147–160.

1. Introduction

The primary aims of this paper are to describe the site characterisation activities that have been conducted at BNFL's Drigg LLW disposal site and to show that such work can, and in many cases must, be undertaken at existing sites as well as during initial site selection activities.

Because Drigg is an operating site, there is a requirement for BNFL to produce a risk assessment for the site which, in turn, is ultimately fed into a site licensing and authorisation process. Through gaining a robust understanding of the geology and hydrogeology of the site, BNFL can be confident in the placement of the statutory monitoring boreholes. Understanding the groundwater flow paths across the site is only one of a number of factors which are built into the risk assessment, but it is the geology and hydrogeology requirements that the site characterisation was designed to underpin.

Situated adjacent to the coast, the site is approximately 1.5 km long and 0.75 km wide and lies about 7 km south of the Sellafield reprocessing plant (*Figure 1*). The land immediately surrounding Drigg consists of a combination of pasture, dunes and grazing land. Forested areas are prevalent on site and the River Irt drains into the Ravenglass Estuary to the south.

The existing geological data set led BNFL to infer that the near surface sediments comprised of a layered sequence of alternating clays and sands, complicated at times by the occasional gravel layer. It was assumed that the clay layers were more or less continuous.

2. Site Operations

Drigg has been in operation for about forty years and accepts most of the U.K.'s LLW which emanates from a number of industries and institutions including hospitals, universities and, of course, predominantly the nuclear industry itself. Historically, waste was tumble tipped into trenches which were subsequently covered with an interim cap incorporating a plastic membrane. Cut-off walls have been emplaced around two sides of the trench area in order to minimise water ingress, and both leachate and surface runoff are collected for analysis prior to appropriate disposal. Waste is now placed in a concrete vault within steel Isofreight containers, having first been compacted in drums and grouted. The photo in *Figure 2* was taken around 1990 and shows the northerly consented area with its trenches and disposal vaults.

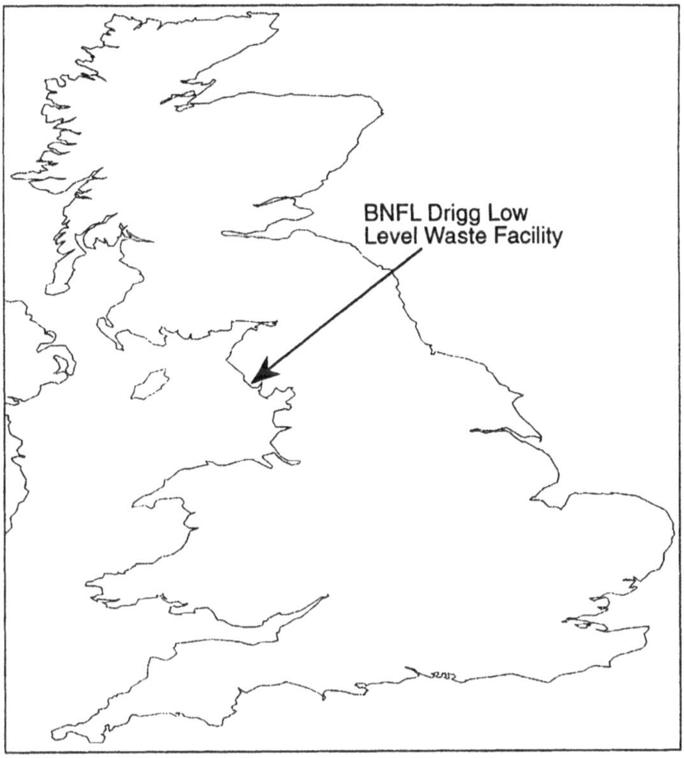

Figure 1. Location of Drigg LLW facility, U.K.

3. Site Characterisation

3.1. PROJECT DRIVER

As previously stressed, one of the main drivers for conducting a site characterisation on what is an existing and operating site is to fulfil regulatory requirements. In addition however, BNFL requires operational monitoring, wishes to further its confidence in the site, and provides information towards a post-closure risk assessment.

3.2. PROJECT STRUCTURE

Effectively a five year programme, BNFL's site characterisation will dedicate the first year entirely to data acquisition. This will essentially be through a combination of drilling, trial trenching and geophysics (*Figure 3*). The following four years will be devoted to collecting a high quality time series data set of water levels, stream flows and meteorological variables and to produce a comprehensive hydrogeological interpretation of the site.

Figure 2. Photograph of Drigg site (ca. 1990) showing trenches and disposal vaults.

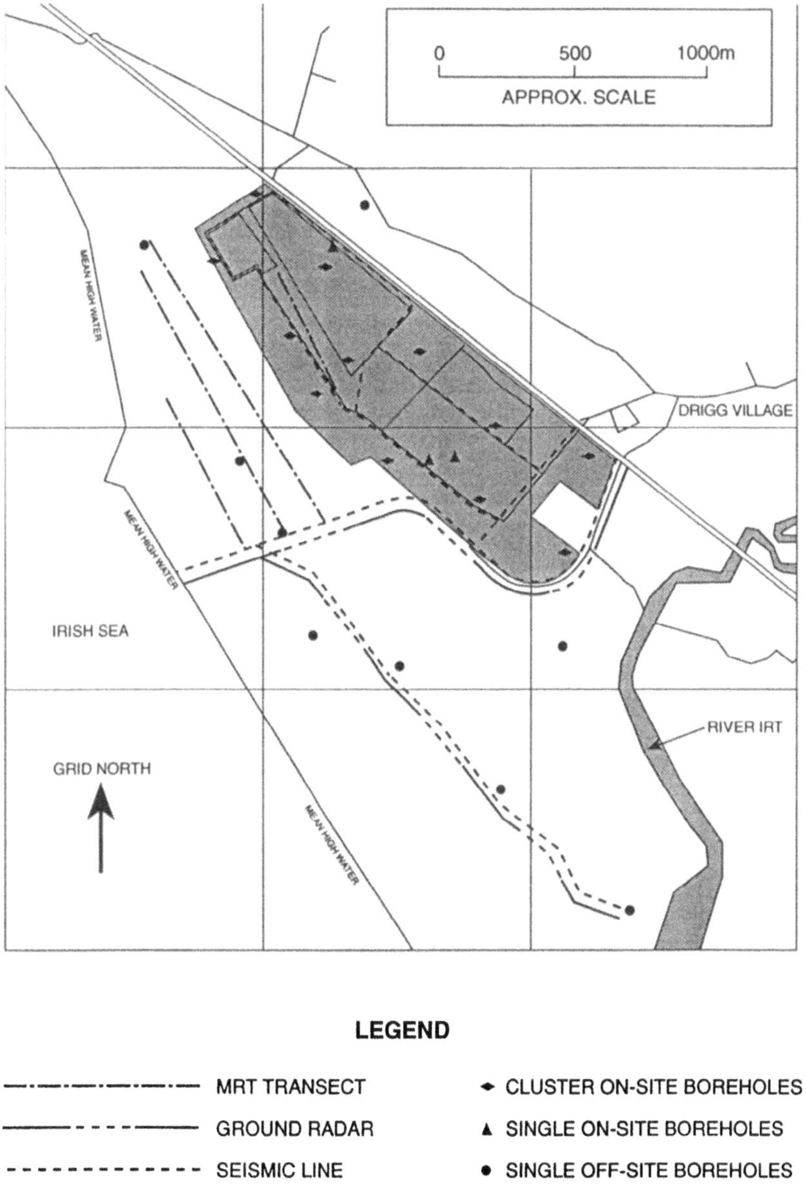

Figure 3. Site characterisation: geophysical techniques.

3.3. SITE CHARACTERISATION METHODOLOGIES

3.3.1. *Hydrology*

The characterisation project inherited a previously initiated high quality surface water monitoring programme. It was deemed important to study the site water balance at Drigg in order to gain a firm understanding of the hydrological regime. This was achieved through the utilisation of the following methods:

- The installation of weirs on the two streams which cross the site.
- Two automatic weather stations were installed which measured meteorological variables.
- Measurement of soil moisture via a network of neutron probes.
- Measurement of soil water tension by way of tensiometers.
- Three forest interception experiments were equipped with rain gauges.

3.3.2. *Geology*

As the Drigg site has been operating for about forty years, BNFL had an existing understanding of the geology in order to underpin their site assessment requirements. The previous geological understanding has now been slightly superseded as a consequence of more detailed site investigations carried out in the West Cumbrian region coupled with the fact that previous drilling methodologies only encountered the near surface sediments. It was important therefore for the Drigg Site Characterisation to consider the geology on a more regional scale and also to select appropriate drilling techniques which could successfully access and sample Quaternary deposits.

Bearing in mind our current understanding of the site geology, we selected the rotary SWF drilling technique, our aspiration being to extract the highest core recovery percentage possible through the Quaternary deposits down to rockhead (Permo-Triassic Ormskirk Sandstone). In all, eighteen rotary boreholes were drilled on, or adjacent to, the site at depths varying from 10 to 55 m.

We decided to excavate a number of trial pits and trenches, thus allowing us to view the geology in two dimensions and to log in greater detail the near surface structures and lithological relationships.

The chosen drilling methodology allowed us to progress through all the geology encountered across the site and obtain a high percentage of core recovery (*Figure 4*). Our work has shown that the simplified stratigraphical sequence consists of interbedded clays and sands, sands, sands and gravels (till), sands and silts, gravelly clays and sands/gravels (till) overlying bedrock.

Figure 4. Quality of core recovery from rotary (SWF) drilling technique.

3.3.3. *Hydrogeology*

A number of shell and auger boreholes were drilled adjacent to the rotary boreholes to form clusters. These boreholes were specifically designed to provide hydrogeological information and were drilled without the addition of water. This allowed water samples to be taken during drilling operations therefore providing a geochemical profile down the borehole. In addition to providing good geochemical data, the dry drilling technique enabled us to record accurate heads and determine water strikes. During drilling a number of field tests were carried out with samples taken for laboratory analysis. These tests are summarised below:

- Rising head tests to measure hydraulic conductivity.
- Standard Penetration Tests (SPT) to measure density.
- Particle Size Distribution (PSD) to measure soil grading.
- Constant head permeameter tests to measure hydraulic conductivity.
- Falling head permeameter tests to measure hydraulic conductivity.
- Bulk density tests.
- Moisture content tests.
- Hydrogeochemical sampling.

A high quality data set was obtained from these boreholes through the utilisation of careful drilling and sampling methodologies which were overseen at all times by a field hydrogeologist. By combining these data with those acquired from the rotary boreholes, it was possible to build up a detailed geological picture at each cluster location. By creating borehole clusters, it was possible to place a completion at each identified aquifer unit at the different cluster locations, rather than opting for multi-level completions within a single borehole.

Thirty boreholes were fitted with pressure transmitters connected to logging units which record water level variations every fifteen minutes. This, in turn, will allow any tidal or atmospheric pressure effect on the water level to be recorded.

3.3.4. *Geophysics*

It is generally recognised that a professionally designed geophysics programme has the potential to provide a quick and cost-effective way of obtaining additional geological and hydrogeological information across a site. Geophysical surveys can be utilised in order to gain information prior to the culmination of a drilling programme, or, where geological control already exists, can assist in interpreting the geology in areas where boreholes are not present. Geophysics is more likely to be successful when a number of techniques are run in tandem or when combined with intrusive methodologies, rather than when utilised as a stand alone tool. There is often a misconception that geophysics will solve all subsurface investigation problems but this is clearly not true. It is imperative that the chosen technique and the survey design is targeted at the specific problem in hand.

Bearing in mind our previous geological knowledge of the site, its topography and accessibility, we chose four geophysical techniques within a programme which allowed close cross correlation. The seismic refraction survey was run over the road and track network and was initially designed to confirm the depth to rockhead (*Figure 5*). However, due to the nature of the underlying strata, the seismic signal was being absorbed by the overlying layers. This allowed us to define three separate velocity layers but not rockhead itself. Ground Penetrating Radar (GPR) was run over the road network and provided structural and stratigraphical information in the near-surface sediments down to depths of about 10 m (*Figure 6*). Micro Resistivity Technique (MRT) was also utilised across ground where we were unable to drill as well as adjacent to some borehole locations (*Figure 7*). This technique was successful in that it allowed us to correlate specific features like channels across the site. All the rotary drilled boreholes have been electrically logged using natural gamma, gamma-gamma and neutron porosity. This has again provided invaluable cross-correlation data against the cores and borehole samples.

4. Developing Techniques

When the site characterisation programme was evolving, we made a decision to try a number of developing technologies at the site in the hope of helping to develop cost effective solutions to site investigation. Two such techniques are briefly described below.

4.1. ELECTROKINETIC SURVEY SYSTEM (EKS)

EKS works through the detection of electrokinetic signals which are generated when seismic impulses vibrate porous rocks containing water or brine. The equipment consists of a seismic source and a signal acquisition system. The impulse can penetrate depths of 100 m and the returning signals are detected via a simple antennae array coupled with electronic circuitry connected to a computer. The computer in turn encompasses signal acquisition, processing and data storage facilities. Specifically designed software allows the processed signals to be converted into aquifer permeability, from which borehole flow can be predicted. (The system utilised by BNFL has been patented by GroundFlow Ltd).

The acquired hydraulic conductivity data can be utilised to discover aquifers, design drilling programmes and compare actual with predicted flows. We were specifically interested in attempting to determine hydraulic

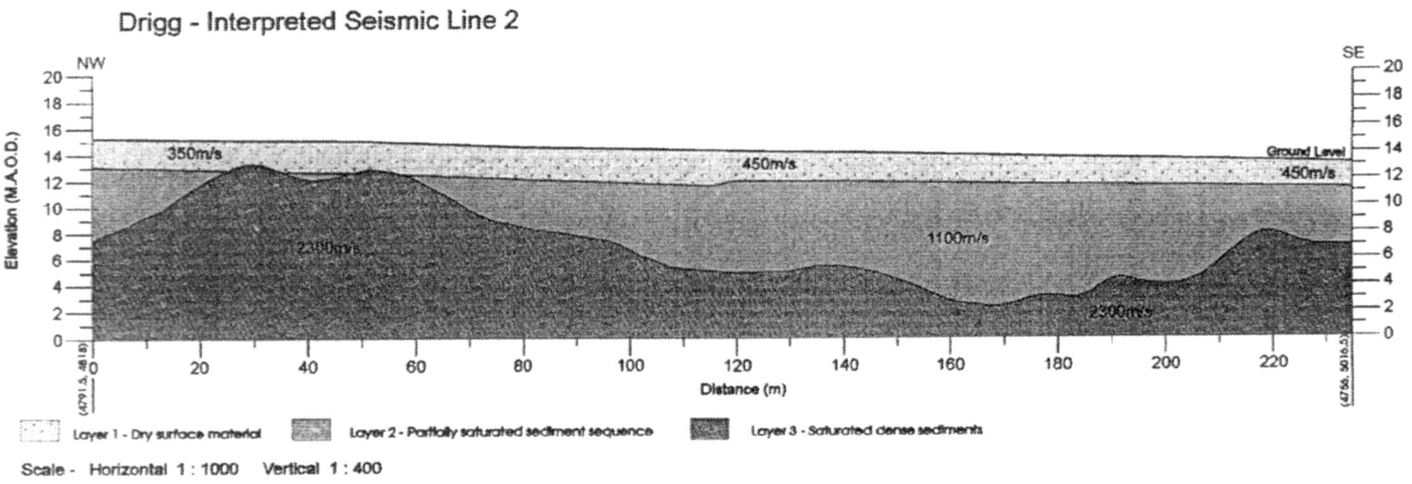

Figure 5. Output from seismic refraction survey.

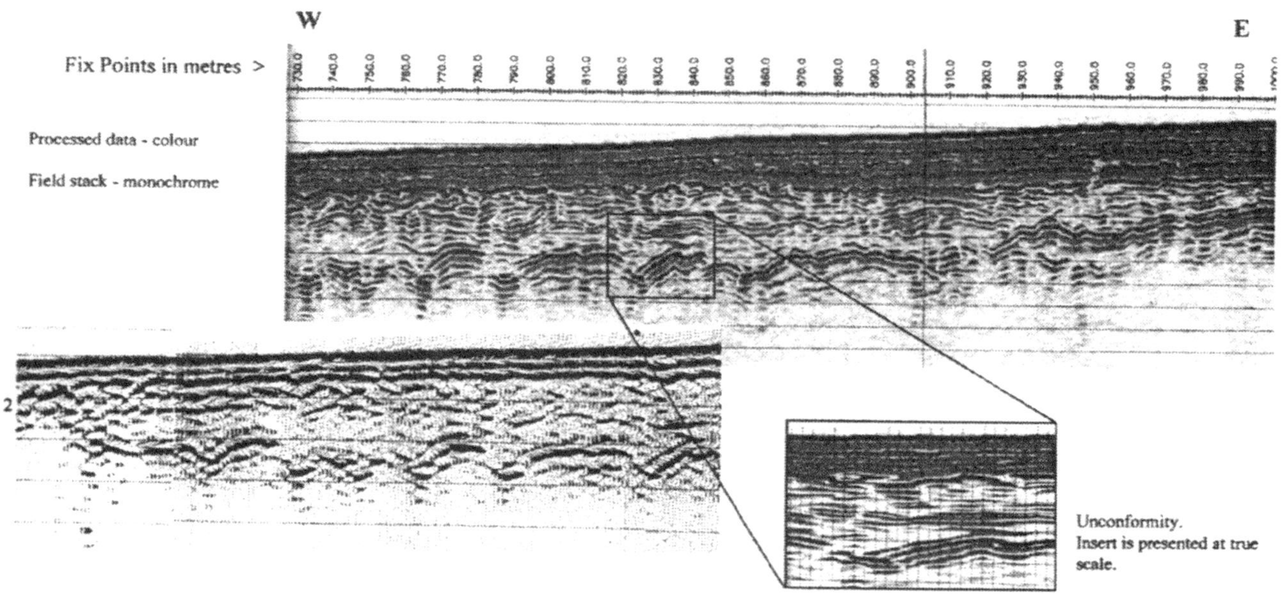

Figure 6. Output from Ground Penetrating Radar (GPR) technique.

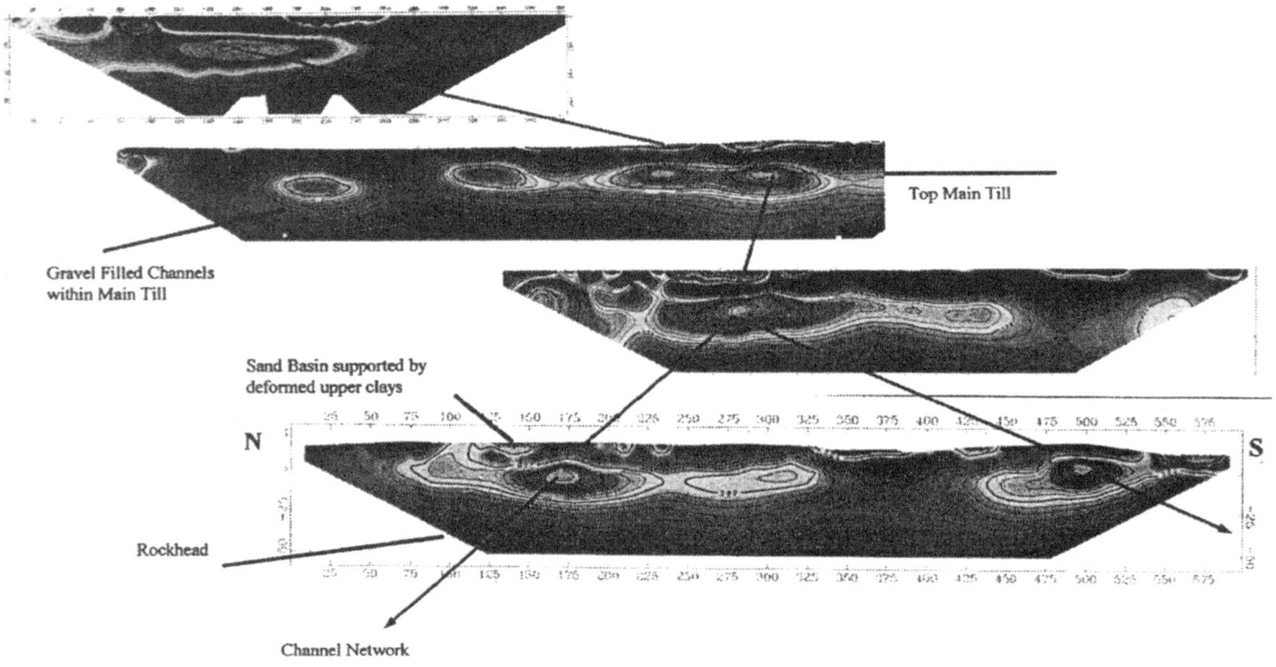

Figure 7. Output from Micro Resistivity Technique (MRT).

conductivity and predict the near surface geology. Results to date have been mixed, primarily because of the heterogeneity of the site which has not allowed the anticipated resolution to be met. In some areas, however, we have produced data which tie in well with borehole information.

4.2. ELECTRICAL RESISTANCE TOMOGRAPHY (ERT)

Currently, substantial world-wide research is concentrating on trying to acquire information on porous media and fluid properties through non-destructive means like tomographic imaging. We have contracted a British university to conduct experiments intent on deriving images of bulk resistivity variation through the utilisation of ERT. This technique is effectively an extension of traditional electrical geophysical profiling whereby four electrode resistivity measurements are made in order to determine the spatial distribution of electrical resistivity within an area of interest.

We were particularly interested in looking to develop cross-hole ERT which may be applied using electrodes placed down pairs of uncased boreholes in an attempt to image the resistivity structure between the boreholes. BNFL is also interested in trying to highlight preferential flow pathways and produce data which could be cross-correlated with existing information.

As this work has only just commenced, there are, unfortunately, no results available yet.

5. Overview of Results

We have now virtually completed the data acquisition aspects of the Drigg Site Characterisation and are starting to interpret the vast data set. An initial appraisal shows that the majority of the acquired data set is of a very high quality. The pertinent points are summarised below.

- The chosen rotary drilling methodology provided us with good core recovery at most locations.
- The various geophysical techniques could be cross-correlated, allowed us to highlight features of interest, and confirmed geological and structural interpretations.
- We now believe that the site has perched aquifers as well as a regional aquifer.
- We have confirmed that faulting is prevalent in the near-surface glacial sediments.

6. Conclusions

Having now performed an initial overview of the site characterisation data we believe that a comprehensive high quality data set has been acquired which will allow us to build on our current geological and hydrogeological understanding of the Drigg site. The success of this can be put down to clearly defining our objectives, appraising all previously acquired information from the site, and then carefully planning the investigation to tackle the anticipated ground conditions, ensuring our objectives were met. The geological model, coupled with the hydrogeological and hydrogeochemical data will now be used as the basis for the production of a conceptual hydrogeological model.

Attempting to obtain substantial information across a site especially in areas where access is arduous, although a prerequisite for a site characterisation, is usually fraught with difficulties. We have, therefore, attempted to acquire additional data through the development of new technologies. Although initial results have been promising, there is a requirement for additional work to ensure these techniques operate successfully under the specific conditions underlying our site.

References

1. Binley, A., Henry-Poulter, S., and Slater, L. (1995) Subsurface imaging using electrical resistance tomography. Lancaster University Report to BNFL.
2. Eaton, G.P., Cooper, S., and Speakman, P. (1996) The geology of the Drigg Site 1996 update. BNFL Report.
3. Sears, R. (1996) The British Nuclear Fuels Drigg Low Level Waste Site Characterisation Programme. BNFL Report.

APPLICATION OF FOCUSING TRANSFORMATIONS TO ENVIRONMENTAL GEOPHYSICAL PROBLEMS

V. A. POZDNYAKOV, V. A. TCHEVERDA & D. V. SAFONOV
Research Institute of Physics and Engineering (NIFTI)
Krasnoyarsk State University P.B. 8678,
Akademgorodok, Krasnoyarsk 660036, RUSSIA

1. Introduction

At present, in the context of the safe final disposal of radioactive waste, scientific opinion favours deep disposal in stable geological formations. Studies on the suitability of specific geological bodies for such a repository requires the investigation of a wide range of problems. This paper deals with an approach to three-dimensional (3D) reconstruction and mapping of a rock mass identified as a potential repository host rock. In our opinion, the first and most important objective in such a process, should be to answer the following questions:

- What is the internal structure of the rock mass?
- Is the rock mass intact and stable?
- Does this rock mass possess tectonic zones (*e.g.* faults and fractures) and if so, what are their orientations and dimensions?
- What are the seismo-acoustic characteristics of the host rock and overlying formations?

The list of questions could be expanded according to the specific problem being investigated.

We believe that the only way to answer the above questions is by the application of advanced geophysical techniques, in particular those based on seismics. The most promising technique in this category for addressing the above topics, appears to be modification of seismic pre-stack migration. Implementation of this technique is rather expensive with respect to data collection and processing, especially given the necessity to deal with rock masses in 3D. However, we suggest that its application is necessary

M. J. Stenhouse and V. I. Kirko (eds.),
Defence Nuclear Waste Disposal in Russia: International Perspective, 161–168.

in order to ensure that the isolation of nuclear waste from the biosphere is maintenance-free and acceptable in the long-term.

This paper is devoted to the treatment of pre-stack migration of seismic data for hi-fi stationary systems created by 2D arrays of sources and geophones, via diffraction and focusing transformations. Such an approach is based on a diffraction model that treats a real geological medium as the sum of two components:

- a smooth velocity model with ordered travel times, with
- diffraction objects placed within it, generating strong scattered waves [1].

Within the scope of this model, regular reflecting boundaries are represented as linearly-ordered arrays of point diffractors.

With regard to the focusing transformation, the technique has a physical basis in the Huygen-Fresnel principle and is carried out by reverse extrapolation of a wave field based on a modification of Kirchoff's integral. In order to provide the focusing of wave fields, grouping is achieved by manipulating the directional patterns generated from the geophones and shots (sources). The dynamic wave image of the medium for each grouping is obtained as a result of focusing scanning over diffraction holograms (hyperboloids) within a specific window with respect to lateral and vertical directions, followed by their weighted stacking. This process generates sections (vertical, lateral or any other spatial orientation) of the medium with selected spatial frequencies (for regular reflecting boundaries, this means their selection with respect to dip)[1].

It should be noted that the image of local diffraction objects will be represented in each section. It should also be emphasised that fractures and faults, particularly their edges, are rather strong diffraction objects. This characteristic allows them to be identified with confidence since they should be mapped through all selected sections. Weighted stacking of these selected sections followed by post-migration filtering gives the complete wave image of the medium. One can refine the velocity model, *i.e.* the smooth component of the geological model, by successive iterations.

2. Focusing Transformation Technique

The focusing transformation of a downward-moving wave field can be treated as a modification of the Kirchoff integral, corrected to minimise the dynamic

[1]Application of diffraction waves in order to obtain a dynamic image of geological structures is discussed in, for example, [2] and [3]). Some focusing transformations of recorded seismic data are described in [4] and [5]. In [6], some basic algorithms are presented of focusing transformations of 3D geosphere plots aimed at generating a dynamic image of a geological medium.

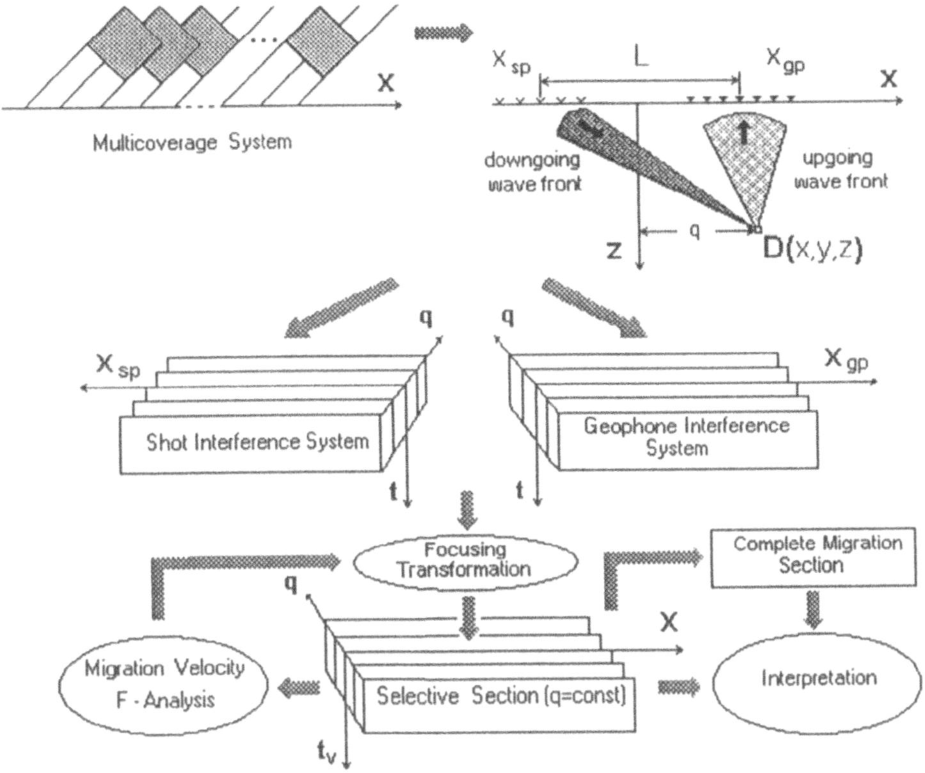

Figure 1. Schematic diagram of the experimental set-up for the focusing transformation technique.

distortion of wave images in accordance with Schnaider's recommendations [7]. Implementation of this technique is achieved via the following scheme (see *Figure 1*):

- creation of pairs of interference systems (sources and geophones) from the complete data set;
- F–analysis of the migration velocity (hereafter,'F' is used to denote 'focusing');
- construction of selective wave images of a geological medium for fixed dips in (x, y, t) coordinates (lateral coordinates and vertical time) as the result of focusing scanning through diffraction hyperboloids;
- weighted stacking followed by post-migration filtering in order to obtain the complete wave image of a medium.

The wave image obtained as the output of focusing transformation for the 3D situation is represented as:

$$R(x,y,t_v) = D(x,y,t_v) \sum_{q_x=-Q_x}^{Q_x} \sum_{q_y=-Q_y}^{Q_y} S(V_m,t_v)$$

$$\sum_{i=1}^{I} \sum_{j=1}^{J} \sum_{k=1}^{K} \sum_{l=1}^{L} K(q_x,q_y,t_{i,j}^{k,l}) A\{t_{i,j}^{k,l} + p(x_{i,j},x_{k,l},V_m,t_v,q_x,q_y)\} \quad (1)$$

where $x, y =$ lateral coordinates of the seismic trace of the section;
$q_x, q_y =$ coordinates of diffraction object with respect to the symmetry axis of the particular interference system;
$t_v =$ vertical time for diffraction objects with $q_x =$ constant, $q_y =$ constant;
$I * J =$ number of elements forming interference system of sources;
$K * L =$ number of elements forming interference system of geophones;
$D(x,y,t_v) =$ postmigration filter;
$S(V_m,t_v) =$ operator that sharpens the array characteristic of the interference systems;
$K(q_x,q_y,t_{i,j}^{k,l}) =$ summation weights;
$p(x_{i,j},x_{k,l},V_m,t_v,q_x,q_y) =$ kinematic correction determined by migration velocity $V_m(x,y,t_v)$ under fixed source $(x_{i,j})$ and geophone $(x_{k,l})$ positions and diffraction object coordinates t_v and q_x, q_y.

All coordinates are given with respect to the symmetry axis of the interference systems formed by sources and geophones [6].

As can be seen from equation (??), focusing transformation of multicoverage data to wave images of a medium can be performed on a step-by-step basis and it is possible to vary, at each step, the parameters which characterise the interference systems formed by sources and geophones (*i.e.* their size and distance, number of sources and geophones, *etc.*) via the processing parameters in the operators D, S, K and p. This process allows the investigator to perform migration velocity analysis, to form wave images of a diffraction medium and regular reflected waves separately, and then to combine these images to obtain the complete wave image.

3. Advantages of Focusing Transformation

The main advantage of the discussed approach is the possibility of obtaining selective wave images. In order to achieve this, one should insert in equation (1) $q =$ constant. Each selective image consists of wave images of dipping reflectors (with a rather narrow range of dips for a given value q) and diffraction objects. Successive variations of q allow the diffraction objects for an arbitrary range of spatial angles to be 'highlighted'.

At present, specialised software based on numerical algorithms of focusing transformations, has been developed and implemented on RISC-platforms. In order to illustrate some of the advantages of the approach described above, we consider an example of its application to the 2D reconstruction of a rock mass containing a fault zone as a highly diffractive object (unfortunately, real field data are not available for 3D reconstruction).

We draw the reader's attention to the fact that seismograms, after the application of focusing transformations, are characterised by regular wave parameters which are more stable compared not only with the raw data, but also with common midpoint (CMP) seismograms. Such comparisons are shown in *Figure 2* in which section (a) represents the raw data, and (b) and (e) are the CMP and focusing transformation ($q_x = 0$) seismograms, respectively. Sections (c) and (d) in *Figure 2* are segments of the CMP and F–transformation sections, respectively.

The higher quality of F–transformed seismograms is explained by two factors:

- the image is the result of the summation (via the interference system of geophones) of the wave field generated by focusing the array of shots;
- for each event, an F–seismogram is provided by the common deep point (focusing point) for a specific migration velocity distribution.

It should be noted that, in accordance with Kirchoff theory, wave-field summation by an interference system of geophones is performed within the first spatial Fresnel zones. As a result, there is a good statistical summation effect with reliable identification of regular (smooth) interfaces for F–sections compared with CMP sections and, consequently, a much more stable spectrum of the signal along interfaces.

As was mentioned above, the main advantage of the approach discussed here is the possibility of forming selective images for fixed values of parameter q (for the 3D problem, q is a vector with components q_x and q_y). *Figure 3* presents a complete wave image of F–transformation and selective images of the chosen window for a range of q varying from -900 m up to 900 m with incremental step, $\Delta q = 300$ m.

Selective image analysis allows the user to interpret complicated interference areas of the wave images, to identify reliably the spatial location of interfaces and estimate their reflectivities, and to locate with confidence diffraction objects as local acoustical inhomogeneities. The key to identifying diffraction objects is the fact that they produce rather strong diffraction waves mapped through selective images for a wide range of values of q, while regular interfaces are mapped only for a rather narrow interval of q.

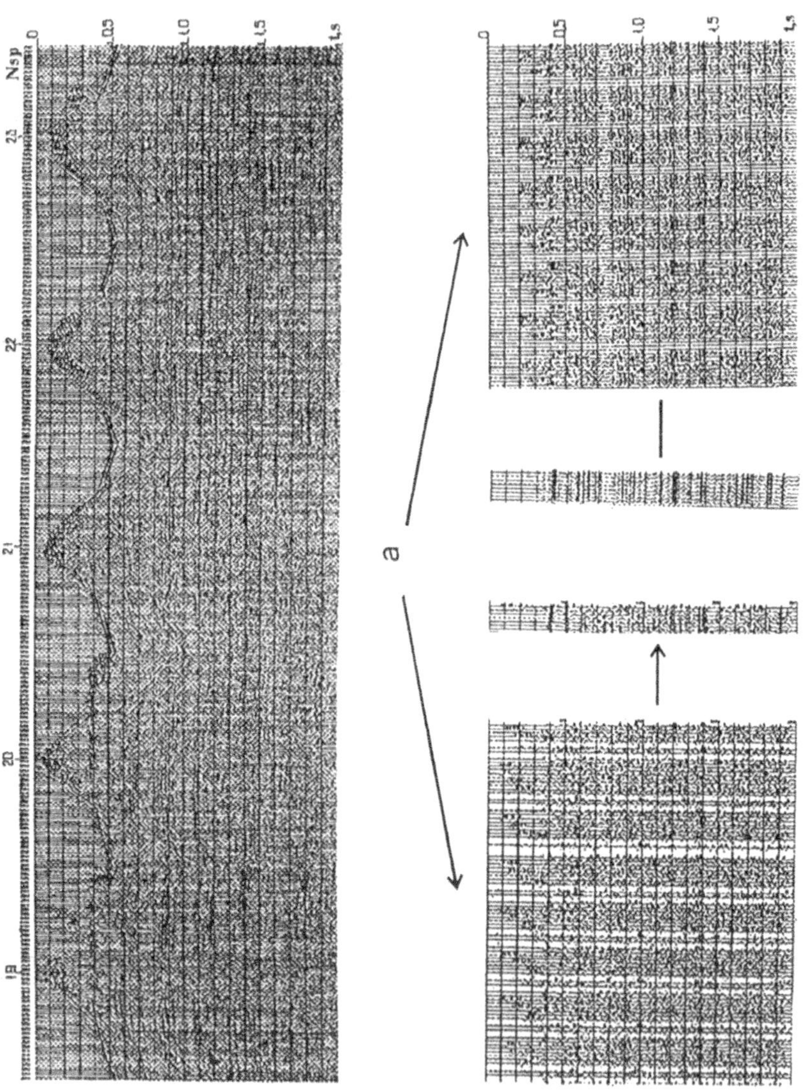

Figure 2. Comparison of seismograms obtained from raw data (a), common midpoint analysis (b), and focusing transformation (e), with selected segments [(c) and (d)] shown.

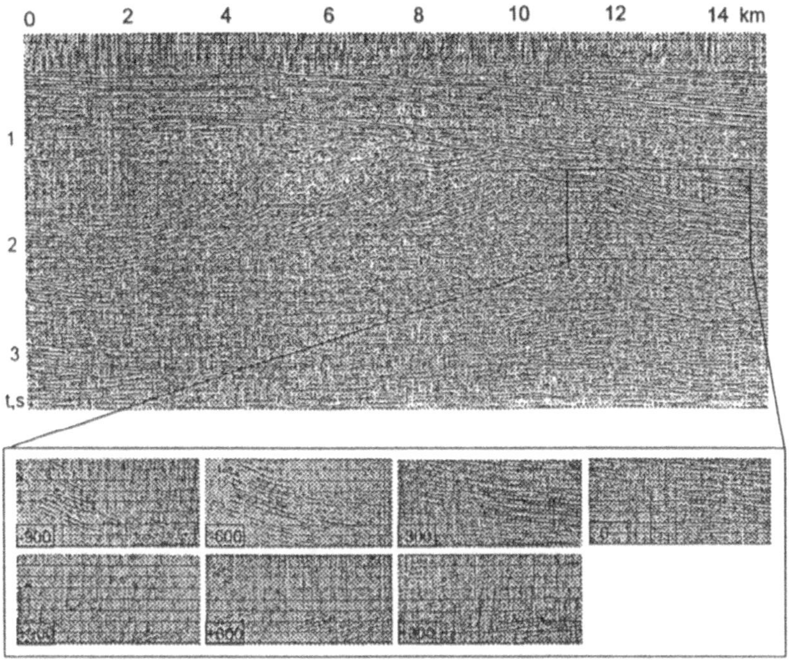

Figure 3. Complete wave image of focusing transformation with selective image analysis of the specified window corresponding to values of q ranging from -900 m to 900 m, with incremental steps of 300 m.

The black 'spots' in *Figure 4* (an F–transformation section) are the result of automatic analysis of a complete set of selective F–transformation images, and identify the locations of highly diffractive objects produced by local inhomogeneities. Such a conclusion is supported by the fact that these objects correspond to a fault zone confirmed independently by geological investigation.

4. Conclusions

To summarise, we conclude that the F–transformation technique allows the investigator:

- to reliably identify interfaces and to improve the quality of wave field dynamic parameters;
- to identify local acoustic inhomogeneities within the medium by means of the resultant diffraction waves produced;
- to perform high-quality migration velocity analysis with an arbitrary lateral step (this is important for a medium with large lateral and vertical deviations of its acoustic properties).

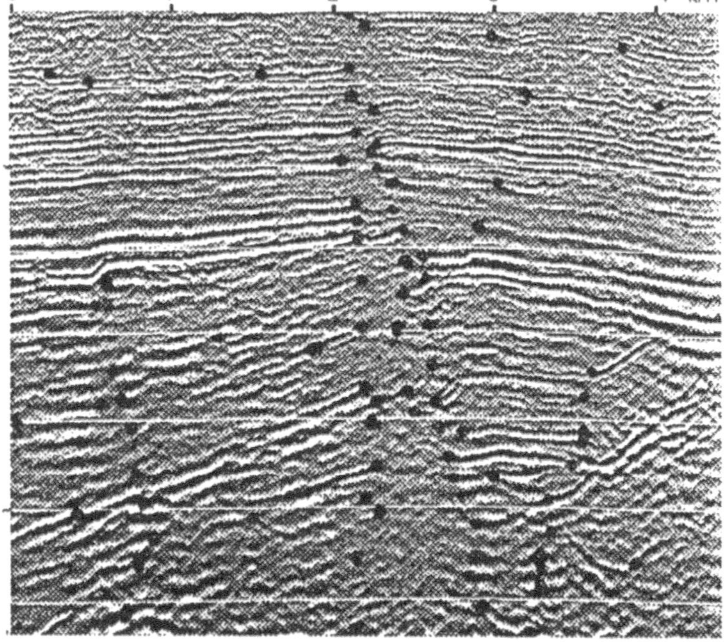

Figure 4. A focusing transformation seismogram with black 'spots' indicative of highly diffractive objects due to local inhomogeneities.

Thus, on this basis, it is possible to answer the questions posed in the Introduction section of this paper.

References

1. DOCHERTY, P. (1991) A brief comparison of some Kirchoff integral formulae for migration and inversion, *Geophysics* **56**, 1164-1169.
2. TROREY, A. (1970) A simple theory for seismic diffractions, *Geophysics* **35**, 1045-1073.
3. STAROBINETS, A. (1986) *Separation and Interpretation of Diffraction and Quasi-Diffraction Waves* (in Russian), Nedra, Moscow.
4. TARASOV, YU. (1968) Some questions of theory and the seismic application of the focusing of downward moving waves (in Russian), *Applied Geophysics* **51/52**, 23-41.
5. DIET, J.-P. AND AUDEBERT, F. (1990) A focus on focusing. Paper presented at the 52nd EAEG Meeting, Copenhagen, May 1990.
6. TARASOV, YU., SHLJENKIN, S., BUSYGIN, I., AND POZDNYAKOV, V. (1985) Focusing wave field transformations (in Russian), *Technical Report of Grozny Oil Institute, No. 3053-85.*
7. SCHNAIDER, W. (1982) Integral formulation for migration in two and three dimensions, *Geophysics* **47**, 49-76.

ACOUSTIC EMISSION MONITORING: A TOOL FOR THE SELECTION AND NONDESTRUCTIVE CHARACTERIZATION OF SITES FOR THE DEEP DISPOSAL OF RADIOACTIVE WASTE

V.A.MANSUROV
Russian Engineering Academy, Siberian Branch
38 Telman St., Krasnoyarsk 660073, RUSSIA

AND

V.A.ANIKOLENKO
Institute of Experimental Geophysics
United Institute of Physics of the Earth
Russian Academy of Sciences
51 Nikoloyamskaya St., Moscow 109004, RUSSIA

1. Introduction

The selection of potential sites for the deep disposal of radioactive waste and the subsequent near- and far-field monitoring is a complex scientific, technological, economical and political issue. The main scientific/technological problem concerns what set of observational methods, and their resultant geological and geophysical data, are considered valid in the context of selecting a site for a radioactive waste repository.

Existing international and national recommendations do not provide specific criteria for performing risk assessments covering extended time periods but can be written in a very general form, *viz.* [1]:

- minimal individual dose/risk should be assessed for a long time period;
- no sudden and dramatic increase in dose should occur over a time period of $>10^4$ years;
- the isolation potential of the repository should be assessed for times $> 10^4$ years;

M. J. Stenhouse and V. I. Kirko (eds.),
Defence Nuclear Waste Disposal in Russia: International Perspective, 169–176.

- assessment of the limiting release of radionuclides to the biosphere over a long period.

Thus, selection and the subsequent development of any radioactive waste repository presents a unique scientific and technological problem. Clearly, the more data relating to site selection which are obtained, the higher the associated cost. A strong correlation between expenditure and information obtained frequently exists, especially in the case of geological research. One of the ways of reducing costs is the use of relatively inexpensive, but reliable, non-destructive methods of characterising rocks at depth. Non-destructive testing using mobile geophysical observatories, for example, helps to reduce deep drilling and transportation costs.

This paper discusses kinetic models [2] and methods of non-destructive characterisation of rock [3] via acoustic emission/ microseismic activity (AE/MA) observations, and their application to the selection of potential sites for radioactive waste disposal and to the monitoring of the repository zone.

2. Basic Mechanical Conditions

The construction of every deep radioactive waste repository involves excavation of the rock mass, thereby perturbing the surrounding rock. This situation also applies to natural underground cavities used for a repository, since the emplacement of radioactive waste also causes perturbation in the surrounding rock, caused in particular by the intense radiation and relatively high temperature of the waste.

The factors which affect the mechanical stability of an entire underground repository system are:

- stresses in the host rock induced by natural tectonic as well as industrial activities;
- shear stresses in the structural components of the repository induced by the same forces;
- fatigue fracture in the host rock and structural components of the repository accumulated during long-term exploitation;
- stresses caused by geochemical factors;
- stresses in the host rocks and structural components caused by the long-term impact of the intense radiation and elevated temperature of the radioactive wastes.

The loss of stability in the host rock caused by excessive stresses and deformation of the rock mass may lead to damage of the natural and technological barriers in a repository and the resultant migration of radionuclides to the geosphere and biosphere. Here, we focus attention only on the first three items above which have a purely geomechanical origin, with a view to using the AE/MA observational method to monitor stress changes.

3. AE/MA Observational Method

Although observation of acoustic emission or microseismic activity is well known, this method has not yet been developed sufficiently for field observations owing to specific limitations, *e.g.* considerable attenuation of AE signals. The AE/MA method could be applied in the course of the selection of a repository site and its subsequent development, to characterise and monitor:

- the presence and locations of fracture zones and active faults;
- hydrological conditions and fluid transport;
- regional tectonics and seismicity, including induced seismicity.

Here, we provide a brief description of the AE/MA method including estimates of its resolution, signal attenuation, working frequency range, and other relevant parameters.

3.1. RESOLUTION OF THE AE/MA METHOD

The working frequency range in the AE/MA observations determines its ability to resolve various blocks in the host rock and/or structural components of the repository.

Figure 1 shows how the average size of a resolved block, calculated as 1/4 of the wavelength of the AE signal, depends on working frequency. The upper curve corresponds to hard rock such as granite, the lower one to relatively soft materials such as a sandstone (see *Table 1*). Thus, the AE/MA observation range lies below these curves.

As can be seen from *Figure 1*, the resolution of rock fragments and specific details in the repository zone which have dimensions typically varying from 0.2 m to 10 m, requires an AE/MA observation frequency in the range 1 to 10 kHz. The velocities of acoustic signals in saturated rock usually differ from those of 'dry' rocks. However, the use of these data to characterise different rocks may be difficult since the difference between 'dry' and saturated is small and its sign varies (see *Table 1*). However, the method is widely used for detecting caverns, fracture and porous zones in ambient rock at the resolution conditions specified above.

Natural and induced geodynamic activity in the vicinity of a radioactive waste repository will cause the substantial emission of elastic waves over a wide frequency band. Consequently, the AE/MA method could be used effectively for early warning purposes because critical geodynamic

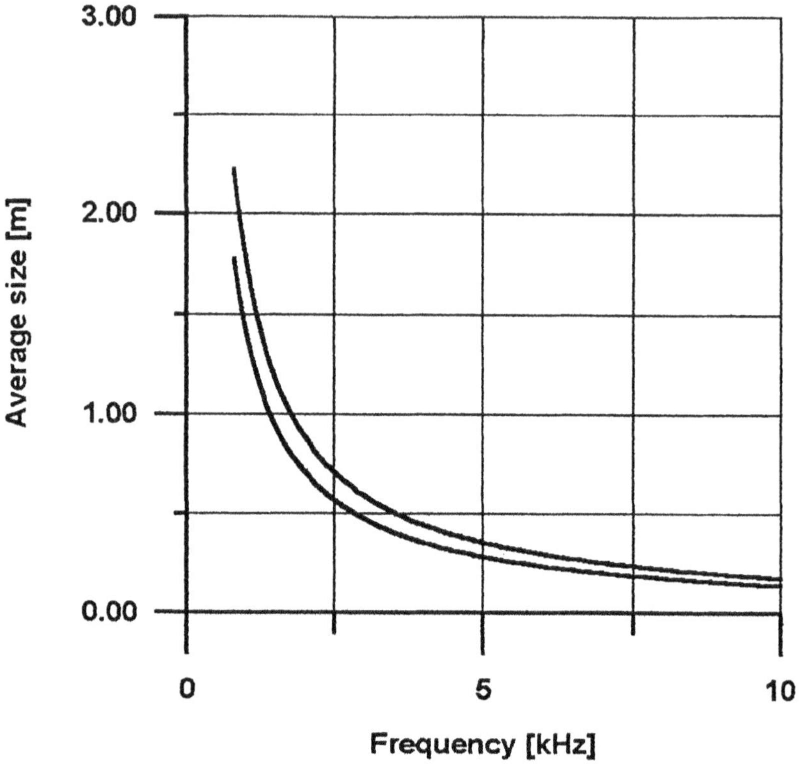

Figure 1. Average size of rock block resolved by AE/MA method plotted as a function of observation frequency.

phenomena are the result of a build-up of energy which starts from microseismic (low energy) events.

3.2. AE/MA SIGNAL-TO-NOISE RATIO AND SIGNAL ATTENUATION

Field observations have shown that an acceptable signal-to-noise ratio could be achieved with the sensor deployment in boreholes at depths below 6–8 m. This avoids the practical problem of surface noise and permits one to carry out AE/MA observations using relatively shallow boreholes. Optimum results for the characterisation of rock can be achieved using inter-borehole sounding combined with the analysis of core samples. As was shown in [4], attenuation factors for most kinds of rock may vary from 0.04 to 0.4 cm^{-1}, decreasing by a factor of 2 when the pressure increases from 8 to 80 MPa. Thus, the attenuation is determined mainly by the structural properties of the rock, rather than their composition. The greater the velocity change, the greater the change in attenuation factor.

TABLE 1. Acoustic velocities at different deformations of rock samples (data from [5]).

ROCK TYPE	V_P (max) (km s^{-1})	ϵ_P (max) (%)	V_S (max) (km s^{-1})	ϵ_S (max) (%)
Gabbro-amphibolite	7.14/7.25	3.0/3.25	3.52/4.06	2.0/2.0
Diorite	6.38/6.94	3.0/2.7	3.35/3.48	1.2/2.5
Granite - I	7.28/6.85	2.0/3.8	3.54/3.40	1.0/1.0
Granite - II	5.70/6.50	1.7/1.8	3.08/3.08	1.8/1.8
Plagiogranite	7.05/ –	2.8/ –	3.40/3.20	3.0/3.4
Granodiorite	5.75/6.06	1.5/2.0	3.14/2.95	1.3/1.5
Gneiss	6.06/6.40	0.7/1.5	3.22/3.30	1.8/2.0
Dolomite - I	4.88/4.80	2.0/3.2	2.58/2.20	2.6/2.6
Dolomite - II	4.92/5.24	1.2/1.2	2.90/2.44	1.2/1.2
Sandstone - I	6.14/6.50	4.2/4.2	3.40/3.18	4.2/4.2
Sandstone - II	6.08/6.08	1.5/1.5	3.08/3.62	1.5/1.5
Sandstone - III	6.70/6.55	1.5/1.5	3.38/3.25	1.5/2.0

The numbers in the columns are the values corresponding to dry / saturated rock. Here V_P (max) and V_S (max) are, respectively, the S and P components of the maximum velocities of elastic waves; ϵ_P (max) and ϵ_S (max) are, respectively, the deformations corresponding to velocities V_P (max) and V_S (max).

3.3. REMOTE EVALUATION OF DILATANCY IN ROCK

The AE/MA method represents a unique technique for the non-destructive characterisation of rock stress state and for the presence of fracture zones [3]. Two experimental curves are shown in *Figure 2* for consideration. Curve 1 shows the dependence of volumetric strain, ϵ_v, on stress, σ, for the rock sample under load. Curve 2 demonstrates the dependence of volumetric strain on the cumulative elastic energy of AE signals. It is readily apparent that Curve 1 remains in the region of negative values of ϵ_v until the load reaches about 0.7 of the strength limit σ^*. According to [4], this effect is associated with the compaction of the sample. The curve then changes to positive values of ϵ_v owing to fracturing of the rock, which starts to overcome the compaction processes as the load increases.

The trend in the curves in *Figure 2* indicates that a decrease in volume caused by compaction of the rock is compensated by a volume increase due to the fracture process. The value of ϵ_{vo} is readily obtained by extrapolating the straight portion of Curve 1 to the point at which it intersects the vertical axis. The right portion of Curve 2 is linear due to the dilatancy caused by fracturing. Its intersection with the horizontal axis also gives the value of

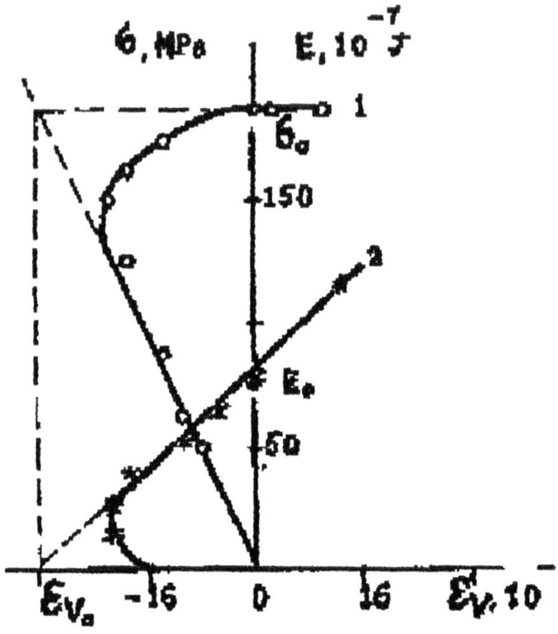

Figure 2. Stress (Curve 1) and released cumulative energy (Curve 2) of AE signals plotted as a function of volumetric strain under uniaxial loading of a granite sample using a laboratory testing machine.

ϵ_{vo}. As shown in *Figure 2*, values of ϵ_{vo} obtained via the two methods are of the same magnitude. Experiments have shown that values of E may vary in the range from 1–100 100 J m^{-3} [5].

3.4. SPATIAL AND TIME UPSCALING

The extrapolation of results of the laboratory rock testing and borehole observations to larger scales forms the basis of long-term monitoring of the geodynamic stability of a radioactive waste repository. Rock testing experiments carried out with the use of samples of volume 1 m^3 and a loading force up to 50,000 tonnes, have not shown any noticeable difference in the behaviour and parameters associated with rock failure. The similarity in failure processes at different scales was demonstrated in [6] for an energy release ranging over 15 orders of magnitude. As shown in [7], timescale correlates with size scale via $B \cdot log\ t$, where B is a constant. Thus, an n-fold extension of the dimension scale is equivalent to a t^n extension of timescale. This correlation forms the basis for predicting the long-term behaviour of a rock mass over larger scales.

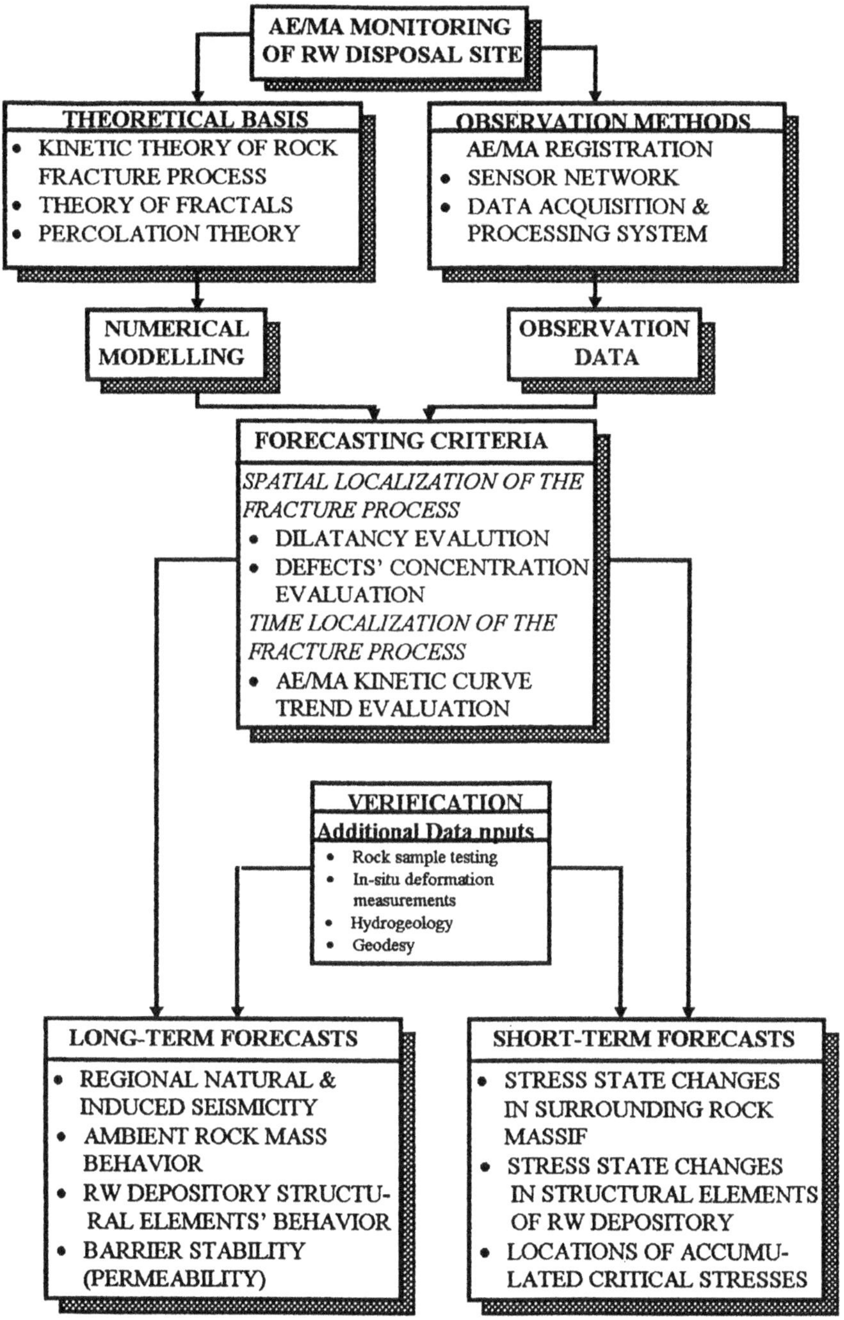

Figure 3. Role of AE/MA Observations for Monitoring a Deep Radioactive Waste Repository.

4. Strategy of AE/MA Monitoring and Site Selection

The main goals and associated methods of a proposed monitoring strategy are shown in *Figure 3*. This strategy is based on the authors' experience of AE/MA observations in coal and bauxite mines as well as on the underground monitoring of a network of 40 m diameter caverns created in compressed salt by leaching, at depths varying from 600 m to 1200 m. The design system permitted the monitoring of a network of individual caverns used as disposal sites for oil wastes, all about 200 m from the only observation borehole.

5. Conclusions

- The results of laboratory and field experiments have demonstrated good performance characteristics of the acoustic emission/microseismic activity method for monitoring the stress state in rocks;
- physical models based on the kinetic concept of the rock fracture process allow the possibility for the monitoring and long-term prediction of the behaviour of a rock mass;
- the monitoring systems deployed in some mines in the former Soviet Union have shown good operational parameters for the efficient prediction of rockbursts and strong motion events;
- the proposed software and system design concept may be applied to the monitoring of stress state in the host rock and structural components of a radioactive waste repository;
- the approach could also be used to monitor fracture processes in the near-field, *e.g.* in blocks of vitrified and solidified radioactive waste.

References

1. SAVAGE, D. (ed.) (1996) *The Scientific and Regulatory Basis for the Geological Disposal of Radioactive Waste*, John Wiley & Sons, Chichester, England.
2. ANIKOLENKO, V.A. (1992) Kinetic study of rock fracture in laboratory testing experiments, *Acta Montana Ser.A* **3** (89), 63-72.
3. ANIKOLENKO, V.A. AND MANSUROV, V.A. (1996) Kinetic approach to the nondestructive monitoring of rock failure, *Mater. Sci. Forum*, **210-213**, 535-540.
4. BIENIAVSKI, Z.T. (1967): The influence of specimen size on compressive strength of coal, *Int. J. Rock Mechanics and Mining Sciences* **4**, 395-430.
5. KARUS, E.V., KUZNETSOV, O.L. AND FAIZULLIN, I.S. (1986) *Inter-Borehole Acoustic Sounding*, Nedra Publishers, Moscow (in Russian).
6. KUKSENKO, V.S., MANZHIKOV, B.T., MANSUROV, V.A AND STANCHITS, S.A. (1986) Burst-prone characterization of rocks using energy release measurement, *Physico-Technical Problems of Natural Resourses Mining*, Novosibirsk **No.5**, 26-32 (in Russian).
7. MANSUROV, V.A. & ANIKOLENKO, V.A. (1996) Evaluation of dilatancy in rock for the forecasting of burst prone zones in mines, *Mater. Sci. Forum* 210-213, 527-534.

KINETIC CONCEPT OF MONITORING THE LONG-TERM STABILITY OF A DEEP REPOSITORY FOR RADIOACTIVE WASTE

V.A. ANIKOLENKO
Institute of Experimental Geophysics
United Institute of Physics of the Earth
Russian Academy of Sciences, 51 Nikoloyamskaya St.
Moscow 109004, RUSSIA

AND

V.A.MANSUROV
Russian Engineering Academy, Siberian Branch
38 Telman St., Krasnoyarsk 660073, RUSSIA

1. Introduction

The long-term monitoring of stressed state is an important tool for monitoring the long-term stability of an underground repository for toxic and radioactive waste. Geodynamic effects caused by natural tectonic and man-made activities may lead to the loss of stability in the surrounding rock as well as the structural components of a repository, with the resultant migration of radionuclides into the geosphere and biosphere.

The only way of predicting the long-term behaviour of the entire repository system is by multi-scale modelling of the process of stress build-up and damage in the near-field. Far-field modelling plays an important role because far-field forces affect the near-field. The modelling has to be based on numerical data obtained from monitoring stressed state. In addition, monitoring also allows one to modify the model to accommodate changes in the near-field stressed state conditions.

One of the approaches to such monitoring is the detection of potentially dangerous regions of excessive stress or deformation in a rock mass, using threshold crack concentration criteria [1]. Another approach is to define invariant quantitative physical criteria concerning the loss of rock

M. J. Stenhouse and V. I. Kirko (eds.),
Defence Nuclear Waste Disposal in Russia: International Perspective, 177–183.

stability in the near-field. To accomplish this task, the overall (instability) development process preceding critical failures in the surrounding rock or structural components of the repository should be considered, as these 'events' are themselves simply stages of the entire, highly stochastic process of instability development.

This paper extends the recent works of the authors [2-4,12] who now apply a rock fracture model based on kinetic theory to the monitoring of stability of radioactive waste repositories constructed in geological formations. The kinetic approach to the failure process was first formulated by Zhurkov (1957) for polymers and later expanded to cover rocks by Regel *et al.* [5], Petrov [6] and others. A similar approach has been successfully applied in the field of solid state physics by Mikhailov [7], and Anikolenko and Mikhailov [8].

2. The Physical Model

It has been established that the rock failure process (fracturing) is, to a large extent, controlled by micro-stresses which exist in rocks owing to the heterogeneity and anisotropy of the geosphere. In a stressed state, micro-cracks or clusters are generated in rocks which then grow in size as deformation develops. During this process, pairs of cracks combine to create higher-level defects by overcoming the potential barriers caused by interatomic interactions. The physical model and kinetic equations showing the cumulative number of crack generation events *vs* time have been described in detail by Anikolenko [2,3]. In its general form, the equation can be written:

$$N(t) = \int_{-\infty}^{\infty} \{1 - exp[-k(r)t]\} dr \tag{1}$$

where $k = k_o exp(-2r/a)$, r is the distance between two joining defects, and k_o and a are constants in time. In the simplest case of a uniform distribution,

$$f(r) = \frac{1}{(r_{max} - r_{min})} = 1/\Delta r$$

(r_{max} and r_{min} are the limits of the distribution, and Δr its width). Hence, we obtain the equation:

$$\frac{N(t)}{N(t_0)} - 1 = \frac{a}{2(r_{max} - r_{min})} ln\frac{t}{t_0} = \frac{a}{2\Delta r} ln\frac{t}{t_0}$$

which, for $t >> t_o$, simplifies to

$$\frac{N(t)}{N(t_0)} = \frac{a}{2\Delta r} ln(\frac{t}{t_0}) \tag{2}$$

where t_o is the time of the initial observation. This approximate isotropic solution corresponds to the common case where the distance between two interacting defects is considerably less than the distance to another pair of interacting defects. The isotropic, three-dimensional (3D) solution to equation 1 has been also derived by Anikolenko [2,3]

$$ln[\frac{N(t)}{N(t_0)} - 1] = (\frac{a}{2r_o})^3[ln^3(k_o t) \; - \; ln^3(k_o t_o)] \qquad (3)$$

Two-dimensional (2D) solutions are not given here for ease of presentation.

The above equations were obtained assuming the Markov process of defect formation, *i.e.* defect size and location depend exclusively on the parameters of the parent defects and are independent of the history of parent defect generation. Solution of the inverse problem for equations 1 and 2 yields the distribution parameter $f(r)$, as well as the value of parameter a, the average cluster size currently involved in the fracture process. Thus, a is a scaling parameter the value of which characterises the size of a typical defect involved in the fracture process at any given time. Useful information about rock failure processes can also be obtained by analysing formal kinetic dependencies.

3. Comparison with Experimental Results

As mentioned above, the monitoring of events at any energy level (which may lead to a loss of stability) requires information about the preparatory process preceding a strong event. In each case, it is important to determine the energy level above which an event will be considered as strong. Once such a level has been defined, all lower-level events occurring prior to the strong event may be considered as part of the preparatory process. Based on this approach, the results of rock test laboratory experiments are considered for comparison with the rockburst data by plotting the cumulative number of registered events *vs* observation time.

Firstly, we analyse the results of rock sample failure experiments where cylindrical rock samples, several cm in length and 3 cm in diameter, were loaded uniaxially using a laboratory testing machine. Defect generation events were registered using the acoustic emission method. Details of the experimental technique are given by Mansurov [4]. *Figure 1* shows two typical differential acoustic emission spectra of failure of sandstone samples. *Figure 2* displays the integral spectra relating to the experiments shown in *Figure 1*, plotted in semilog coordinates according to equation 2. Thus, these plots show the cumulative number of acoustic emission events *vs* the logarithm of the time of observation, *log t*.

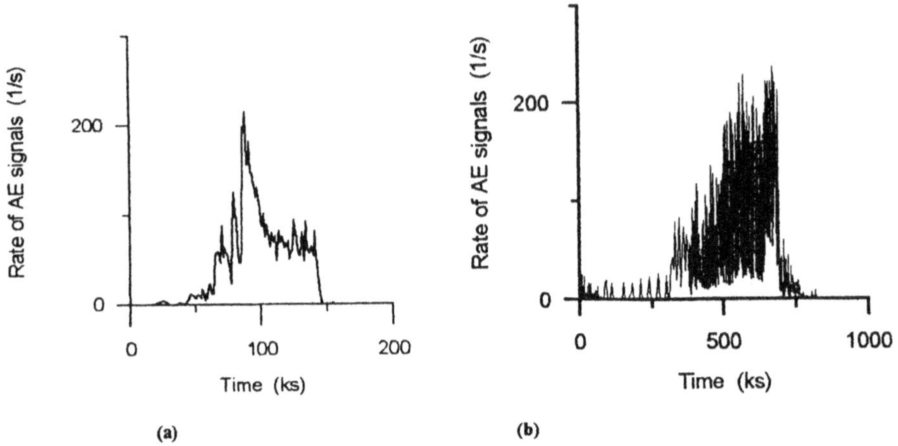

Figure 1. Rate of acoustic emission (AE) signals vs elapsed time at failure of two different sandstone samples using a laboratory testing machine.

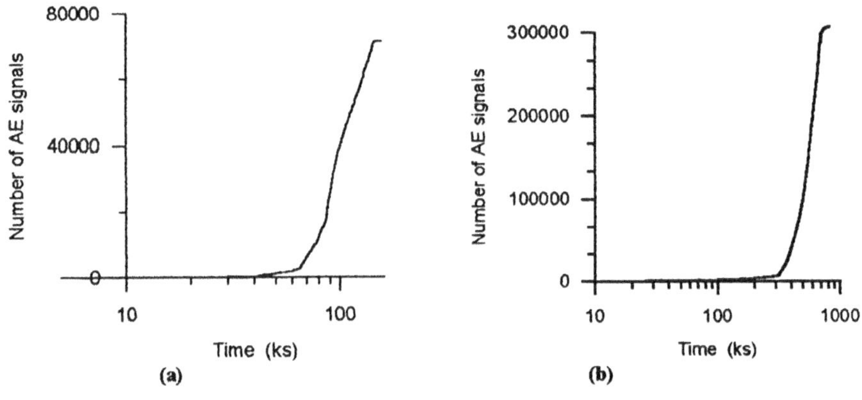

Figure 2. Dependencies of Figure 1 plotted as a function of the logarithm of elapsed time.

The two plots shown in *Figure 1* relate to sandstone testing. The size of the sandstone sample for *Figure 1b* was several times greater than that used in *Figure 1a*. The rate of loading for experiment (b) was lower than in experiment (a). Regardless of the substantial differences in shape, the curves in *Figure 1*, when plotted in the semilog coordinates of equation 2, look similar. Such a similarity means that these invariant curves allow the opportunity to study the process of failure in terms of the factor a/r, *i.e.*

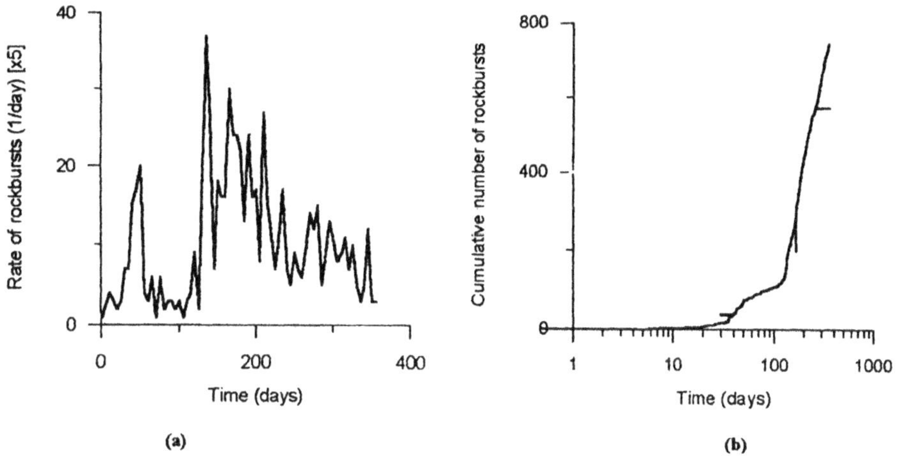

Figure 3. Rockbursts at Gottwald II - Vinarice mines in 1985:
(a) Rate of rockbursts vs time; (b) Cumulative number of rockbursts plotted as a function of logarithm of elapsed time

in terms of the qualitative parameters of energy and space distributions.

Figure 3 shows the rockbursts registered in the Gottwald II - Vinarice mines during 1985 as a function of observation time [10]. Here, the initial time t_0 is taken as zero on January 1, 1985. As seen from *Figure 3b*, the kinetic curve plotted in semilog coordinates consists of two linear segments similar to the curves in *Figure 2*. *Figure 3a*, similar to *Figure 1*, has a stochastic nature and can yield no data for considering trends in the failure preparatory process.

The observed kinetic dependencies describe a whole set of parallel failure processes which are taking place at different levels in the sample. The absence of sharp deviations in the curves of *Figure 2* is evidence of compaction of rock samples under load due to the shear movement of rock grains. Grains may slide over one another without total failure of the sample (see [9]). In the case of scaling effects, one obtains a kinetic curve consisting of several linear segments when plotted in the coordinates of equations 2 or 3 (see *Figure 3b*).

4. Discussion

The above analysis shows good agreement with the mechanical theory of rock failure proposed by Sadovsky [11]. From a physical perspective, this theory means that in the course of time the system has to overcome potential barriers of larger widths, the average waiting time for higher energy events increases, and rock failure processes seem to slow down. With regard to earthquake or rockburst forecasting, a seismic gap is a reliable predecessor of a strong seismic event in the case of formation of a rupture crack (see [12]). In addition, the deviation of the kinetic curve from linearity in semilog coordinates is evidence of an approaching earthquake or rockburst. During such an event, full or partial energy release follows and the kinetic dependence follows a similar trend to that shown in *Figure 2*. The ratio $a/2r$ remains invariant for each separate linear segment of the kinetic curve.

Such kinetic dependence is shown in *Figure 3b*. The plot indicates at least three linear segments during 1985: January – March – May – December. The onset of each period corresponds to a new stage of energy accumulation by the rock mass after partial energy release. These time periods can most probably be explained by the action of tectonic forces since there is no evidence of strong rockbursts occurring at the end of any of the three periods. Unfortunately, no data relating to seismic events of a lower-energy level, are available for 1985.

5. Conclusions

Thus, the proposed strategy of rock mass monitoring includes the following key elements:

- real-time or quasi real-time observations of the cumulative kinetic dependencies in the far-field. The registered cumulative parameters in kinetic dependencies could be stress and/or strain deformations, the number of acoustic emission / microseismic signals, mass transfer, underground fluid flow *etc.*
- selection of critical values of parameters and their threshold values for a given type of rock or structural material corresponding to the specific conditions of a radioactive waste repository;
- subsequent analysis of the kinetic curves plotted according to the coordinates of equations 2 or 3, for detecting non-linear trends in these curves;
- analysis of physical processes which can lead to non-linear deviations with a view to revealing trends in rock mass behaviour, and to forecasting possible geodynamic events which could cause a loss of stability in the repository.

The simplest monitoring system could detect in real time only the cumulative number of seismic events for which the energy level is lower than that of an expected critical event. Such a monitoring system would use very few sensors but would be capable of forecasting the time of occurrence of a strong event such as a rockburst or rock failure. In order to predict the location of expected events, the system should have a relatively dense network of seismic sensors.

The simultaneous processing of signals coming from different cells of this network via the same software also permits one to forecast the location of a strong event with an accuracy corresponding to the cell size. Thus, analysis of the cumulative kinetic curves provides new information about the trends and the mechanism of rock failure processes and rock mass behaviour.

References

1. MANSUROV, V.A. AND ANIKOLENKO, V.A. (1996) Evaluation of dilatancy in rock for forecasting burst prone zones in mines, *Mater. Sci. Forum* **210-213**, 527-534.
2. ANIKOLENKO, V.A. (1992) Kinetics of rock failure, in (G. Sobolev ed.), *Physical and Seismo-geological Fundamentals of Rock Failure Prediction*, pp. 41-46, Nauka, Moscow, (in Russian)
3. ANIKOLENKO, V.A. (1992) Kinetics of rock failure in laboratory experiments, *Acta Montana* **89** (No. 3), 63-72.
4. MANSUROV, V.A. (1994) Laboratory experiments: their role in the problem of rock burst prediction, *Comprehensive Rock Engineering*, **3**, 745-771.
5. REGEL, V.R., SLUTSKER, A.I., AND TOMASHEVSKI, A.E. (1974) Kinetic nature of strength in solids, *Uspekhi Fizicheskikh Nauk*, **106**, 193-228 (in Russian).
6. PETROV, V.A. (1981) Principles of kinetic theory of microfailure forecasting, *Fizika Tverdogo Tela*, **23**, 33-72 (in Russian).
7. MIKHAILOV, A.I. (1971) Kinetics of electron transfer in solids, *Doklady AN SSSR*, **197**, 136-140 (in Russian).
8. ANIKOLENKO, V.A. AND MIKHAILOV, A.I. (1976) Post-radiation decay of ion radicals in a solid nonpolar matrix, *Doklady AN SSSR*, **230**, 102-105 (in Russian).
9. STANCHITS, S.A. AND TOMILIN, N.G. (1983) Investigation of time parameters of acoustic signals during tear-off crack generation, *Prediction of Earthquakes*, Dushanbe, USSR, **4**, 31-45 (in Russian).
10. RUDAEV, V. (1989) Catalogue of rockbursts in the Gottwald II – Vinarice mines in 1985, D.Sc. Thesis, Prague.
11. SADOVSKY, M.A. (1984) Geomechanical theory of rock failure, *Prediction of Earthquakes*, Dushanbe, USSR, **4**, 268-273 (in Russian).
12. ANIKOLENKO, V.A. AND MANSUROV, V.A. (1996) Kinetic approach to the non-destructive monitoring of rock failure, *Mater. Sci. Forum* **210-213**, 534 - 540.

IDENTIFYING CONTAMINANTS OF CONCERN FOR POSTCLOSURE ASSESSMENT

KISHOR K. MEHTA & BRUCE G. GOODWIN
Atomic Energy of Canada Limited
Whiteshell Laboratories
Pinawa Manitoba ROE 1LO Canada

Abstract

The nuclear fuel waste disposal concept currently being studied in Canada involves isolation of the waste in long-lived containers emplaced in a sealed vault located deep in plutonic rock of the Canadian Shield. An important element of the study is the evaluation of potential impacts that may occur far into the future. These impacts are examined in the postclosure assessment, a long-term performance and safety assessment that extends for thousands of years, beginning from the time that the disposal facility is permanently sealed.

The impacts of concern include exposure to both radiotoxic and chemically toxic materials, collectively referred to as "contaminants". There are several hundred contaminants present in nuclear fuel waste. However, many are present in very small quantities and cannot lead to any significant detriment. In addition, many radionuclides have very short half-lives and quickly decay to very small quantities, and thus would not pose a significant hazard. Thus, preliminary screening provides a quantitative rationale for eliminating many contaminants from further detailed consideration.

Such a preliminary screening was carried out for a case study [1] of deep geological disposal of used fuel from CANDU^R nuclear generating stations. The screening started from a large list of contaminants present in used-fuel bundles (irradiated UO_2 fuel and zirconium alloy sheaths). Our first objective was to identify contaminants that require detailed assessment. Our second objective was to simplify the four actinide decay series, each of which contains 15 or more members linked together through radioactive decay. Our analysis used simple and conservative mathematical models;

M. J. Stenhouse and V. I. Kirko (eds.),
Defence Nuclear Waste Disposal in Russia: International Perspective, 185–186.

data representing contaminant inventories and properties such as half-life, sorption on minerals and groundwater velocities in the rock; and criteria for radiotoxicity and chemical toxicity impacts.

From a starting list of 200 nuclides, we identified 72 radionuclides that required detailed evaluation in the postclosure assessment for their potential radiotoxicity impacts. The initial 200 nuclides are from 85 different chemical elements, but the screening showed that only about 25 required evaluation for their potential chemical toxicity impacts. The four actinide decay series can be shortened to chains containing four or fewer radionuclides.

These decreased numbers of contaminants are particularly important for postclosure assessment studies that follow a probabilistic systems assessment approach. In this approach, we use a comprehensive mathematical model to describe the disposal system (including the biosphere), and probability distributions to represent uncertainty and variability in the model parameters. Typically, tens of thousands of simulations are performed using SYVAC (SYtems Variability Analysis Code), In each simulation, values for the model parameters are randomly sampled from their associated distributions, and the set of simulations yields distributions of potential impacts that reflect the underlying parameter uncertainty and variability. The computational requirements of SYVAC, and the associated data requirements, are substantially reduced because of the simplifications afforded by the preliminary screening.

The conclusions and recommendations from the screening study are specific to the nuclear fuel waste and the case study that were considered. However, the general approach can be applied in other assessments that may involve different wastes and different disposal concepts.

References

1. GOODWIN, B.W. AND MEHTA, K.K. (1994) Identification of contaminants of concern for the postclosure assessment of the concept for the disposal of Canada's nuclear fuel waste. AECL Reoprt AECL-10901, COG-93-265.

EXPERIENCE OF SAFETY ASSESSMENT FOR THE DEEP DISPOSAL OF INTERMEDIATE-LEVEL WASTE IN THE U.K.

J.E. SINCLAIR
AEA Technology plc
Harwell, Oxfordshire
United Kingdom

AND

G.E. HICKFORD
United Kingdom Nirex Limited
Harwell, Oxfordshire
United Kingdom

1. Introduction

United Kingdom Nirex Ltd is investigating Sellafield as a potential site for the deep geological disposal of solid intermediate-level waste (ILW) and some low-level radioactive wastes (LLW). Attention is currently focused on a potential repository located in volcanic rocks (the Borrowdale Volcanic Group, BVG) at a depth of 650 m below ordinance datumOD. This paper summarises Nirex 95 [1], a preliminary analysis of the groundwater pathway for this repository. Nirex 95 has focused on the calculation of annual individual risk, which is a principal safety requirement as set out by the U.K. regulatory bodies [2]. It is not a comprehensive safety assessment, which would require the systematic analysis of a wider range of features, events and processes. However, it does illustrate a number of key components of the methodology employed in the Nirex programme.

Data and underlying research to support assessments of the repository are provided by the Nirex Safety Assessment Research Programme (NSARP). The assessment was based on a hydrogeological conceptual model of the site [1] constructed on the basis of the extensive characterisation programme involving a series of deep boreholes [3], associated testing and geophysical surveys. Based on this conceptual model, effective hydrogeological

M. J. Stenhouse and V. I. Kirko (eds.),
Defence Nuclear Waste Disposal in Russia: International Perspective, 187–197.

properties were generated [1] that were used in calculations of groundwater flow and radionuclide transport.

The uncertainties in the values of the parameters used in the assessment models are addressed through the use of probabilistic safety assessment (PSA). For many parameters, probability density functions (PDFs) that characterise the uncertainty have been obtained in structured data elicitation sessions, which are held with appropriate expert groups [4]. An important component of the assessment is the propagation of these uncertainties through the calculations in a systematic manner.

2. Modelling Approach

Calculations of risks arising from the transfer of radionuclides from the repository through the geosphere to the marine environment, to streams and rivers and to the deep soil, were based on calculations of radionuclide fluxes to the biosphere using the program MASCOT [5]. In addition, preliminary scoping calculations of risks arising from the abstraction of water from agricultural wells were based on calculations of radionuclide concentrations in the geosphere using the program NAMMU [6].

2.1. NEAR FIELD

The reference repository design used in the assessment consisted of seven vaults, which would be constructed in the BVG, five containing unshielded ILW, and two containing LLW and shielded ILW. Waste packaged in steel or concrete containers would be emplaced within each vault, and surrounded by a cementitious backfill designed to sustain alkaline conditions in the groundwater within the vaults. Above the waste and backfill there would be an unfilled crown space, within which much of the groundwater flow through the vault would be concentrated.

Hydrogeological models were developed that identified a number of distinct hydrogeological units and different categories of features. Of particular significance was a set of features in the BVG that were more transmissive than the background rock ('transmissive features'). These features were considered to consist of organised structures of fractures. A major aspect to be taken into account in near-field modelling was the assumed intersection of the transmissive features with the repository vaults. Several of the features were judged to intersect each vault, significantly affecting groundwater flow through the vault (see *Figure 1*).

The near-field part of the MASCOT model (the radionuclide source term) was constructed from several submodels, representing different sections of the repository. A single representative model of each vault type was used, and for each of these, separate submodels represented:

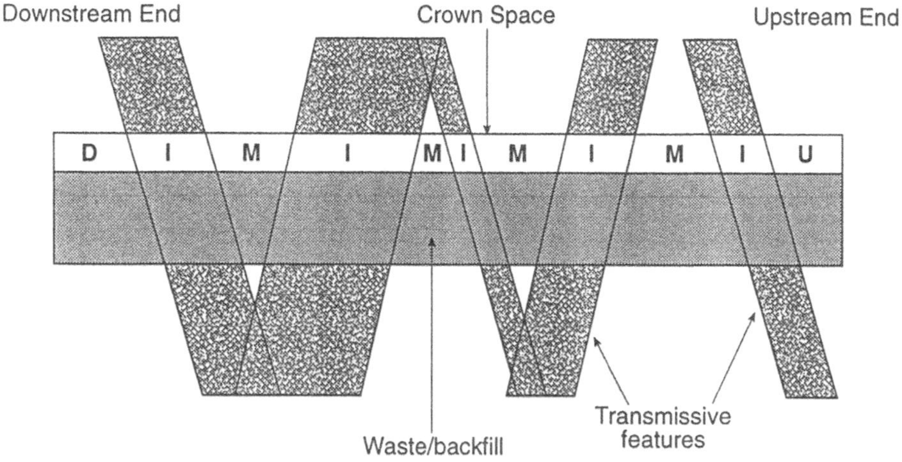

Figure 1. Relationship between the disposal vaults and the planar transmissive features in the host rock.

D (downstream), M (midstream), U (upstream), I (intersection) regions.

(a) the intersection regions with transmissive features, where groundwater flow was assumed to flow through the waste/backfill region;

(b) 'midstream sections' between two transmissive feature intersections, in which flows are mainly along the crown space;

(c) 'upstream' and 'downstream' sections between a transmissive feature intersection and a vault end, in which flows are lower, and mainly along the crown space.

In submodels (b) and (c), the transport (diffusive and advective) of radionuclides from the waste/backfill region to the crown space was taken into account, limiting the impact on the overall source term. The values for the groundwater flows in transmissive features and through the vaults were obtained from separate continuum and fracture-network hydrogeological modelling (see Section 2.2).

Within each source-term submodel, the effects of primary package containment, sorption onto backfill materials, solubility limitation, and radioactive decay and ingrowth were taken into account.

2.2. GEOSPHERE

Detailed models of groundwater flow were used, not only to gain an understanding of the flow behaviour and the pathways for radionuclide transport through the geosphere, but also to provide data to calibrate the parameter PDFs used in the MASCOT submodels for radionuclide transport.

A hierarchy of groundwater flow models was used in support of the assessment. These ranged from models used in the interpretation of data and the derivation of effective properties, to regional flow models. Two approaches were used to calculate regional groundwater flow. Coupled groundwater flow and salt transport was calculated in a two-dimensional (2-D) vertical cross section, on the basis of using a continuum porous-medium approach, using the program NAMMU [6]. A pathline calculated with NAMMU from the potential repository location is shown in *Figure 2*. In addition, a three-dimensional (3-D) discrete fracture-network model was used to address flow through the network of transmissive features within the repository host rock, the Borrowdale Volcanic Group (BVG). This model was implemented using the program NAPSAC [7]. Calculations of radionuclide transport in two dimensions were also undertaken with NAMMU in order to obtain concentrations for use in calculations for the agricultural wells.

A network of one-dimensional geosphere transport submodels representing segments of the flow path was then created within MASCOT. The NAMMU and NAPSAC models were used to obtain parameters characterising the groundwater volume flux through the repository and the travel times and path lengths in different hydrogeological units. These calculated parameters were then used in the MASCOT calculations.

Most calculations with MASCOT were undertaken for a base-case hydrogeological model [1]. However, a number of hydrogeological issues were identified that were not addressed by the base-case model. Some of these issues were investigated by means of a number of variant calculations. These calculations ensured that there was not undue focus on the base-case hydrogeological model of the site, which was not considered to be the only possible hydrogeological interpretation.

Most of the water movement in fractured rocks takes place in the fractures. Radionuclides in solution can diffuse into immobile water in the rock matrix away from the fractures. This effect can considerably retard transport. Experiments have been conducted within the NSARP in order to measure the diffusivity into rocks of interest [8]. The extent of rock-matrix diffusion was explicitly calculated in the radionuclide transport calculations undertaken using the program MASCOT. The idealised model considers a system of uniformly spaced, parallel fractures, all carrying similar, spatially uniform flows and radionuclide concentrations. Retardation by sorption is considered to modify the transport both within the fractures and in the matrix. Sorption was treated as a simple linear equilibrium process with parameters based on batch and intact experimental measurements of sorption [8].

2.3. BIOSPHERE

The MASCOT calculations used an equilibrium biosphere submodel, with the factors relating radionuclide fluxes from the geosphere to doses being based on separate, more detailed calculations.

Factors were calculated, based on a simple model, relating dose to a unit flux of radionuclide to the biosphere or unit concentration of the radionuclide in abstracted groundwater. The radionuclide fluxes to the biosphere were taken to be divided among a number of soil types and surface water bodies within the resource area of the community under consideration. Radionuclides were taken to be transported directly from each soil type to surface water bodies. The radionuclide concentrations in each soil type were calculated using an analytical technique. Assessment calculations were undertaken for a number of climate states that were assumed not to change with time.

Radiological risks are largest when radionuclide fluxes are discharged to terrestrial rather than marine environments (see *Figure 3*). Some component of terrestrial discharge of radionuclides is expected with a sea-level stand similar to the present and this component would be expected to increase with falling sea level, as would occur in colder climatic conditions. A key factor in assessing the radiological risks arising from terrestrial discharge is the fraction of activity that is discharged to surface soils compared to the fraction that enters surface water bodies. This issue is being addressed using the program SHETRAN-UK, developed to undertake surface water catchment simulations [9]. The studies undertaken to date indicate that the fraction of activity that is discharged to surface soils is no more than 0.1; further studies are being undertaken.

For natural discharges of radionuclides to soils or to surface water bodies, the most significant radionuclide over the first 10^5 years has been found to be ^{36}Cl. For this radionuclide, the dominant pathway is retention in soil, uptake by pasture plants and transfer to animal products, notably milk. The main uncertainties relating to the biosphere concern the dynamics of ^{36}Cl in soils and uptake in plants. These matters have been studied as part of the NSARP, in lysimeters and associated column experiments [9].

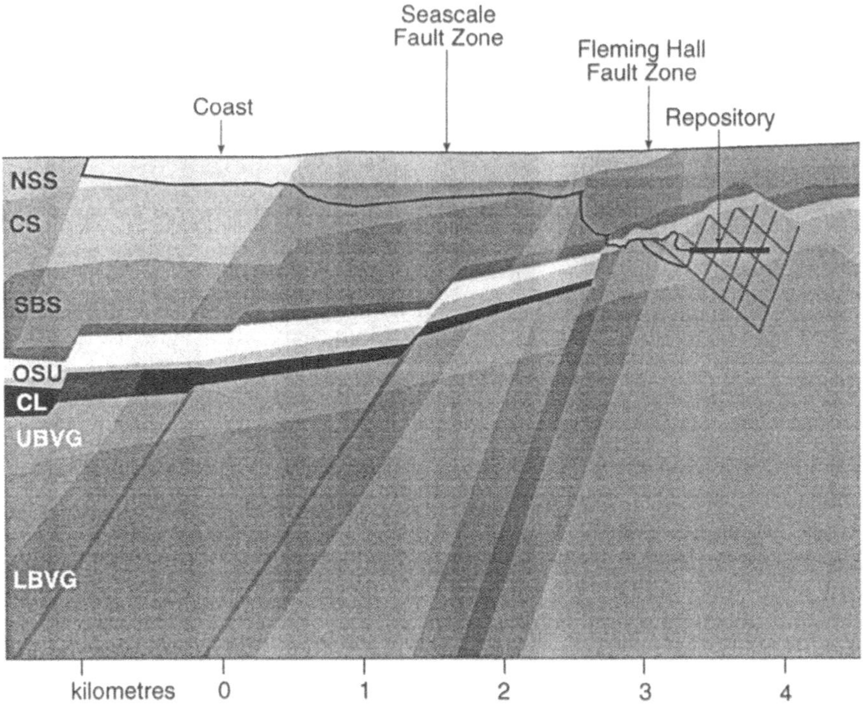

Figure 2. Pathlines from the potential repository (same vertical and horizontal scales).

Key to geological units: LBVG: Lower BVG; UBVG: Upper BVG; CL: Carboniferous Limestone; OSU: Other Sedimentary Units; SBS: St Bees Sandstone; CS: Calder Sandstone; NSS: Near-surface Sandstone.

3. Risk Calculations

A probabilistic MASCOT calculation was performed using base-case distributions for the hydrogeological parameters, and with biosphere treatment corresponding to 'natural' discharge of radioactivity to the marine environment, streams, rivers and deep soil. Supplementary deterministic calculations were performed to investigate alternative hydrogeological assumptions, and for exposure resulting from abstraction of water from agricultural wells.

The probabilistic MASCOT calculation contained 500 sampled realisations. The mean dose was calculated as a function of time for each steady-state biosphere model, and converted to annual individual risk values using the conversion factor 0.06 Sv^{-1}. The calculations were extended up to 10^8 years after closure although quantitative assessment beyond 10^6 years is

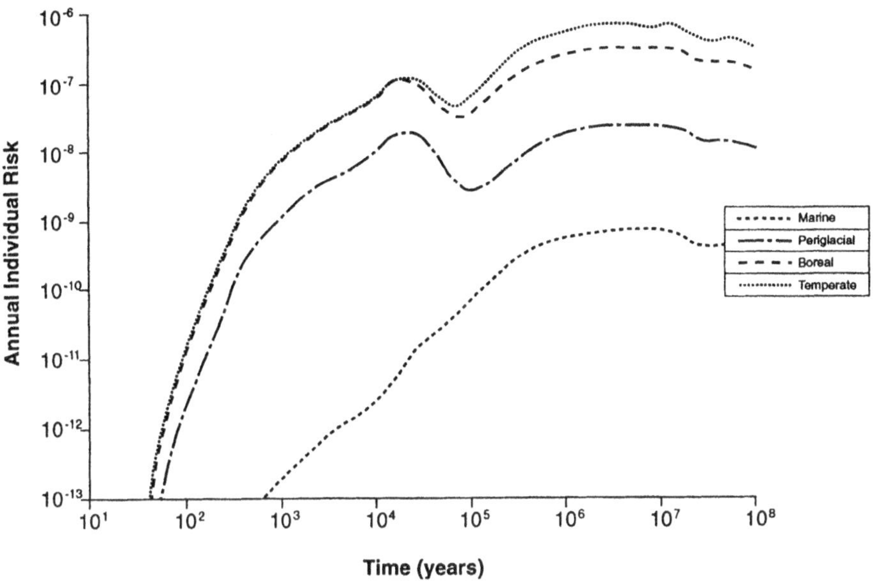

Figure 3. Base-case probabilistic calculation: risk plotted against time for each of the climate states with terrestrial discharge, and for the temperate climate state with marine discharge.

The horizontal dotted line shows the regulatory target for risk. The most important set of parameter values derived from other detailed calculations were hydrogeological parameters that affect the flux of groundwater through the repository, and the travel times and path lengths in different hydrogeological units. The calculations undertaken to obtain these parameters were explained in Section 2.2.

highly speculative.

The annual individual risk as a function of time, obtained from the base-case probabilistic calculation, is shown in *Figure 3.* Results are shown for four different biosphere models, corresponding to different climate states and discharge environments. The risk peak at around $2{\cdot}10^4$ years arises mainly from ^{36}Cl, whereas the broader peak beyond 10^6 years arises mainly from ^{226}Ra, one of the daughters of ^{238}U. Many of the other radionuclides present in the wastes made no significant contribution to risk, because they were effectively contained as a result of one or more of the system components acting to keep the radionuclides out of the human environment long enough that decay reduces the released amounts to insignificant levels.

Risks for the agricultural well were obtained from calculated radionuclide concentrations in near-surface waters, these concentrations having first been obtained from a 2-D radionuclide transport calculation undertaken with NAMMU. These preliminary deterministic calculations for the agricultural well exposure pathway indicated a peak risk of $1.7 \cdot 10^{-6}$ at 4,000 years after closure, with a dominant contribution from ^{129}I.

4. Understanding the Results

The numerical calculations of risk were supplemented with simple analyses aimed at providing a clear understanding of why it is that certain radionuclides are dominant contributors to risk, and how different components of the system contribute to performance.

The extent of containment was investigated using simple analyses of the contributions to containment time of the primary physical containment, the chemical near-field conditions, and the geosphere. The containment time contributed by the geosphere is strongly affected by the degree of retardation by sorption of radioelements by the rocks. Comparison of the containment times with the half-life of each radionuclide showed how much each of these 'barriers' contributes to containment for each radionuclide. These comparisons showed that the vast majority of the disposed inventory decays in the repository near field or in transit through the geosphere.

The contribution to risk of those radionuclides that are too long-lived to be effectively contained by any component of the system is determined by the extent of the spreading (in time) and the dilution (into a larger volume of groundwater) that they experience during release from the repository and transport through the geosphere. To a large extent, spreading and dilution are two aspects of the same phenomenon.

It was found that the total time over which the arrival in the biosphere of a long-lived radionuclide was spread could be estimated from a source-term contribution, σ_s, and a geosphere-transport contribution, σ_g. For such long-lived radionuclides, decay in the near field and geosphere may be considered to be unimportant. The total spreading time, σ, is given by $\sigma^2 = \sigma_s^2 + \sigma_g^2$ [1].

Simple analyses of the source-term and geosphere processes showed that for the release of ^{36}Cl, the most important physical parameter affecting σ_s is the groundwater volume flux through the repository, Q_r. For other radionuclides (*e.g.* ^{238}U) the solubility limit can also be important, as can the extent of near-field sorption. The geosphere spreading time, σ_g, is affected by the total groundwater travel time, the retardation of the radionuclide and the hydrodynamic dispersion length.

The distributions that were found for σ_s and σ_g for ^{36}Cl as a result of model parameter uncertainty for the base-case probabilistic calculation are shown in *Figure 4*. The spread for σ_s is greater, resulting from the

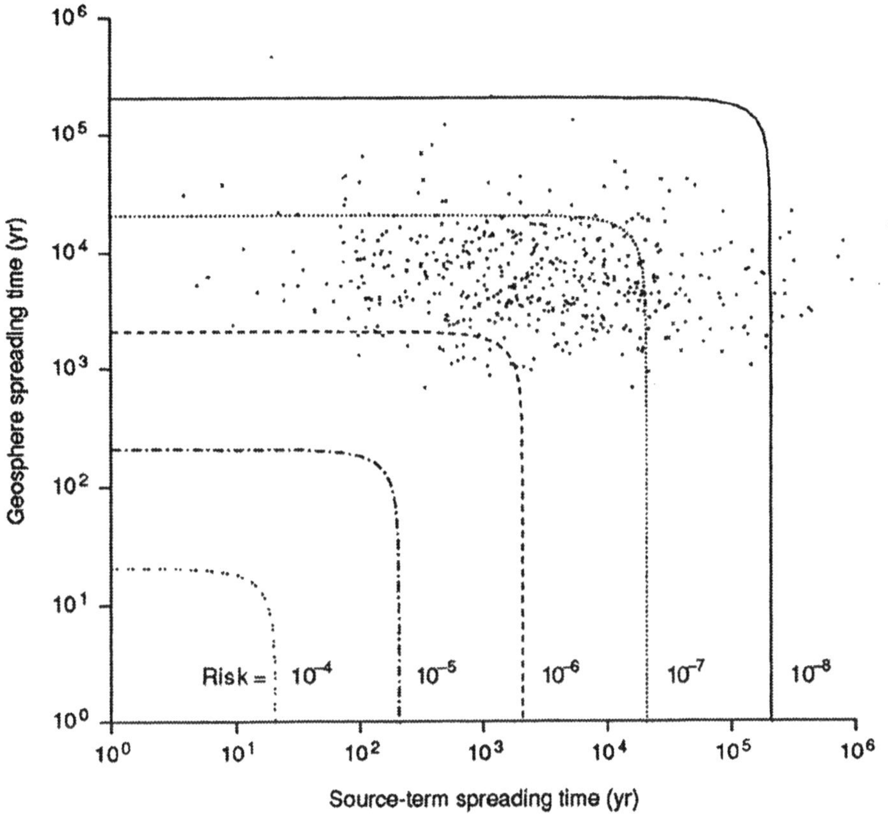

Figure 4. Distributions for the source-term and geosphere-transport contributions to spreading time, σ_s and σ_g, together with contours of estimated peak risk contribution, for the base-case probabilistic calculation, 'natural' discharge of ^{36}Cl.

uncertainty in Q_r. Superimposed on the diagram are contours of risk, as estimated by a simple formula (whose agreement with the MASCOT calculations has been shown to be good). Using this simple formula, it is possible to understand how the risk for each radionuclide varies as a function of parameters such as the travel time through the geosphere. Demonstration of such understanding is considered to be an important feature of any safety case.

5. Key Issues

Considerable progress has been made in identifying the range of hydrogeological models that are consistent with observations of head and salinity at the site and in the calibration of hydrogeological models to ensure a better fit with observations. In the future, continuing validation will ensure that the range of models consistent with site data is identified and the acceptable parameter ranges within these models constrained.

The development of a good understanding of factors controlling risk is regarded as a key component of a safety case. This understanding may be accomplished by careful analysis of the results from models of varying complexity. A particularly powerful approach has been developed in which the performance of various barriers in the repository system can be understood on the basis of very simple models. This has served to focus research priorities.

As noted earlier, this assessment is of a preliminary nature. Detailed assessments will involve a fuller treatment of many aspects of the system. Certain issues are currently the subject of study with the aim of developing robust and appropriate treatments. More sophisticated treatments of time dependence and spatial variability are under development, and 3-D groundwater flow models are likely to be increasingly used. The implementation of a systematic scenarios-based approach will provide a sound basis for the assessment calculations.

A number of assumptions have been made in the current assessment which it is considered may in due course prove to be unduly pessimistic. Work is in hand to address the basis of these current assumptions, which may lead to revised models whose incorporation in future assessments will tend to lower calculated risks.

Acknowledgment

The Nirex 95 assessment was carried out by AEA Technology under contract to United Kingdom Nirex Ltd.

References

1. NIREX (1995): Nirex 95: A preliminary analysis of the groundwater pathway for a deep repository at Sellafield, *Nirex Science Report* S/95/012 (3 Volumes), U.K. Nirex Ltd., Harwell.
2. DoE (1995) *U.K. Review of Radioactive Waste Management Policy: Final conclusions*, DoE Document Cm 2919, Her Majesty's Stationery Office.
3. NIREX (1994a) The geology and hydrogeology of the Sellafield area: interim assessment: December 1993, *Nirex Report* No. 524, U.K. Nirex Ltd., Harwell.
4. NIREX (1994b) Post-closure performance assessment: information management, *Nirex Science Report* S/94/002, U.K. Nirex Ltd., Harwell.
5. SINCLAIR, J.E. AND AGG, P.J. (1994) MASCOT and MOP programs for probabilistic safety assessment, Part A: Overview, *Nirex Safety Series Report* NSS/R336 Part A Version 3C/3A.
6. HARTLEY, L.J., JACKSON, C.P., AND WATSON, S.P. (1994) NAMMU (Release 6.2) User Guide, *AEA Technology Report* AEA-ESD-0138, AEA Technology, Harwell.

7. GRINDROD, P., HERBERT, A.W., ROBERTS, D.L., AND ROBINSON,P.C. (1992) NAPSAC Technical Document, *AEA Technology Report* AEA-D&R-0270, AEA Technology, Harwell.
8. JEFFERIES, N.L., LEVER, D.A., AND WOODWARK, D.R. (1993) NSARP Reference Document: radionuclide transport in the geosphere: January 1992, *Nirex Safety Series Report* NSS/G118, U.K. Nirex Ltd., Harwell.
9. THORNE, M.C. (ed.) (1995) Nirex Biosphere Research: Report on Current Status in 1994, *Nirex Science Report* S/95/003, U.K. Nirex Ltd., Harwell.

DEEP-WELL INJECTION OF LIQUID RADIOACTIVE WASTE IN RUSSIA: PRESENT SITUATION

A. RYBALCHENKO
Head, Research Laboratory
All-Russia Design and Technology Institute (VNIPIPT)
33 Kashirskoje shosse, Moscow 115409, RUSSIA

1. Introduction

At present there are 3 facilities (polygons) for the deep-well injection of liquid radioactive waste in Russia, all of which were constructed in the mid-60's. These facilities are operating successfully, and activities have started in preparation for decommissioning.

Liquid radioactive waste is injected into deep porous horizons which act as 'collector-layers', isolated from the surface and from groundwaters by a relatively thick sequence of rock of low permeability. The collector-layers (also collector-horizons) contain salt waters or fresh waters of no practical application, lying beneath the main horizons containing potable waters.

Construction of facilities for the deep-well injection of liquid radioactive waste was preceded by geological surveys and investigations which were able to substantiate the feasibility and safety of radioactive waste injection, and to obtain initial data for facility design. Operation of the facilities was accompanied by monitoring which confirmed that the main safety requirement was satisfied *i.e.* localisation of radioactive waste within specified boundaries of the geologic medium. The opinion of most specialists in the atomic power industry in Russia favours deep-well injection as a solution to the problem of liquid radioactive waste management; during the period of active operation of defence facilities (atomic power industry of the former U.S.S.R.), this disposal method prevented the impact of radioactive waste on man and the environment.

The experience accumulated concerning the injection of liquid radioactive waste in Russia is of interest to scientists and engineers engaged in problems of protection and remediation of the environment in the vicinity

M. J. Stenhouse and V. I. Kirko (eds.),
Defence Nuclear Waste Disposal in Russia: International Perspective, 199–217.

of nuclear industry facilities; an example of the utilisation of the deep subsurface for solidified radioactive waste and the disposal of different types of nuclear materials.

Information on the scientific principles and background for the development of facilities for the injection of liquid radioactive waste in Russia is provided in the report "Deep-well Injection of Liquid Radioactive Waste in Russia - General View of the Situation and Scientific and Technical Principles". This report gives a brief description of the operating facilities.

2. Deep-well Injection Facilities, Dimitrovgrad

At the Research Institute for Nuclear Reactors (State Scientific Centre in the city of Dimitrovgrad, Ulyanovskaya region), porous horizons in the depth range 1130–1550 m are used for radioactive waste disposal.

The development of a liquid radioactive waste injection facility was preceded by seismic investigations, the drilling of exploratory wells and a study of their stratigraphy using geophysical techniques, and the analysis of rock samples and groundwaters. The results obtained showed that the geological structure of the region is favourable for liquid radioactive waste injection. Within the boundaries of the territory, there are no tectonic disturbances (faults/fractures) which could connect deep horizons with the surface.

In the geological section, porous water-saturated horizons and the low-permeability layers separating them, were identified. *Figure 1a* shows a stratigraphic sequence of the facility.

Two horizons were chosen for radioactive waste injection:

- upper layers at a depth interval of 1130–1410 m, composed of limestones, and
- lower layers at a depth interval of 1440–1550 m, composed of clayey sandstone with interlayers of fractured limestoney sandstones.

The collector-horizons contain salt waters with a salt concentration up to 260 g l^{-1}. The rate of groundwater movement is extremely small, and waters of the collector-horizons are essentially isolated from the surface.

Above the collector-horizons is a low-permeability horizon (*Vereisky* horizon) consisting of clays which are essentially impermeable. In the Ulyanovskaya territory and contiguous regions, this horizon plays the role of a 'cap' for oil accumulations, thereby testifying to its high isolation property. Above the Vereisky horizon are porous horizons containing salt waters, separated from shallow horizons containing fresh waters by rocks of low permeability lying at a depth interval of 50–260 m.

Investigation of the groundwater compositions from a series of aquifer layers confirmed the isolation of deep horizons from fresh waters and the surface.

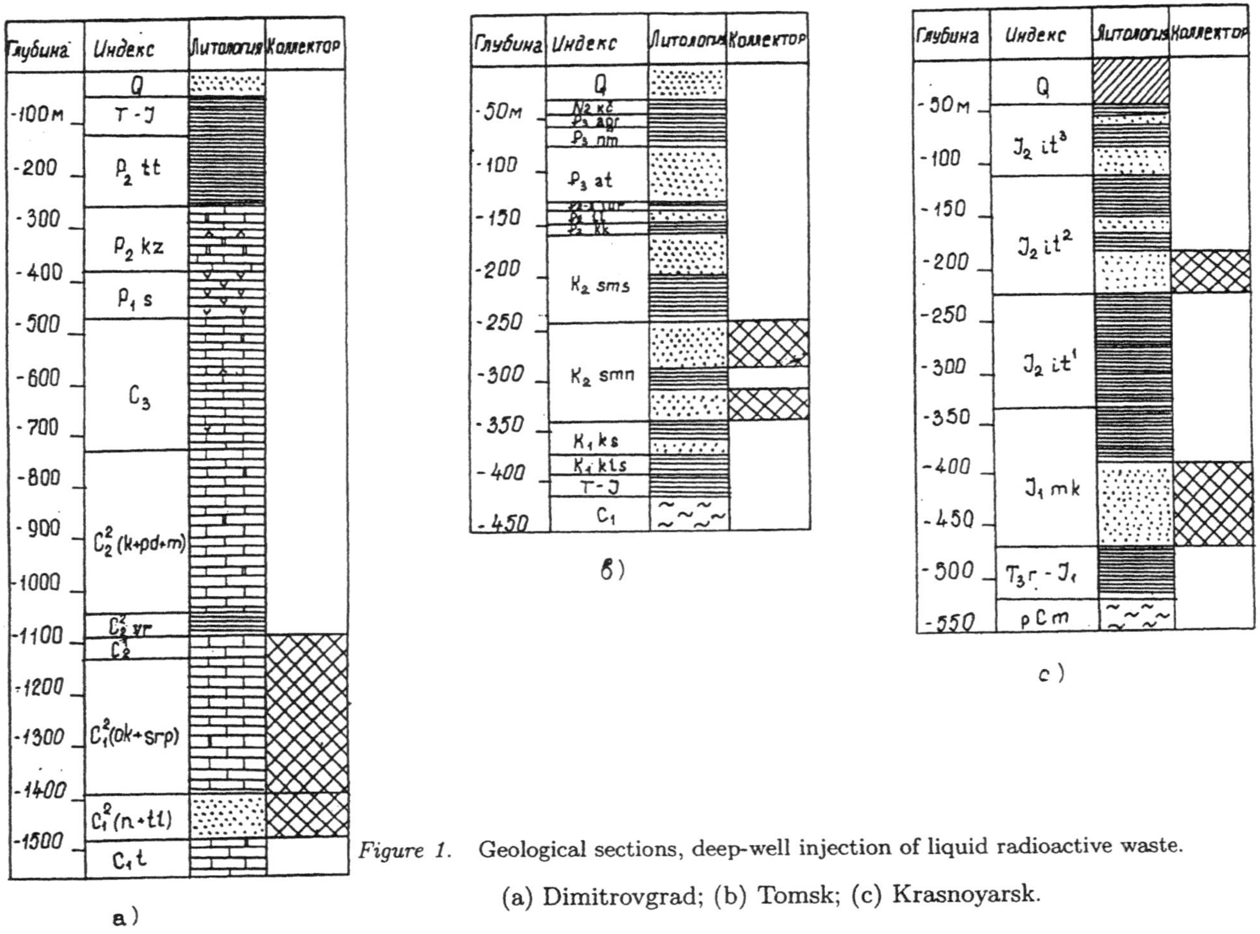

Figure 1. Geological sections, deep-well injection of liquid radioactive waste.

(a) Dimitrovgrad; (b) Tomsk; (c) Krasnoyarsk.

Based on the results of filtration tests which included water pumping and injection, the porosity, ϕ, of horizons was determined, *viz.* total porosity ($\phi_T = 0.26$) and effective porosity ($\phi_e = 0.1$). The filtration coefficient was in the range 0.1–1 m d^{-1} (0.08–0.8 darcy).

While predictions of collector-layer performance were being performed, injection of liquid radioactive waste in the collector-layers under investigation was authorised to be carried out under limited pressures not exceeding 5.0 MPa. It was assumed that the distribution of injected radioactive waste in the collector-layers would lie within a radius of 1,000-2,000 m from the injection well. Analysis of the experience of oil field production utilising a flooding technique indicated that the disposal of liquid radioactive waste under such conditions would be rather efficient.

Hydrogeological investigations led to the conclusion that the subsequent migration of nuclides would be negligible over a time period long enough for radionuclides to decay to safe levels. Studies concerning the compatibility of the liquid radioactive waste with its geological surroundings were able to verify that the requirements for radioactive waste disposal had been met, and to allow the technology for radioactive waste disposal to be developed.

Based on the results obtained, subsequent activities successfully completed the design of:

- the facility,
- injection and observation wells,
- transport pipelines,
- pump stations, and
- operation and monitoring methods.

The injection wells are located directly on the facility land in the vicinity of the sub-station for the collection and processing of radioactive waste. The wells have a multiple-inclination design, the upper part of the well, in the geological section corresponding to the fresh water horizons, being cased with a minimum of two strings. The annular space beyond the outer pipe, and the space between the tubes, were grouted along the entire length of the casing, thereby providing reliable protection of fresh groundwaters from radioactive waste and salt waters from deeper horizons. The lower part of the well, at a depth interval corresponding to the collector-horizons, was equipped with a filter column. The supply column is located along the axis of the well, positioned to direct waste into the collector-layer.

Since the liquid radioactive waste directed for disposal was not corrosive, the casing pipe was manufactured from ordinary steel. Both supply and filter columns were manufactured from corrosion-resistant materials over the maximum length which could be subjected to the influence of chemicals added to the layer, in order to prolong its receiving ability. *Figure 2* shows the design of the well.

The observation wells have a similar design to that of injection wells. In all, the facility includes 5 injection wells and 32 observation wells. *Figure 3* shows the layout of the wells and a schematic geological cross-section.

The waste directed for injection consists of solutions derived from:

- decontamination of research plants,
- decontamination of equipment and special clothes,
- perimeter water discharges and
- "cooling" ponds for nuclear reactor spent fuel.

These solutions contain phosphates, oxalates, sulphates, nitrates, and oils. The salt content is up to 3.5 g l^{-1}, and the specific activity of waste reaches $3.7 \cdot 10^2$ kBq l^{-1}. The radionuclides include isotopes of caesium, strontium, ruthenium, zirconium and the rare earth elements. The long-lived, transuranic nuclides (TRU) are present in negligible amounts.

Volumes of waste up to 550 m^3 d^{-1} (150,000 m^3 a^{-1}) are injected under an injection pressure of 5.0 MPa. Waste injection is accompanied by monitoring observations and measurements. The monitoring of the injection process includes the determination of the rate of flow and pressure of injection, as well as the composition of the injected radioactive waste.

3. Monitoring Requirements

Monitoring of the state of the geological formations includes measurements of formation pressures or groundwater levels, analysis of water samples taken from the wells, and geophysical measurements made within the well (thermometry, radioactive logging, electric logging). As a result, information is received concerning the development of the repressuring dome in the collector-layer and how the radioactive waste components are distributed throughout the layer's area and thickness.

Monitoring of the technical conditions of the wells includes determination that the pipes of the casing string are leak-tight, evaluation of the isolating cement in the angular space behind the pipe, and determination of the acceptance capability of the well. Acoustic cementometry, thermometry and other measurements are carried out for this purpose. The monitoring results are used for preventive maintenance in order to maintain the necessary technical condition of the well.

Results of measurements of formation pressures or groundwater levels provide information on the trends in waste distribution long before radionuclides are detected in any of the observation wells, in order to confirm isolation of the collector-layer from the above-lying horizons. Geophysical monitoring (radioactivity logging and thermometry) allows the determina-

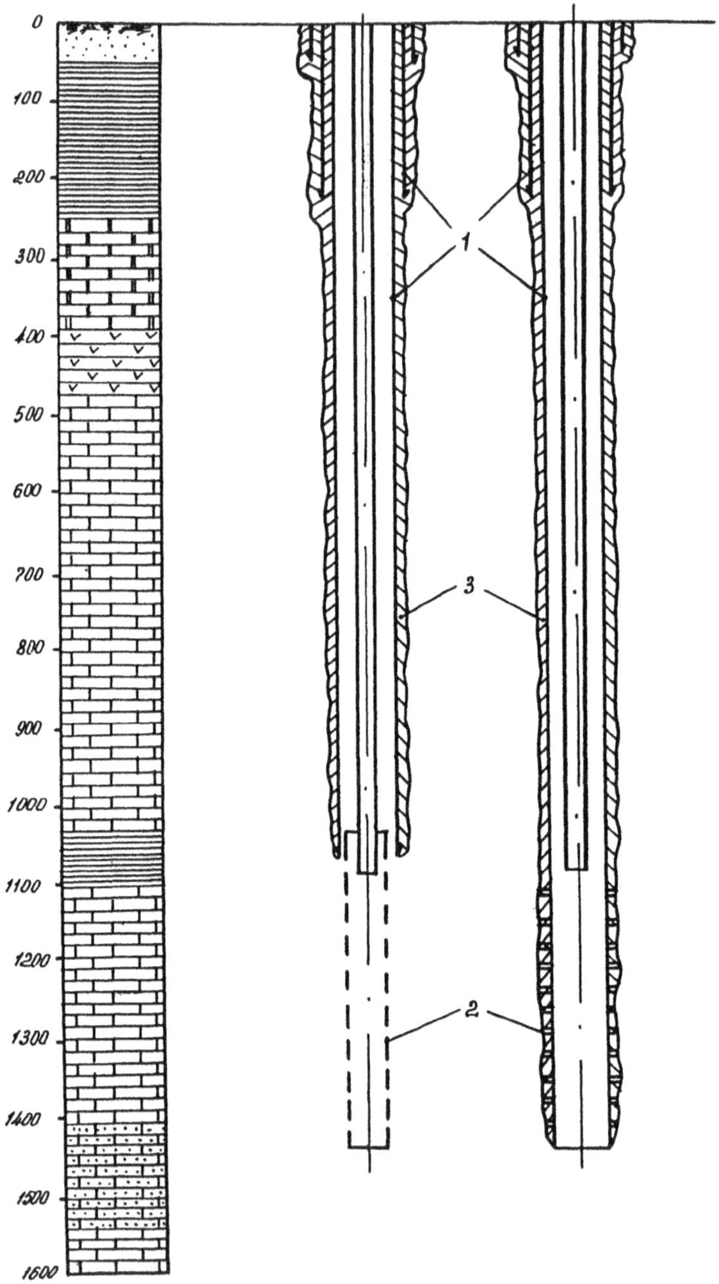

Figure 2. Scheme of well construction.

1 – casing; 2 – cementation; 3 – filter casing.

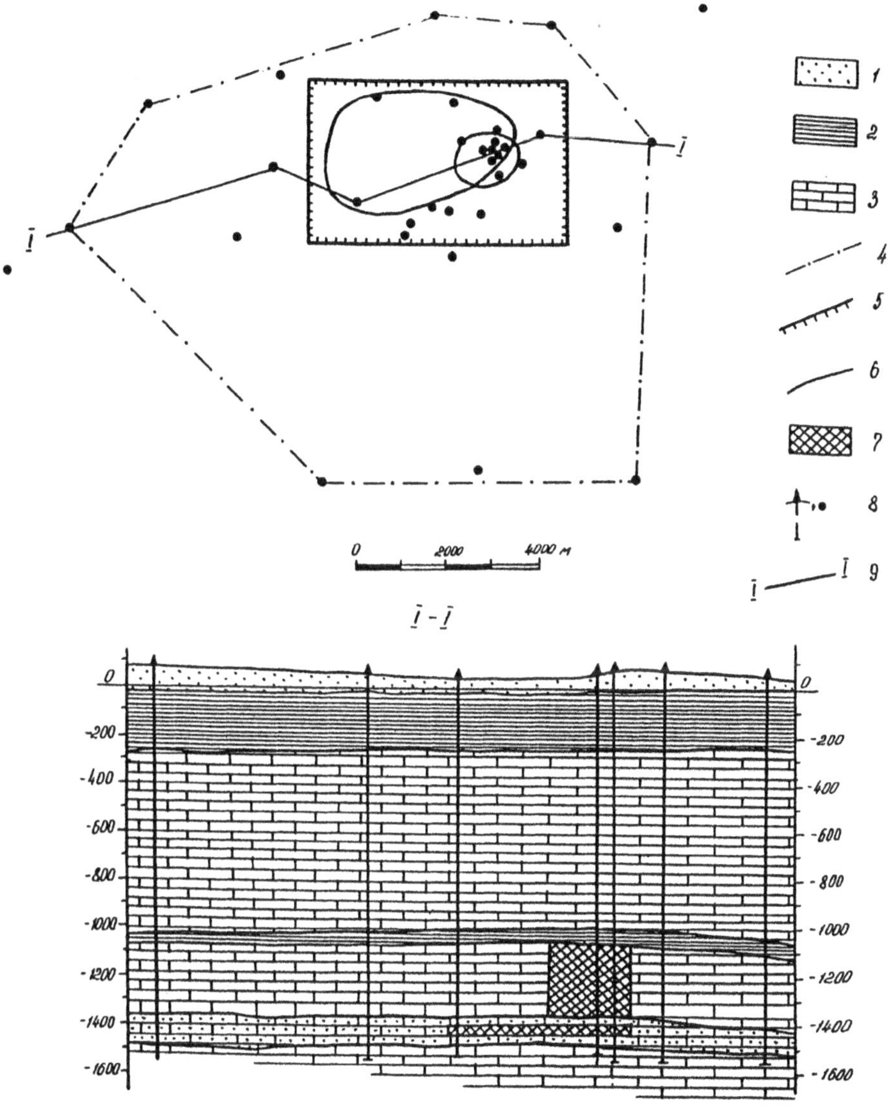

Figure 3. Deep-well injection facility, Dimitrovgrad.

1,3 – permeable rocks; 2 – impermeable rocks; 4 – allotment boundary; 5 – boundary of facility; 6 – outline of waste; 7 – waste in collector layer or horizon; 8 – well; 9 – section line.

tion of the waste distribution over the thickness (height) of the collector-layer, to assess the technical condition of the well. Sampling and analysis of groundwaters provide data concerning the presence of radioactive waste constituents in the area around the observation wells.

Based on the monitoring data, conclusions are made concerning:

- the level of agreement between actual and predicted data on radioactive waste distributions;
- implementation of safety requirements while injecting radioactive waste;

In this way, procedures are developed and implemented to optimise the radioactive waste injection process (*Table 1*).

3.1. MONITORING SCHEDULE

Decisions on the periodicity and scope of observations depend on the location of a well, on the specific tasks to be solved, and on the results of previous observations. Monitoring conditions at Dimitrovgrad are:

- pressures and the rate of flow are recorded continuously,
- waste composition are determined daily,
- groundwater levels are measured in different wells on a weekly or monthly basis, and
- geophysical measurements and groundwaters sampling are performed quarterly or yearly.

Inspection of the technical condition of wells is carried out every year or once every 3 years.

The area of the facility for radioactive waste injection represents an 'allotment' (*Figure 3*). The main requirement for liquid radioactive waste injection is to limit the distribution of radioactive waste constituents to within the boundaries of this allotment, or allocated area. Within the boundaries of the allotment, utilisation of the deep subsurface zones of the earth not associated with radioactive waste injection, is restricted.

Originally, while conducting experimental and test activities, injection of radioactive waste was carried out in a deep horizon, in the depth interval 1z40-1520 m, which consisted of clayish sandstone, with interlayers of limestoney sandstone. The monitoring data indicated a marked lack of uniformity in the waste distribution; in particular, filling of the limestoney sandstone interlayers occurred. It was accompanied by a decrease in permeability of the clayish sandstones while injecting radioactive waste with a lower salt content compared with formation waters. For this reason, the subsequent injection of liquid waste was carried out in the upper horizon (depth interval 1130-1410 m), resulting in a marked uniformity of

TABLE 1. Control over deep-well injection of liquid radioactive waste.

NN	Types of control	Defined indices	Periodicity	Application of the results
1	Injection control:			Control over schedule, determination of sum volumes for injection
	– pressure	Pressures at well mouths	continuous	
	– Rate of flow	Volume of injection per time unit	– " –	
	– Radwaste composition	Content of main components	periodically (daily, weekly)	
2	Control over conditions of geological formation:			Determination of repressuring dome in collector-layer, position of contour for radwaste components, revelation of possible sources for seismic vibrations, confirmation of the collector-layer isolation from overlying horizons
	– hydrodynamic measurements	Position of underground water levels, formation pressures	periodically (daily, monthly)	
	– hydrogeochemical	Composition of underground water samples	periodically (daily, monthly)	
	– geophysical measurements	Radioactivity, temperature, electrical resistance seismic signals	periodically (monthly, yearly)	
3	Control over technical state of wells:			
	– state of filtration zone	Degree of mud grouting	periodically (yearly, once per 3-year period)	Preventive maintenance for maintaining technical condition
	– casing string	Integrity	– " –	
	– behind the pipe cementation	Isolating properties	– " –	

distribution of waste in the limestone layers. The deeper horizon is reserved for the injection of waste with increased salt content. The generalised contour of radioactive waste constituents in the collector-layers of the facility is also shown in *Figure 3.*

According to model calculations, the radionuclides will remain within the boundaries of the allotment for a time period sufficient for the nuclides to decay to safe levels. Deep-well injection of liquid radioactive waste does not prevent the utilisation of fresh groundwater, or oil-field development, beyond the boundaries of the allotment.

The utilisation of deep-well injection has avoided the need for liquid radioactive waste discharge to surface water systems, as well as the creation of storage ponds and complex facilities for waste processing, which may not provide the necessary protection of the environment.

4. Deep-well Injection Facilities at Tomsk, Krasnoyarsk

Facilities for the deep-well injection of liquid radioactive waste from the Siberian Chemical Integrated Plant (city of Seversk, Tomsk region) and the Integrated Mining and Chemical Plant (city of Zheleznogorsk, Krasnoyarsk territory) differ significantly from the facility in Dimitrovgrad in terms of the type of formations, the depth of wells, and the volumes and compositions of injected waste.

Figures lb and 1c show the stratigraphic sequences of the injection facilities at Tomsk and Krasnoyarsk, respectively. Sand horizons are used as the collector-layers, which are isolated from the surface and shallow waters by layers of low-permeability clay. Construction of the facilities was preceded by detailed geological exploration and characterisation which included geophysical investigations of the area (electrometric and seismic), drilling of wells and their characterisation, and the study of rock and groundwater samples. As a result, the necessary information was obtained on the geological structure and rock properties of both areas, their tectonic structure, hydrogeology, as well as their isolating properties. *Figures 4 and 5* show the layout of the facilities at Tomsk and Krasnoyarsk, respectively, and their corresponding geological cross-sections.

Special attention was given to the determination of the filtration and isolation properties of the rock, and to the characteristics of the natural movement of groundwaters. For this purpose, hydrogeological investigations, long-term test-filtration studies, indicator tests, and a helium survey were conducted.

It was determined that the porosity of the collector-horizons is characterised by values of 0.3–0.4 (total ϕ_T) and 0.08–0.15 (effective ϕ_e), with filtration coefficients in the range 0.8–2.2 m d^{-1} (0.6-1.8 darcy). The filtration

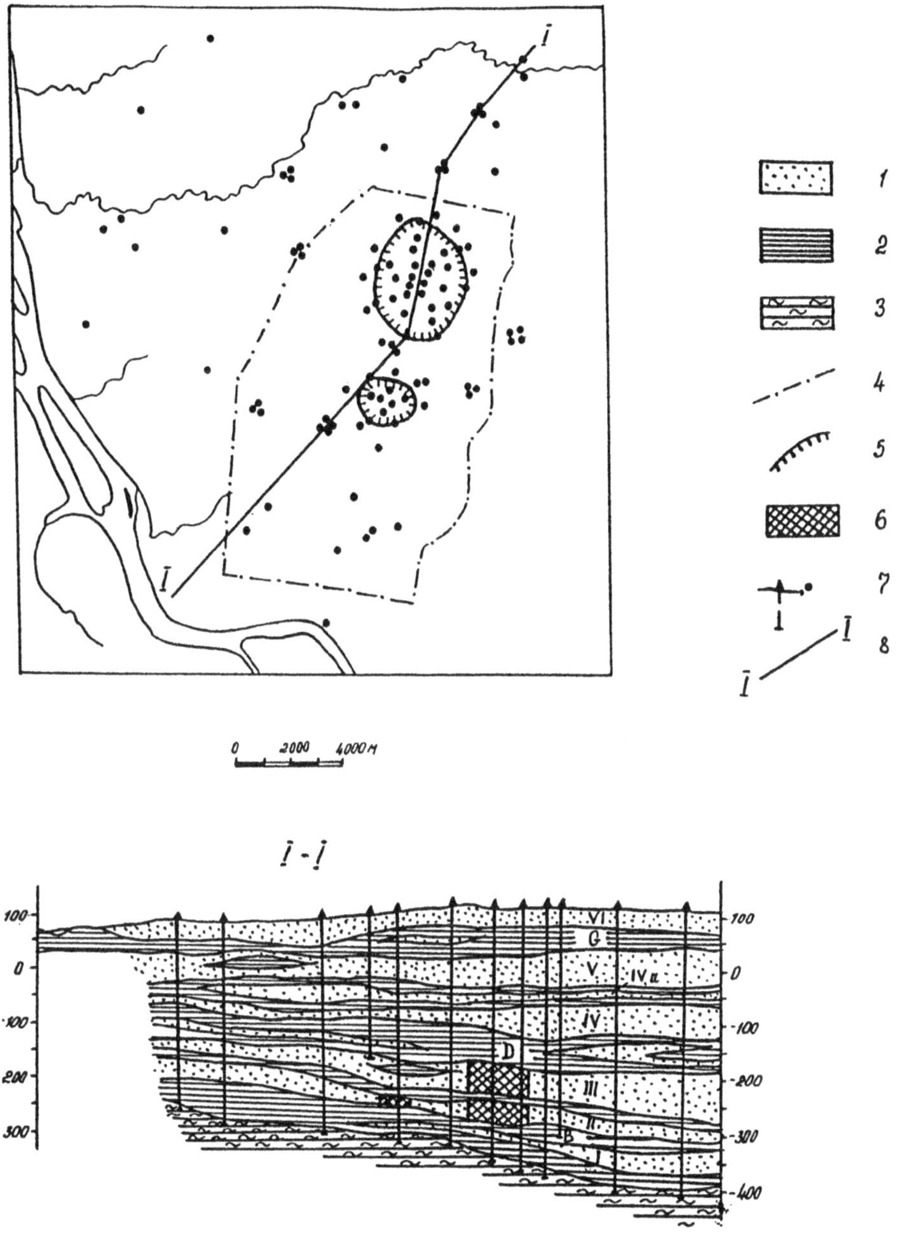

Figure 4. Deep-well injection facility, Tomsk.

1 – permeable rocks; 2 – impermeable rocks; 3 – rocks of basement; 4 – allotment boundary; 5 – boundary of facility; 6 – waste in collector layer or horizon; 7 – well; 8 – section line.

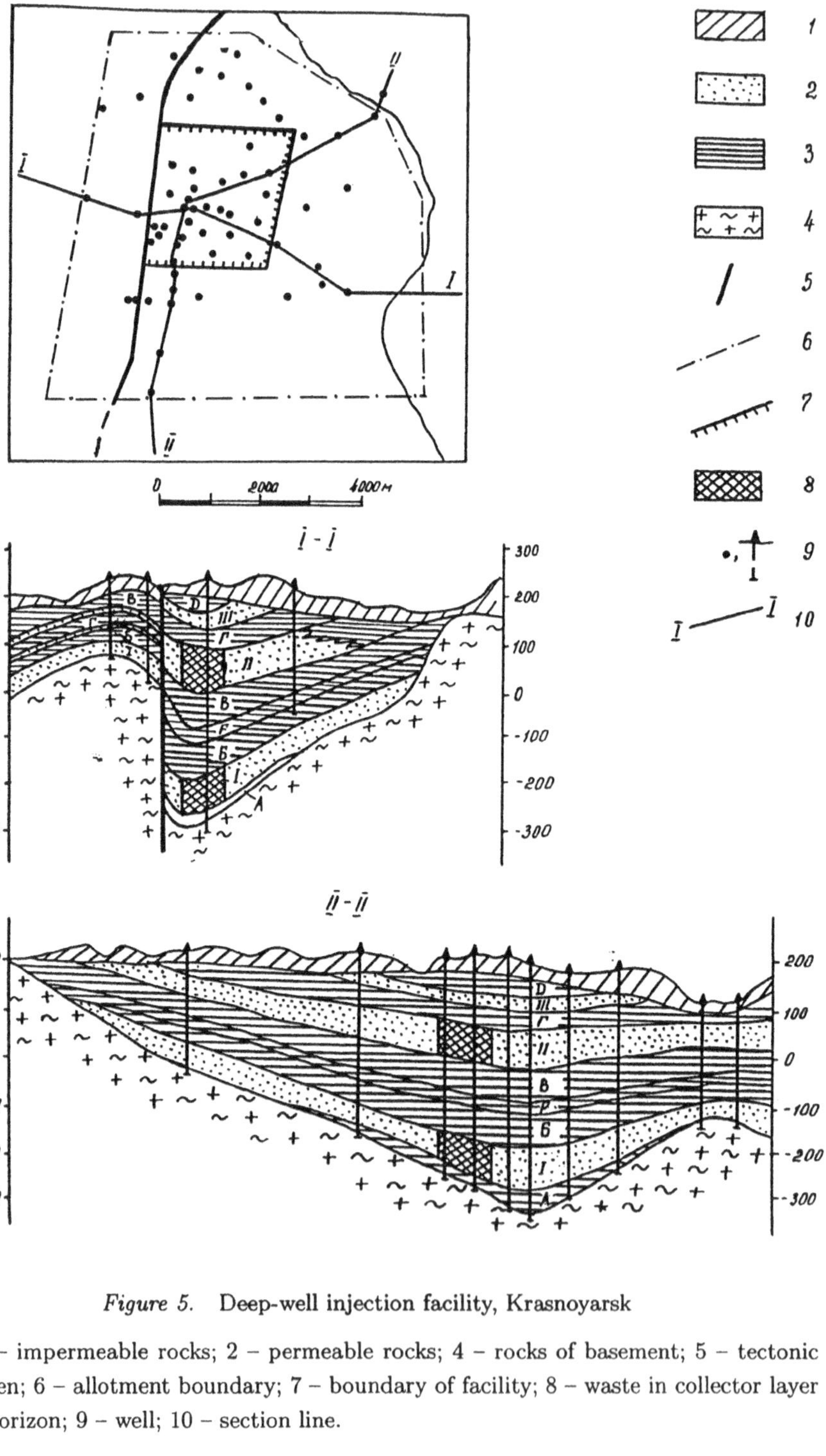

Figure 5. Deep-well injection facility, Krasnoyarsk

1,3 – impermeable rocks; 2 – permeable rocks; 4 – rocks of basement; 5 – tectonic screen; 6 – allotment boundary; 7 – boundary of facility; 8 – waste in collector layer or horizon; 9 – well; 10 – section line.

coefficient of low-permeability clay is characterised by a value of 10^{-4} m d^{-1} (0.1 mdarcy). The rate of natural movement of groundwaters is in the range 3–5 m a^{-1}. According to the data obtained from groundwater compositions, helium measurements, and age determinations, the collector-horizons used for injection appear to be isolated from the surface and shallow fresh-water horizons.

In the region where the radioactive waste injection facility for the Siberian Chemical Integrated Plant is located, there is no tectonic structure which would disturb the collector-layers and confining layer. In the western part of the facility at the Integrated Mining and Chemical Plant, there is a fault, the plane of which is displaced, thereby providing a hydraulic barrier. The isolation properties of this barrier were determined by special investigations of the wells located on both sides of the barrier. This tectonic 'shield' prevents the movement of radioactive waste westwards.

A characteristic and unusual feature of the geology of the facility area at the Integrated Mining and Chemical Plant is the location of the collector-horizons and their isolation by low-permeability clay layers within the boundaries of the depression in the foundation. The density of radioactive waste is higher than that of groundwaters, resulting in the localisation of radioactive waste in the deep part of the depression.

The design and construction of facilities for radioactive waste injection was preceded by mathematical modeling of the conditions for radioactive waste injection and subsequent migration. It was found that the migration of radionuclides which do not interact with the rock over a period of 1,000 years, taking into account hydraulic dispersion and other factors, will amount to 5–6 km in the direction of natural groundwater movement. The distribution of radionuclides which are retarded by rock will be significantly less. During this 1,000-year period, radionuclides (fission products) will decay to safe concentrations.

Based on these predictions, the boundaries of the 'allotments' were determined. Within the limits of these allotments, utilisation of the deep subsurface zones of the earth which are not associated with radioactive waste injection, was restricted.

Based on the investigations carried out, design of the facilities for liquid waste injection was accomplished. Alternative methods for liquid radioactive waste management were also considered. It was found that storage of radioactive waste in tanks, open ponds or reservoirs can lead to contamination of the environment and, hence, to an impact to the population. Management of radioactive waste in such a manner was used during the initial stages of operation of the facilities. Objective data concerning the drawbacks of radioactive waste demanded long-term studies which could not, therefore, be realised in the short term.

4.1. SIBERIAN CHEMICAL INTEGRATED PLANT, TOMSK

At the Siberian Chemical Integrated Plant, there are two facilities for deep-well injection of radioactive waste: Site N18 for injection of LLW, and Site number 18a for ILW.

4.1.1. *LLW injection*

In the facility for LLW injection, two collector sandstone horizons are available at depth intervals of 270-320 m and 350-390 m (*Figure 1.4* [??]). The volume of injected radioactive waste amounts to 3000 m^3 d^{-1} (10^6 m^3 a^{-1}), with injection pressures up to 2.0 MPa. The radioactive waste includes salts, surfactants, and radionuclides (fission products), with a total activity of not more that $3.7 \cdot 10^4$ Bq l^{-1}.

The facility comprises 11 injection wells and 106 observation wells. In addition, outside the facility but within the boundaries of the allotment, there are 108 observation wells to monitor the condition of groundwaters in all aquifers of the territory in which the production complex of the Siberian Chemical Integrated Plant is located.

4.1.2. *ILW injection*

One collector-horizon is used at the facility for injection of ILW, which lies at a depth interval of 315-340 m. The volume of injected waste originally amounted to 550 m^3 d^{-1}, and currently has decreased to 100-150 m^3 d^{-1} (50,000 m^3 a^{-1}), with an injection pressure no greater than 1.2 MPa.

The waste contains salts (up to 350 g l^{-1}), ions of heavy metals, and radionuclides (fission products), and the activity of the injected waste is up to $3.7 \cdot 10^8$ Bq l^{-1}. At the same facility, the injection of liquid radioactive waste with a greater concentration is carried out periodically (1,000–7,000 m^3, several times a year). This ILW facility has 10 injection wells and 45 observations wells.

4.2. INTEGRATED MINING AND CHEMICAL PLANT, KRASNOYARSK

Injection facilities for LLW and ILW are combined at the Integrated Mining and Chemical Plant. Injection wells for waste of the types described are located close to each other, but they feed into different collector-horizons. This facility is called *Severny* polygon.

For the injection of ILW, the sand collector-horizon at a depth interval of 355-500 m is used, and for LLW, the collector-horizon at 180-280 m is used.

The bulk of the waste is similar in composition to, but less in volume than, that injected at the facilities of the Siberian Chemical Integrated Plant:

- LLW - up to 800 m^3 d^{-1} (up to 250,000 m^3 a^{-1}), and
- ILW - up to 300 m^3 d^{-1} (up to 100000 m^3 a^{-1}).

The volumes of injected waste have decreased several times until the present time. The facility includes 13 injection wells and 70 observation wells.

The wells at both the Siberian Chemical Integrated Plant and the Integrated Chemical and Mining Plant are similar in design to, but shorter in length than, the wells at the Research Institute for Atomic Reactors (city of Dimitrovgrad).

The injection of liquid radioactive waste at the facilities associated with both plants is accompanied by a wide range of monitoring observations. The scope of observations is similar to that used at the facility in Dimitrovgrad and is presented in *Table 1*.

In connection with the large number of observation wells, a special system was created and equipped with the necessary instrumentation, equipment, and a means of transportation. At the Siberian Chemical Integrated Plant, all wells are incorporated in a unified system for monitoring the depths of the earth. This system embraces all objects in any event which affects the environment.

The radioactive waste injection facilities at both the Siberian Chemical Integrated Plant and the Integrated Mining and Chemical Plant are located in a zone of low seismicity. However, owing to a fault in the western part of the facility at the Integrated Mining and Chemical Plant, a seismic observation system was established in this region. This system includes 12 monitoring points, and a device for the transmission and processing of data. According to the monitoring data, no seismic effects resulting from waste injection have been registered.

Figure 6 shows the results of monitoring observations relating to the transport of radionuclides through collector-horizons of the Siberian Chemical Integrated Plant facility. Distribution contours for nuclides, based on the analysis of water samples as well as on calculations, show that radionuclides did not reach the boundary of the facility.

Figure 7 shows the distribution contours for radionuclides in the collector-layer of the first (upper) horizon at the Integrated Mining and Chemical Plant facility. The unusual feature of the first horizon's operation lies in its unloading, *i.e.* the pumping out of pure water from unloading wells to decrease the formation pressure. Unloading of the collector-layer makes it possible to regulate the distribution of injected radioactive waste by selecting the rates of flow and pumping.

The radionuclide distribution contour has not reached the unloading wells. While pumping out pure water, continuous monitoring of the water composition is conducted.

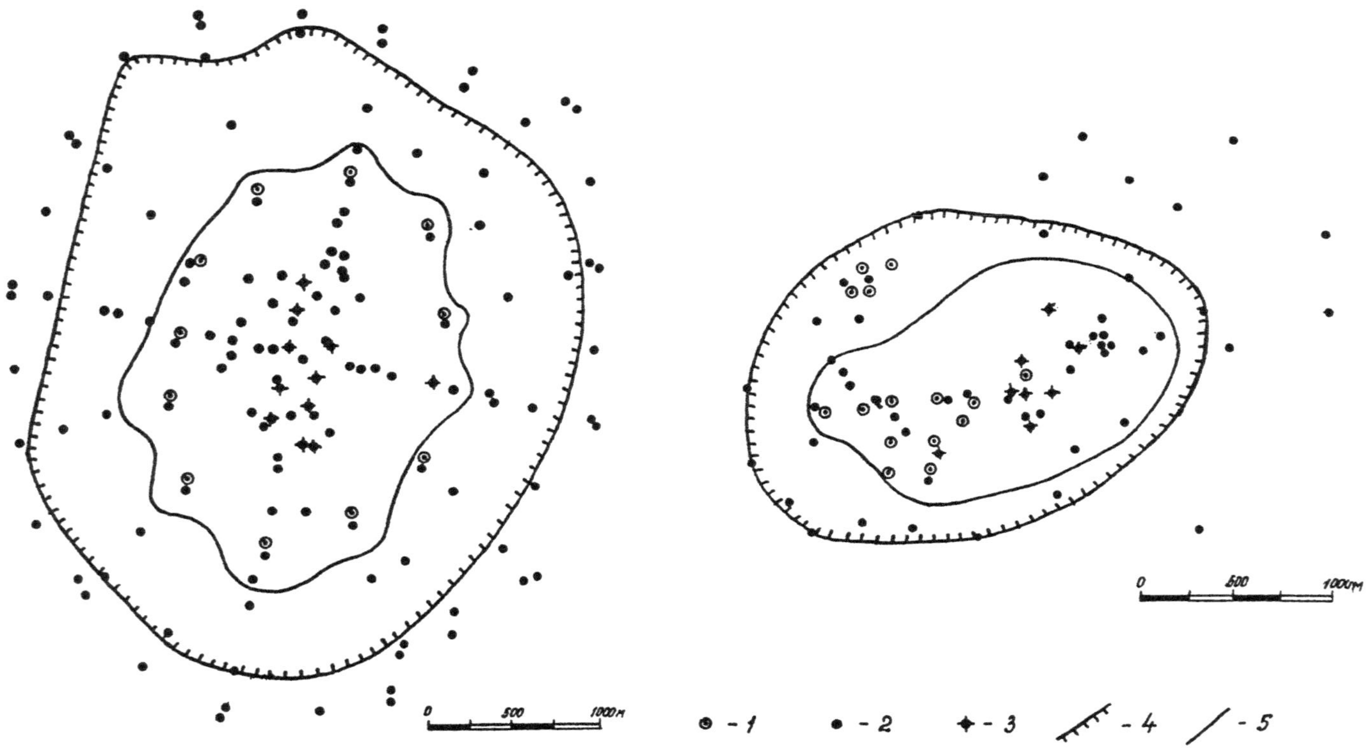

Figure 6. Wells layout and outline of radioactive waste in collector-layer, Tomsk.

(a) low-level activity waste; (b) medium-level activity waste.
1,2,3 – respectively, injection wells, observation wells, and wells after conservation;
4 – boundary of facility; 5 – outline of radioactive waste.

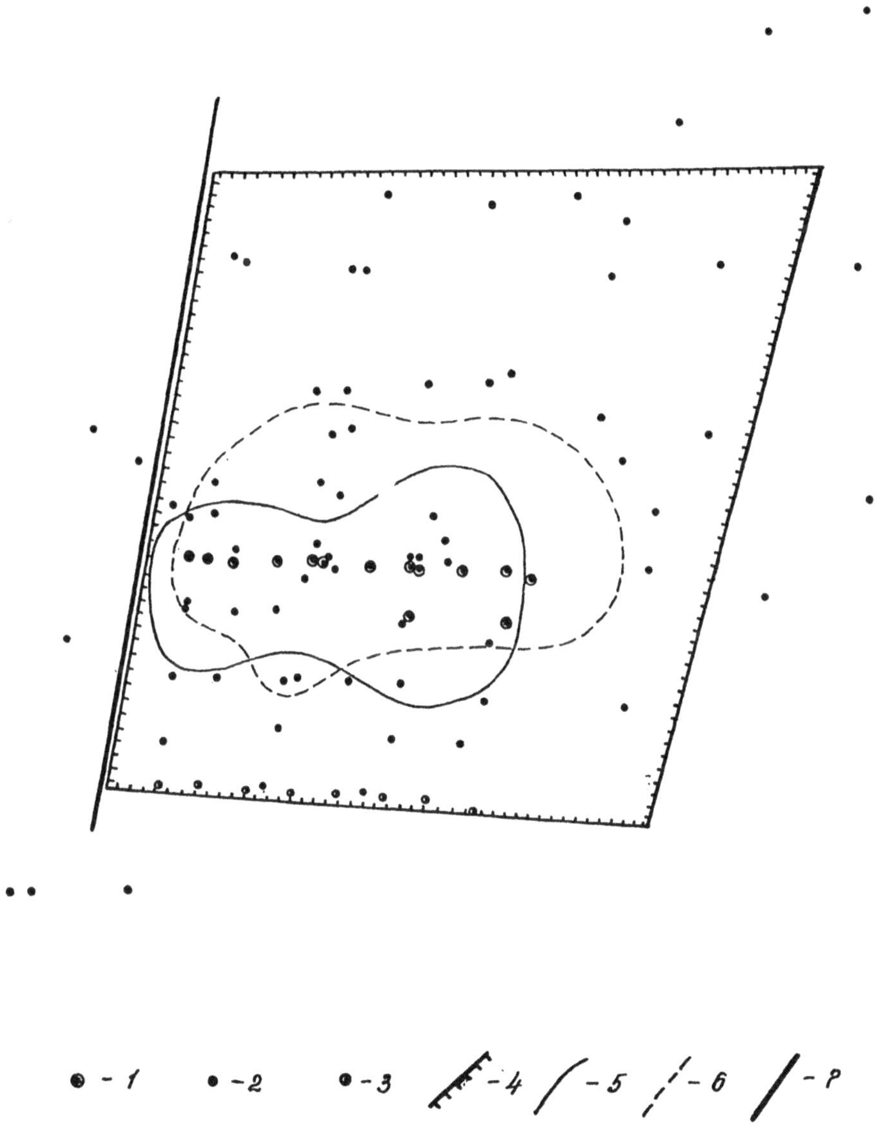

Figure 7. Wells layout and outline of radioactive waste in collector-layer, Krasnoyarsk.

1,2,3 – respectively, injection wells, observation wells, and wells after conservation; 4 – boundary of facility; 5 – outline of medium-level activity waste; 6 – outline of low-level activity waste; 7 – tectonic screen.

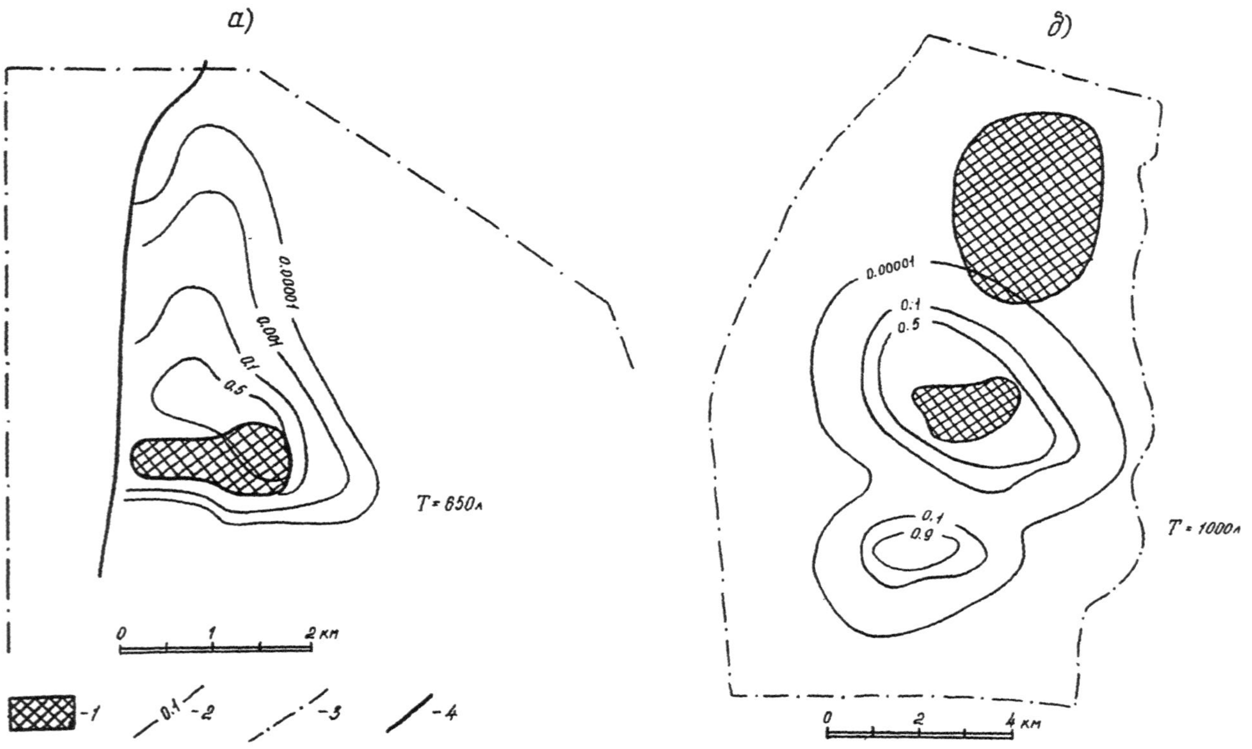

Figure 8. Radioactive waste migration.

(a) Krasnoyarsk, 650-year period; (b) Tomsk, 1000-year period.
1 – boundary of waste, time = 0; 2 – isoline of relative concentration;
3 – allotment boundary; 4 – tectonic screen.

Figure 8 shows the results of predictions of radionuclide migration after decommissioning of the facilities at Krasnoyarsk and Tomsk. Calculations were carried out for an arbitrary radioactive waste indicator (nuclide) which does not interact with the rock and does not decay with time, thereby establishing the maximum possible radionuclide distribution. Figure 8 shows that migrating radionuclides will remain within the boundaries of the allotment.

Thus, the main requirement for the safe deep-well injection of liquid radioactive waste is satisfied, *i.e.* radionuclides are, and will continue to be, located within the boundaries of the geologic medium-allotment.

The final stage in the liquid radioactive waste injection process involves the decommissioning of facilities at the 3 companies identified above. At present, technology is being developed for decommissioning, and enaction of a number of practical measures is underway.

The focus of decommissioning activities is on the closure of wells at particular locations where isolation, covering periods of tens and hundreds of years, should be provided for all horizons intersected by a well and the surface of a collector-horizon containing waste.

At the same time, no large-scale activities will be conducted directly within deep geological formations, *e.g.* the construction of anti-filtration or anti-migration screens, the extraction and decontamination of groundwaters, *etc.* Thus, the geology, hydrogeology and properties of the rock will provide for the localisation of radionuclides within the boundaries of the allotments for a period of time sufficient to allow decay of waste radionuclides to safe levels.

DEEP-WELL INJECTION OF RADIOACTIVE WASTE IN RUSSIA

JAAP HOEK
AEA Technology plc
Harwell, Oxfordshire
United Kingdom

1. Introduction

In the Russian federation, deep borehole injection of liquid radioactive waste has been established practice since at least 1963. The liquid is injected into sandy or other formations with high porosity, which are isolated by water-tight layers. This technique has also been used elsewhere for toxic liquid waste and residues from mining operations. Deep-well injection of radioactive waste is not currently used in any of the European Commission (EC) countries.

In this paper the results of a EC-funded study were presented. The study is entitled 'Measurements, modelling of migration and possible radiological consequences at deep well injection sites for liquid radioactive waste in Russia', COSU-CT94-0099-UK. The study was carried out jointly by AEA Technology, CAG and the Research Institute for Nuclear Reactors (NIIAR) at Dimitrovgrad. Many scientists have contributed to the results reported here. A full list can be found in [4].

The aims of the study are:

- Provision of extensive information on the deep-well injection repositories and their use in the former Soviet Union.
- Provision of a methodology to assess safety aspects of deep-well injection of liquid radioactive waste in deep geological formations. This will allow evaluation of proposals to use deep-well injection techniques in other regions.
- Support for Russian regulatory bodies through evaluation of the suitability of the sites, including estimates of the maximum amount

M. J. Stenhouse and V. I. Kirko (eds.),
Defence Nuclear Waste Disposal in Russia: International Perspective, 219–230.

of waste that can be safely stored in them.
- Provision of a methodology to assess the use of deep-well injection repositories as an alternative disposal technique for EC countries.

The site selected for this study was the repository near Dimitrovgrad, operated by NIIAR. Though the potential environmental impact of the Siberian sites is much greater, it proved difficult to obtain the data required for the assessment of one of these sites. However, the core methodology developed is not very site-specific, and can be extended to the repositories in Tomsk and Krasnoyarsk.

The approach taken is based on the analysis of Features, Events and Processes (FEPs) relevant to the repository, as advocated by the Sandia [1]. This technique is applied by AEA Technology to the Sellafield site in the United Kingdom, which was under investigation as a potential site for the construction of a repository for the U.K. low- and intermediate-level radioactive waste (L/ILW) [2]. The development of the conceptual model for the Dimitrovgrad site can be divided into a number of steps. Firsly, the current list of NEA FEPs ([3], version 0.5) has been structured into a Directed Diagram, clarifying their interrelations and their influence on the risk to humans from exposure to radionuclides originating from the repository. A number of extra FEPs were introduced in this phase. Subsequently this general Directed Diagram has been made specific to the Dimitrovgrad site, by screening out FEPs irrelevant to that location, and developing some FEPs further to clarify various contributing factors. A number of scenarios were identified, and these have been modelled with existing AEA codes. Because of the size of the project, it was not possible to perform the FEP analysis in an exhaustive way. However, the methodology lends itself very well to refinement.

The codes used to perform chemical modelling, groundwater flow calculations (NAMMU), biosphere calculations (BIOS 3A), and a Probabilistic Safety Assessment (MASCOT) have all been previously verified, through international validation exercises such as PSACOIN, HYDRACOIN and BIOMOVS. The results of the calculations are presented here.

For the NIIAR site the assessment has not only been done for the existing inventory, but also for an artificial inventory. The reason is that the actual inventory consists mainly of short-lived isotopes, such as Cs-137. This inventory decays rapidly to safe levels, and many of the scenarios which result in the return of the nuclides back to the biosphere would not be applicable. Since the study is aimed at leading to a methodology that can be applied to the other Siberian sites as well, an artificial inventory was introduced which includes a number of long-lived isotopes. Only the results on the actual inventory are presented here.

2. FEP Analysis

The various steps in the process of analysis of FEPs are summarised below.
The main objectives of this approach are:

- to identify all the exposure scenarios that can lead to radiological risk to humans as a result of disposal of radioactive waste;
- to provide a framework for registering all decisions made on the irrelevance of some FEPs to a particular disposal site (given a particular assessment basis);
- to be able to prove completeness of the analysis to regulatory bodies.

The stages in this process can be summarised as follows:

1. Development of a Directed Diagram (expansive phase);
2. Development of a conceptual model for a disposal site (screening phase);
3. Identification of exposure scenarios;
4. Development of mathematical models to model the migration of radionuclides from the disposal site to humans;
5. Use of computer codes to calculate the migration of radionuclides and their effect on humans in the various scenarios, as part of a probabilistic safety assessment.

The first two steps were summarized in the introduction. A full account can be found in [4]. In order to identify the possible scenarios to be considered in the assessment, possible routes by which radionuclides may return from depth to the biosphere are considered. The analysis of FEPs in the construction of the conceptual model has led to the following cases to be considered:

2.1. BASE CASE

The base case features the natural groundwater pathway, *i.e.* return of radionuclides with the normal groundwater flow in the region. The confining layers are taken to be continuous and unbroken (as they are according to NIIAR);

In addition the following scenarios and scenario representations have been identified:

1. *Undetected features* [NEA 0.5 FEP 2.2.12]
   Transport via natural groundwater pathway, but with breaks in the confining layers due to faulting (it is accepted that these cannot be too frequent or the layer would not act so effectively as a confining layer, however transport through a small number of breaks could be radiologically significant). It is accepted by NIIAR that the basement

is faulted but it is thought that these faults do not propagate through the sediments.

2. *Borehole pathway* [NEA 0.5 FEP 2.1.05]
In which an existing borehole, subsequently abandoned, is imperfectly sealed and acts as a conduit for transfer of radionuclides from depth to shallower layers.
3. *Intrusion pathways* [NEA 0.5 FEP 1.4.04]
Two scenario representations are considered:
    - a case in which a water supply well acts as the final part of the return pathway;
    - *an intrusion scenario* leading to exposure of a geotechnical worker as a result of prospecting activities.

A full account of all scenarios can be found in [4]. In this paper we concentrate on the base scenario.

3. Description of the Dimitrovgrad repository site

NIIAR performs research and development in the fields of nuclear energy, nuclear materials and radiation damage, chemistry and production technology of transuranic elements, technological aspects of the different stages of the nuclear fuel cycle, and research on the utilisation of achievements in nuclear science and technology in other sectors of the economy.

The research activities generate liquid waste of complex chemical and radionuclide composition. The waste arrives through a common collector in three mixing storage tanks of 2,000 m^3 each. When a storage tank is nearly full, the mixed waste is sampled to determine its chemical and radionuclide composition. Once the storage tanks are filled, the waste is conditioned and the next cycle of waste injection into the underground repository is performed. The annual discharge rate is 55,000 - 65,000 m^3.

The available data suggest that the Dimitrovgrad site (also referred to as the 'pilot proving ground' [PPG]) lies in the central part of an extensive sedimentary basin, the so-called Melekesskian depression. The basin has the general shape of a northward stretched triangle. The region of interest thus has the hydrogeological features of an artesian basin. A hydrogeological cross-section running through the repository site is presented in *Figure 1.*

The geological sequence at the site (from the surface down) and significant features of each layer are as follows:

- **Water-Bearing Complex 7**
Quaternary sands, clay *etc.*, fresh water, water supply wells;
- **Confining Layer**
Permian, up to 200 m of clay;

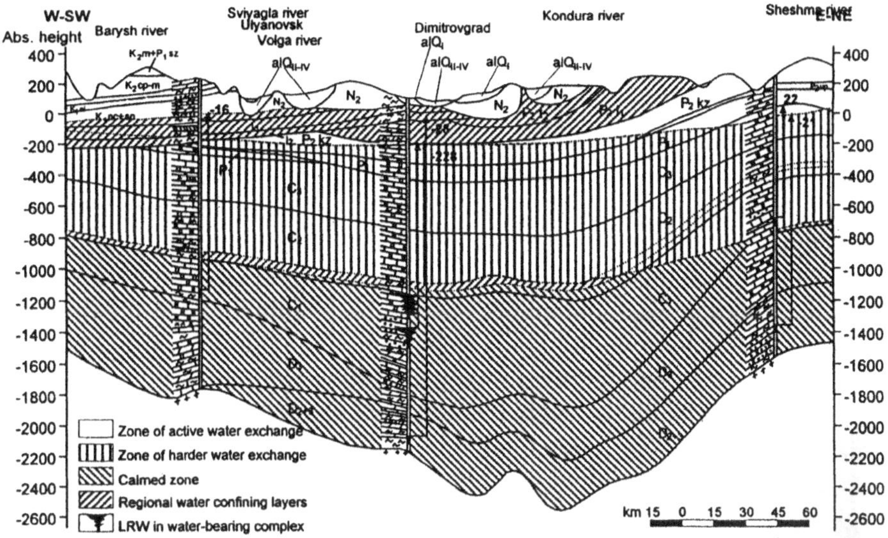

Figure 1. Vertical cross section passing through the repository site.

- **Water-Bearing Complex 6**
  Permian limestone, dolomite, saline water (density 1.03 g cm^{-3});
- **Water-Bearing Complex 5**
  Middle / Upper Carboniferous limestone, dolomite, brine (density 1.13 g cm^{-3});
- **Vreskian Confining Layer**
  Middle Carboniferous, argillaceous, brine;
- **Water-Bearing Complex 4**
  Lower Carboniferous, fractured limestone, brine (density 1.14 g cm^{-3}), disposal horizon;
- **Low permeability carbonate / clay horizon**
- **Water-Bearing Complex 3**
  Lower Carboniferous, coal measures, brine (density 1.15 g cm^{-3}), disposal horizon;
- **Compact limestone horizon**
- **Water-Bearing Complex 2**
  Lower Carboniferous / Upper Devonian, limestone /dolomite, brine (density 1.16 g cm^{-3});
- **Confining Layer**
  Upper Devonian, Argillaceous, brine;
- **Water-Bearing Complex 1**

Devonian, limestone / dolomite, brine (density 1.19 g cm^{-3});
- **Confining Layer**
  Devonian, argillaceous;
- **Crystalline Basement.**

Disposal has occurred by injection into two of the more permeable horizons in the sequence (Water-Bearing Complexes 3 and 4). Injection has been either into coal measures (Complex 3) at a depth around 1500 m or into a limestone/dolomite formation (Complex 4) at depths of around 1100 m. It is notable that the pressures in these formations imply static water levels at depths ranging from 27 - 87 m (depending on the formation and well under consideration). It is clear that all of the 'Water-Bearing Complex' layers 1-6 are confined and have significant formation pressure in the site area. This would be expected as the site appears to be at the low point of an extensive sedimentary basin. Another notable point is the existence of the water supply boreholes in the shallowest 'water bearing complex' (Complex 7).

3.1. INJECTION HISTORY

The waste injected into the repository layers consists mostly of short-lived isotopes, *viz.* Co-60, Sr-90, Ru-106, Cs-134, Cs-137, Ce-144, Eu-152, Eu-154, Pu-238, Pu-239, Pu-240, Am-241, Cm-244.

The injection history is sketched in *Figure 2.*

4. Groundwater flow modelling

The first step in the development of a numerical model of the groundwater flow in the region around the NIIAR site was to develop a conceptual model of the geological and hydrogeological structure of the region. The basis for this conceptualisation of the system was the information supplied by NIIAR about the site.

After consideration of the available information, it was considered that a groundwater flow model of a two-dimensional vertical cross section through the site was the most appropriate type of model to construct for this illustrative assessment.

A base-case numerical model of the groundwater flow in the region around the NIIAR site was developed. The model is used to estimate the quantities related to the groundwater flow system that are required for the PSA calculations of risk from the groundwater pathway. The numerical model was developed using Release 6.3 of the NAMMU software package [5]. NAMMU can be used to model groundwater flow and solute transport in porous media. It can be used to model steady-state and time-dependent

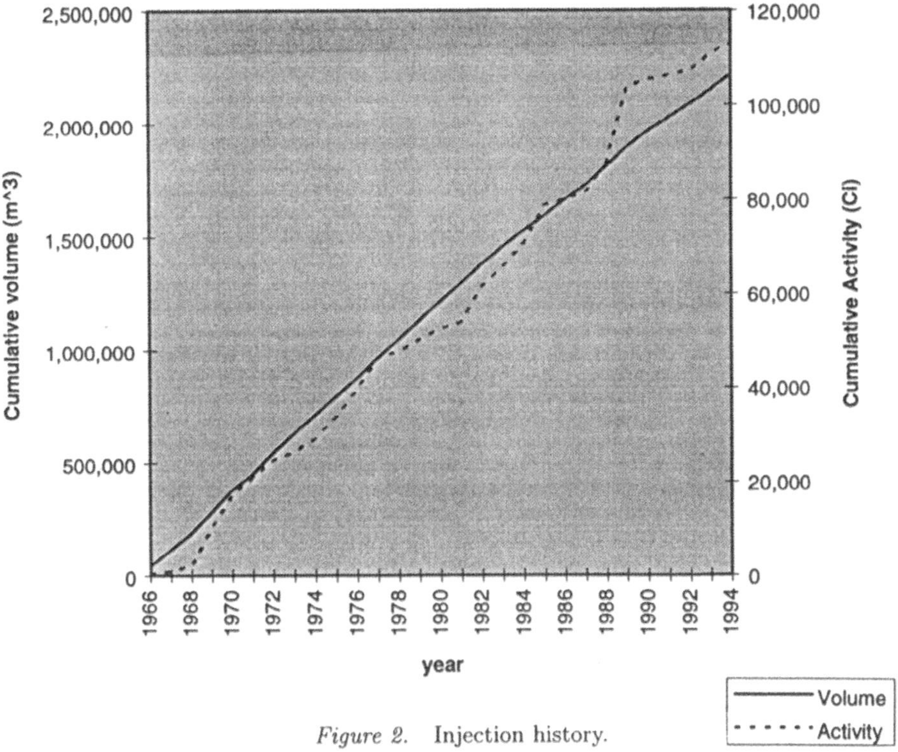

Figure 2. Injection history.

behaviour, including unsaturated flow and the transport of heat and mass. An option is available for modelling radioactive decay and the transport of chains of radionuclides. The software is based on an efficient implementation of the finite-element method that provides many options for modelling complex geological strata. Of particular relevance for the case considered here is the fact that NAMMU can be used to calculate coupled groundwater flow and solute transport with the fluid density strongly dependent upon concentration. Thus NAMMU can take account in a self-consistent way, of the significant salinity variations observed in the vicinity of the NIIAR site and the effects that these salinity variations exercise on the groundwater flow through the corresponding variations in groundwater density that they cause.

The artesian groundwater conditions observed at depth at the NIIAR site suggest that the normal groundwater return pathway will be roughly subvertical, at least as far as the groundwater flow system in the shallower layers. Radionuclides reaching these near-surface layers will be transported to the discharge zones associated with the near-surface groundwater flow system (most likely rivers and low lying ground) or may be captured by the

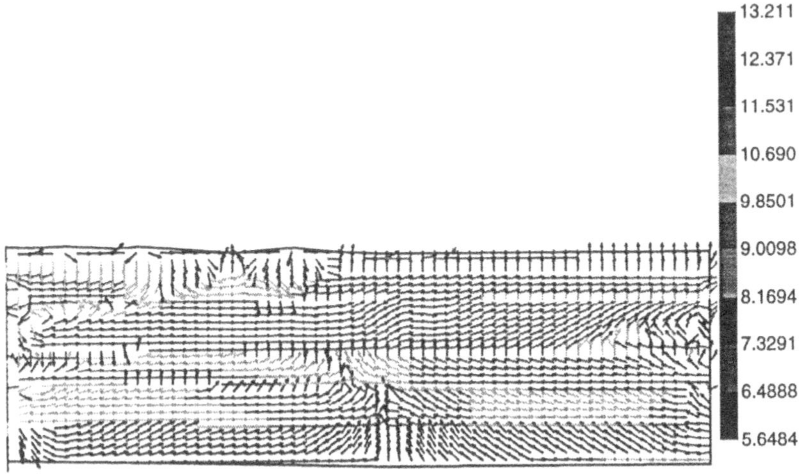

Figure 3. Velocity vectors from the base-case NAMMU model.

water supply wells that are located in the uppermost layers. This behaviour is confirmed by the results of the numerical model. The velocity field is displayed in *Figure 3*.

Pathline calculations were performed in order to generate the type of information required by the MASCOT PSA code [6]. The pathlines were started from the centre of the layers used for waste injection (Water-Bearing Complexes 3 and 4) and at a position directly below the NIIAR site. These paths are shown in *Figure 4*. It can be seen that the paths travel upwards from the Water-Bearing Complexes until they reach the upper regional flow system (that with NW-SE flow) above the Vereiskian layer. The paths are then carried to the right (SE) by the regional flow. It can be seen that the paths do not actually reach the ground surface within the model domain but exit the model on the SE boundary with Water Bearing Complex 6. This means that the travel time to the biosphere will actually be longer than that calculated by the NAMMU model. For this illustrative assessment, the conservative assumption was made that the calculated travel time was the travel time to the biosphere. For a more comprehensive assessment it would be necessary to extend the groundwater flow model so that the pathlines reach the biosphere within the region that is modelled. However, it should be borne in mind that over such a long sub-horizontal path the action of transverse dispersion could play a significant role in returning radionuclides

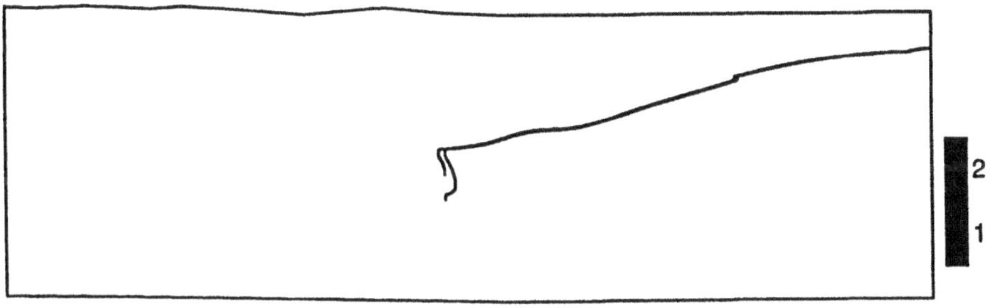

Figure 4. Pathlines from the base-case NAMMU model.

to the biosphere. Two-dimensional calculations of radionuclide transport would be required to scope this effect and would therefore be an important part of a comprehensive assessment of a site with this type of behaviour.

5. Modelling results on the actual inventory

The groundwater calculations reported above consisted of a base-case calculation, in which all the uncertain hydrogeological parameters were assigned their base-case values, and a suite of variant calculations. In each of the variant calculations one of the hydrogeological parameters was varied relative to its base-case value with all the other parameters being held at their base-case value.

For each of the NAMMU calculations, pathline calculations were carried out starting at the two injection points. These pathline calculations lead to values of the path length and travel time for each of the units through which the pathline passed. The representation of the geosphere within the MASCOT calculations was based on the results of the NAMMU calculations.

The first step in setting up the MASCOT model was to study the travel times in the various hydrogeological units. If the travel time in a given unit was not significant, there was no justification for representing that unit in the MASCOT model. This analysis lead to a network of submodels in which the paths starting at the two injection points are treated separately until Water Bearing Complex 5, where they join. This is because above Water Bearing Complex 5 the travel times in the various units are equal for both pathlines.

A simple biosphere model is included in the network of MASCOT submodels. Biosphere modelling takes the flux of radionuclides discharged from the geosphere and addresses transfer through the environment to humans. The BIOS computer program [7] was used to calculate rates and modes of transport through biosphere environments, and the doses to exposed individuals. Two critical groups have been represented in the biosphere model: a 'resident', and a 'farmer'. Dose conversion factors were calculated for these two groups. There is very little difference between the results for the two critical groups, and only the one for the 'farmer' group are presented in *Figure 5.*

A number of comments can be made about the results given in *Figure 5.* First, the estimated doses are extremely low. Secondly, it is seen that there is no significant dose prior to 10^6 years. Taken together these observations indicate, as expected, that the injection of the radionuclide inventory supplied by NIIAR poses no significant risk from the groundwater pathway. Note that all the radionuclides that are found to be important contributors to dose are daughter products of parents in the actinide decay chains. For instance, between 10^6 and 10^7 years, there is a contribution arising from Ac-227 that is the result of decay of the disposed Pu-239.

5.1. OTHER SCENARIOS

A detailed assessment was made of the risks associated with the exposure of a geotechnical worker as a result of prospecting activities. Assessment of the radiological risks arising from intrusion has been undertaken using the AEA Technology code HINDSITE. The other scenarios identified above were discussed in detail in [4], but none changes the conclusions presented below.

6. Conclusions

The deep-well injection repository operated by NIIAR in Dimitrovgrad, was studied in detail. The operation is licensed until 2001.

The assessment of the deep-well disposal technique employed in Dimitrovgrad has led to the development of a methodology that can be applied to all deep-well injection sites. The results will be different in detail at different sites because of the sensitivity to many parameters such as chemical properties, and in particular on the geology and the properties of the confining layers. Also, the inventories at the various sites are quite different.

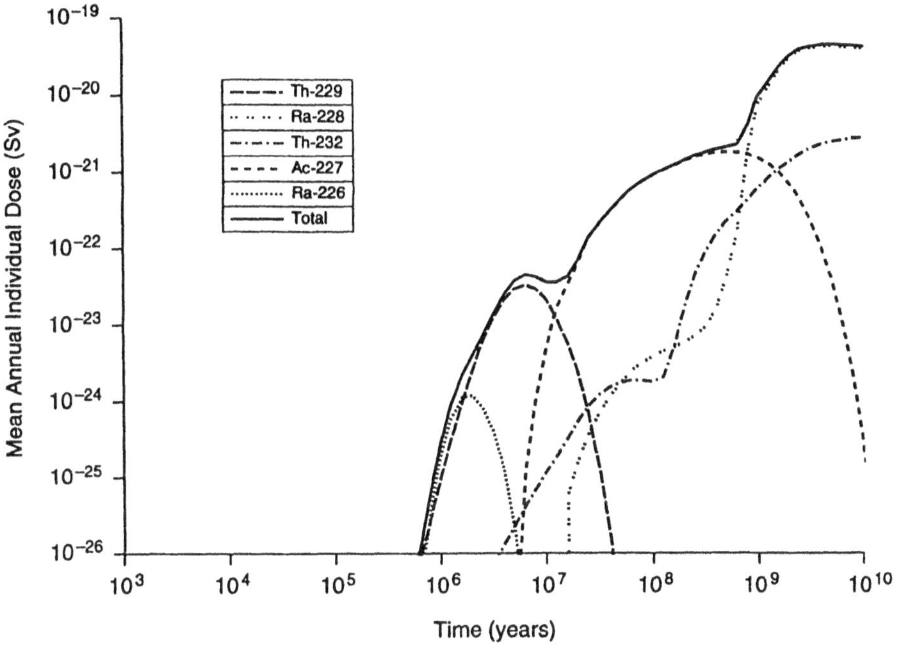

Figure 5. The mean total annual individual dose and the mean annual individual doses from the most important radionuclides (in Sv) as a function of time (in years). Results for disposal of the NIIAR radionuclide inventory, doses calculated for exposure of the 'farmer' critical group.

However, provided that the general disposal methodology is similar, the qualitative features of an assessment approach should be the same.

The assessment was based on data provided by NIIAR, describing in detail the geology, the properties of the host rock and groundwater, and the composition and properties of the injected waste. The assessment was carried out using computer codes, developed by AEA Technology, which have been validated through international validation exercises.

Subject to some limitations discussed in [4] and the data provided, the calculations demonstrated that deep-well disposal of low level, short-lived, radioactive waste, using the methodology applied in Dimitrovgrad, is acceptably safe within the regulatory requirements. On a 300-year timescale, the waste is very well confined. This was the result of Russian assessment calculations, and is corroborated by the travel time results of the methodology proposed by the current study. On a very long timescale, some very small dose is expected to reach the biosphere between 1 million and 100 million years, but the risk at that time to humans is negligible. It is more than sixteen orders of magnitude below the level of natural radiation sources, which is typically of the order of 2.2 mSv per year.

References

1. CRANWELL, R.M., *et al.* (1982) Risk methodology for geological disposal of radioactive waste: scenario selection procedure, *U.S. NRC Document* NUREG/CR-1667, SAND80-1429.
2. NIREX (1994) Post-closure performance assessment, *Nirex Science Report* S/94, U.K. Nirex Ltd., Harwell.
3. NEA (1996) Nuclear Energy Agency FEP Database Working Group.
4. HOEK, H., *et al.* Measurements, modelling of migration and possible radiological consequences at deep well injection sites for liquid radioactive waste in Russia, Final Report COSU-CT94-0099-UK, European Commission, DG XI.
5. HARTLEY, L.J., JACKSON, C.P., AND WATSON, S.P. (1996) NAMMU (Release 6.3) User Guide, *AEA Technology Report* AEA-ES-0138, AEA Technology, Harwell.
6. SINCLAIR, J.E., ROBINSON, P.C., COOPER, N.S., WORGAN, K.J., AND WILLIAMS, M.J. (1994) MASCOT and MOP Programs for Probabilistic Safety Assessment Part D: MASCOT (Version 3C) User Guide, *U.K. Nirex Report* NSS/R336, Version 3C Part D, Issue 1.
7. MARTIN, J.S., BARRACLOUGH, I.M., MOBBS, S.F., KLOS, R.A., AND LAWSON G. (1991) User Guide for BIOS 3A, *U.K. National Radiological Protection Board Report* NRPB-M285, U.K. NRPB, Didcot.

PROBABILISTIC SAFETY ASSESSMENT FOR HIGH-LEVEL WASTE TANKS AT HANFORD

L.H. SULLIVAN, D.R. MACFARLANE & D. W. STACK
Probabilistic Risk and Hazard Analysis Group
Technology and Safety Assessment Division
Los Alamos National Laboratory
Los Alamos, New Mexico
U.S.A.

1. Introduction

Los Alamos National Laboratory has performed a comprehensive probabilistic safety assessment, (PSA), including consideration of external events, for the 18 tank farms at the Hanford Tank Farm (HTF). This work was sponsored by the Department of Energy /Environmental Restoration and Waste Management Division (DOE/EM).

The Hanford tank farms are divided into east and west quadrants. The 18 tank farms are given one– or two–letter designations and contain either double–shell tanks (DSTs) or single–shell tanks (SSTs).

The DSTs consist of three concentric structures: (1) an outer, reinforced concrete tank designed to sustain induced loads from soil and seismicity, (2) a secondary, carbon-steel tank that lines the concrete tank and is designed to serve as a barrier to primary tank leaks, and (3) a free-standing carbon-steel primary tank that rests on an insulating concrete pad within the secondary tank. The primary tank contains the waste material; the secondary tank, which is 5 ft (1.5 m) larger in diameter, encloses the primary tank to create a surrounding annular space. The annulus is ventilated and monitored constantly for evidence of primary tank leakage. Each of the 28 tanks has a capacity of $1.2 \cdot 10^6$ gallons ($4.5 \cdot 10^3$ m^3). No leaks have occurred from the DSTs. The active induced-draft ventilation system for the DST farms has two completely separate sub-systems: a primary tank ventilation system and an annulus ventilation system. Tanks are connected to the two sub-systems by manifolds, which maintain a slightly negative pressure with the tanks and annulus. The ventilation sub-systems have no redundancy; the ductwork is above ground in some cases and underground in others. The ductwork routes the ventilation air from the primary tank and annulus of each tank to the respective filter trains and exhaust fans.

M. J. Stenhouse and V. I. Kirko (eds.),
Defence Nuclear Waste Disposal in Russia: International Perspective, 231–262.

SSTs are composed of a reinforced concrete enclosure with an inner steel liner on the bottom and sides. The 149 SSTs range in capacity from 55,000 to 1,000,000 gallons ($\sim$ 200–4,000 m^3) The tanks have a history of leaking into the surrounding soil. Most SSTs are vented through passive filter systems.

An earlier Los Alamos study focused only on the risks from Tank SY-101 [1]. This tank, which periodically undergoes sudden releases ("burps") of a mixture of gases that includes hydrogen (H_2), nitrous oxide (N_2O), ammonia (NH_3), and nitrogen (N_2), was analysed first because of public safety concerns associated with the potential release of the radioactive tank contents should this flammable gas mixture be ignited during one of the burps. The Tank SY-101 releases have been mitigated by the insertion of a large mixer pump into the tank to promote a slow, continuous release of the gases.

The HTF PSA involved three distinct tasks. First, the accident-sequence analysis identified the frequencies of those potential accidents whose consequences result in the release of tank material to the environment and quantified them. Second, radionuclide source terms for the airborne and liquid radioactive releases were determined. Finally, the consequences, as measured by on-site and off-site potential health effects and cleanup costs resulting from radionuclide release, were estimated, and overall risk curves were constructed. This PSA did not consider risk reductions from tank remediation activities (*e.g.* the Tank SY-101 mixer pump) or the risks from deliberate sabotage.

The accident-sequence identification task began with the construction of a master logic diagram (MLD) to identify the potential initiating events, which then were grouped into categories according to their effect on the tanks. These initiator groups included external events, such as earthquakes and airplane crashes, and internal events, such as gas releases ('burps') and liquid leaks.

At the next stage, event-sequence diagrams, whose events represent physical phenomena, hardware responses, and emergency operator responses, were constructed for initiating-event groups. Accident sequences were defined for quantification by event trees developed for each initiator. The event trees were developed in such a way to allow dependencies between top events and initiators to be identified. Finally, the frequencies of accident sequences were determined by combining initiating-event frequency estimates with the branch-point probabilities, or split fractions, for the occurrence of each event on the event-tree paths. An important aspect of this process was quantification of the branch-point probabilities. This quantification used a combination of tank farm historical operating databases and occurrence reports, generic component/system failure data, and specific deterministic analyses for flammable-gas-releasing tanks. This effort involved considerable interaction with Westinghouse Hanford Company (WHC) tank farm operations personnel and analysts at Los Alamos and WHC who had performed other related safety analyses.

The airborne source-term characterisation task involved identifying factors that influence the magnitude and timing of a radionuclide release and defining release categories for accident-sequence grouping. Both deterministic and probabilistic analyses were necessary for modeling material release mechanisms for the various accident sequences, thereby providing estimates for the quantity of material and energy involved in each case. Core sample analyses of the contents of tanks were used to characterise the radionuclide composition of the source terms.

The consequence analysis provided estimates of radiological health risks for both co-located workers and off-site residents via the airborne pathway and for off-site residents only for the groundwater pathway. Because of the large worker population in relatively close proximity to the 200 West and 200 East areas, the airborne dose consequences to this group were estimated as well. The airborne-transport population doses were calculated with AP-RISK, a computer code recently developed at Los Alamos to calculate dose consequences and surface contamination resulting from waste tank accidents.

The discussion of the analysis is divided into four sections: defining the initiating events and their frequencies, accident sequence modeling and quantification, source terms, and consequence assessments. These areas are discussed in the following sections. Results and conclusions are provided in the last section.

2. Initiating Events

The first step in developing a risk model is to define a set of initiating events. For an accident sequence to occur, an event must first perturb the steady-state condition of a waste tank or its contents. Subsequent events may (or may not) result in a release of radionuclides or chemicals.

The primary objectives of the initiating event exercise are

- to provide a comprehensive list of initiating events and adequate assurance that all possible events are taken into account,
- to account for unique tank design and operational features,
- to provide a way to categorise events in all of the unique ways that the event may affect the entire tank population, and
- to group events that present similar threats to safety functions for quantification.

Candidate initiating events were identified for the HTF PSA using several different analytic approaches, including

- MLD development,
- hazard and operability (HAZOP) study analysis, and
- external events analysis.

Applications of the MLD and HAZOP techniques to the HTF are described later. In addition to the information from the above analyses, new insights often are obtained from other PSA tasks that lead to identification of new initiating events or that alter the judgments made in finalising the list. This list is a highly iterative process and is difficult to predict. For example, the results of event-sequence quantification for one initiating event may be used to eliminate some others from the list. A review of operating experience often reveals new initiating events and becomes an important source of data and information.

2.1. MASTER LOGIC DIAGRAM

The MLD is similar to a fault tree and provides a deductive approach for directly answering the question "How can a significant release of radioactivity, chemicals, or toxic gas occur?" The first page of the HTF MLD is shown in *Figure 1.* A key objective of developing the MLD is to identify all possible types and sources of

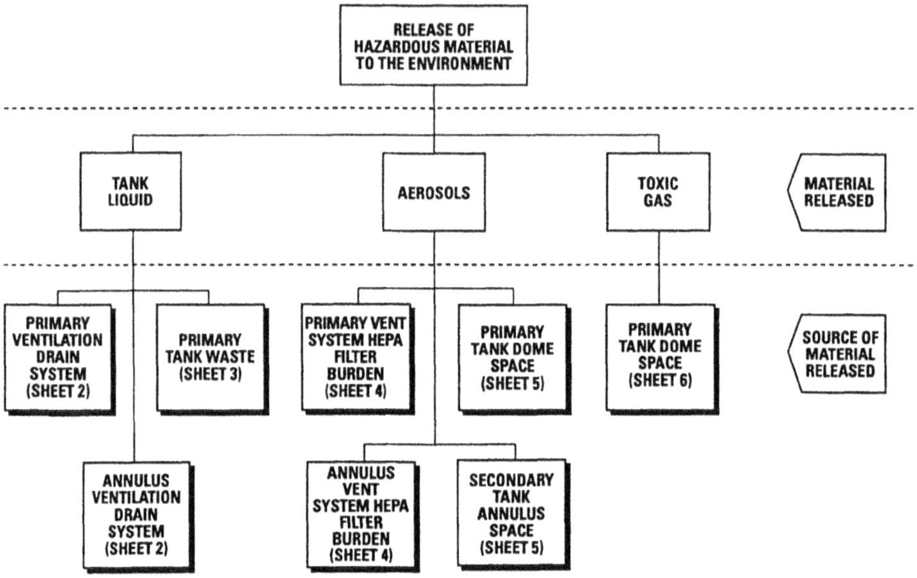

Figure 1. Hanford Tank Farm master logic diagram (MLD).

the hazardous materials and the pathways by which the top event can be satisfied down to a level of detail at which all important safety functions and barriers have been taken into account. When this is accomplished, specific causal events that can threaten a safety barrier or function can be listed. The question of completeness then is reduced to an assessment of the total frequency of all causal events that could produce any of the conditions shown on the MLD. It should be noted that it is not the objective of the MLD to delineate all possible accident scenarios that could result from the initiating event. For example, safety function failures could result in the failure of one or more fission-product barriers. Such scenarios are not shown by the MLD but are addressed later in the accident-sequence model.

Many of the initiator events shown at the bottom of the MLD may be subdivided to reveal more specific causal events. *Table 1* contains a list of events that could threaten the safety barriers or functions shown at the bottom of the MLD. The events listed also are matrixed against the MLD initiator events to help identify common-cause initiators that can threaten multiple safety barriers or functions simultaneously. However, not all of the basic events shown on the HTF MLD qualify as initiators. For example, events such as 'Flammable Gas Ignition' do not by themselves initiate the release and, therefore, are not true initiators, but they occur afterwards to an initiator in an accident sequence. This and other subsequent events typically are found under an "AND" gate in the MLD (*Figure 1*) and are shown for completeness by rounded rectangles, but they are not included in *Table 1*.

2.2. INITIATING-EVENT GROUPS

The concept of grouping initiating events by similarity of expected response is common to most PSA models and helps to limit the number of event-sequence models to be developed. It is necessary and practical to analyse only those initiating events that are expected to make appreciable contributions to risk. Given knowledge of the approximate frequency of the initiating events and the relative effect of these events on the tanks, it is possible and desirable to group and screen initiating events to simplify the quantification of risk but without introducing large errors into the risk estimates.

The causal events listed in *Table 1* are put into initiating-event groups in *Table 2*. Different causal events that affect the tanks in a similar way are grouped together. Where a causal event could be applied to multiple groups, the event was assigned to the more severe initiator group. This grouping results in a one-to-one correspondence between a causal event and an initiating-event group. However, care has been taken to keep common-cause initiators and initiators with special dependencies separate. The initiating-event groups listed in *Table 2* are generally applicable to all storage tanks. However, because of differences in tank design, tank status, and waste characteristics, some of these general initiating-event groups have been subdivided to account for these differences. Evaluations of the susceptibility of individual tanks to each initiating-event group are discussed under the accident sequence modeling section (Section 3).

3. Accident-Sequence Modeling and Quantification

The HTF accident-sequence model serves two primary purposes: to document the PSA team's understanding of how radionuclide and/or toxic gas releases from the HTF facility could occur and to create a logic model describing the potential release scenarios that can be used to quantify the likelihood of releases. The general approach used to develop accident-sequence models is shown in *Figure 2* and is described in detail in the Tank SY-101 PSA report [1]. For this analysis, the modeling approach from the Tank SY-101 analysis was expanded to address all 177 tanks.

Generic initiating-event groups that had the potential to lead to material releases from one or more tanks, were identified in the initiating-events section (Section 2). The accident-sequence modeling process began by examining each generic initiator to identify any characteristics of the tanks or waste materials that could influence the assessment of the frequency of occurrence of the generic initiator for a particular tank or group of tanks. If segregation of the 177 tanks into subgroups or families was justified, then initiator frequencies were calculated for each of the types. This calculated frequency represents the total per year for the initiator occurring over all of the tanks belonging to the type. For example, if the frequency of a generic initiator was determined to be 10^{-4} per tank per year and the event is equally applicable to all tanks, then the total type-initiator frequency for this analysis is 177×10^{-4} or $1.8 \cdot 10^{-2}$ per year.

Table 1. Initiator Events

EXAMPLE CAUSAL EVENTS	MLD BASIC EVENTS**																
	Primary Tank Shell Breach	Secondary Tank Shell Breach	Tank Dome Failure	Annulus Vent Line Breach	Riser Breach	Primary Exhaust Vent Line Breach	Exhaust HEPA Filter Breach or Bypass	Vent. Drain Line Leak	Vent. Seal Pot Leak	Waste Transfer Event	Waste Transfer System Boundary Failure	Inadequate Primary Ventilation Flow	Inadequate Level Control	Criticality Event in Waste	Bound Flammable Gas Release	Flammable Gas Accumu-lation	Uncon-trolled Waste Heatup
Drilling Contact with Tank	X	X															
Excavation Contact with Tank	X	X	X		X												
Tank Thermal Stress	X																
Tank Liner Corrosion	X																
Vehicle Overloads Dome			X														
Load Dropped Over Tank			X	X	X	X											
Vehicle Impact w Above Ground Equip.				X	X	X	X	X	X								
Vent. Duct Corrosion				X													
Human Error, Equip. not restored after maint.				X	X	X	X	X	X		X	X					
Vent. Drain Line Corrosion								X	X								
Vent. Drain Line Freezing								X									
Excessive Moisture in Vent System							X										
Ventilation Exhaust Filter Blockage												X					
Ventilation Inlet Blockage												X					
Vent Fan Failure												X					
Loss of Power to Vent Fan												X					
Loss of Air Supply to ALC's																	X
Dryout of Waste																	X
Tank Inundated by Transfer Spill			X										X				
Tank Inundated by Raw Water Leak			X										X				
Tank Inundated by Heavy Precipitation			X										X				
New Waste Transfers From Other Facilities										X			X	X	X	X	X
Salt Well Transfers to Collector Tank										X				X	X		
Liquid Transfers to 242-A Evaporator										X			X	X	X	X	
Slurry Transfers from 242-A to DST Storage										X			X	X	X	X	X
Radiolysis of Water in Waste																X	
Bound Gas Release	X	X	X	X	X	X	X	X	X			X				X	
Lightning Strike on Tank												X					
Seismic Event	X	X	X	X	X	X	X	X	X		X	X					
Aircraft Crash	X	X	X	X	X	X	X	X	X		X	X					
Range Fire*												X					
High Wind & Dust Storm*				X		X						X					
Volcanism*												X					
Dam Break*			X										X				

Table 2. HTF Initiating-Event Groups

EXAMPLE CAUSAL EVENTS	HTF INITIATING EVENT GROUPS [a]												
	Tank Shell Breach	Tank Dome Failure	Riser or Vent Line Breach	Exhaust HEPA Filter Breach or Bypass	Vent. Drain System Leak	Waste Transfer Event	Inadequate Primary Ventilation Flow	Uncont. Waste Heatup	Flammable Gas Accumu-lation	Bound Gas Release	Water Intrusion to Tank	Seismic Events	Aircraft Crash
	TSB	DOMEF	RVB	FB	VDL	WT	LOTV	HEATUP	FGA	BURP	WI	SEIS	AC
Drilling Contact with Tank		X											
Excavation Contact with Tank		X											
Tank Thermal Stress	X												
Tank Liner Corrosion	X												
Vehicle Overloads Dome		X											
Load Dropped Over Tank		X											
Vehicle Impact w Above Ground Equip.			X										
Vent. Duct Corrosion			X										
Human Error, Equip. not restored after maint.			X										
Vent. Drain Line Corrosion					X								
Vent. Drain Line Freezing					X								
Excessive Moisture in Vent System				X									
Ventilation Exhaust Filter Blockage							X						
Ventilation Inlet Blockage							X						
Vent Fan Failure							X						
Loss of Power to Vent Fan							X						
Loss of Air Supply to ALC's								X					
Dryout of Waste								X					
Tank Inundated by Transfer Spill											X		
Tank Inundated by Raw Water Leak											X		
Tank Inundated by Heavy Precipitation											X		
New Waste Transfers From Other Facilities						X							
Salt Well Transfers to Collector Tank						X							
Liquid Transfers to 242-A Evaporator						X							
Slurry Transfers from 242-A to DST Storage						X							
Radiolysis of Water in Waste									X				
Bound Gas Release										X			
Lightening Strike on Tank							X						
Seismic Event												X	
Aircraft Crash													X

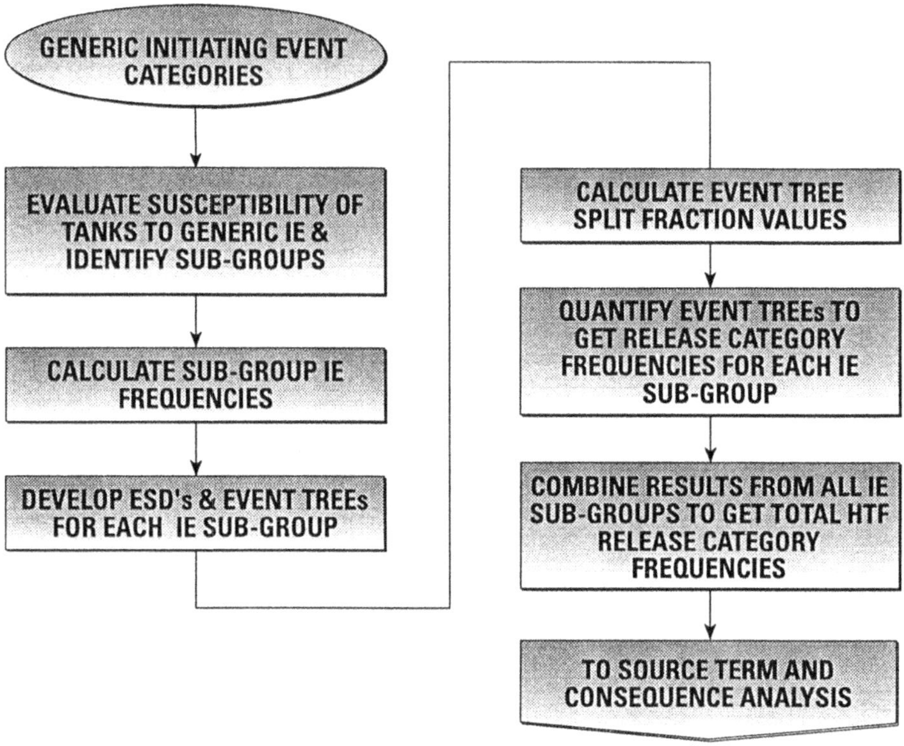

Figure 2. Accident-sequence modeling steps.

TABLE 3 Hanford Tank Farm PSA release categories.

Table 3. HTF PSA Release Categories

RELEASE PATHWAY		RADIONUCLIDE CONTENT	ENERGY OF RELEASE	RELEASE CATEGORY CODE
To Atmosphere	Unfiltered Release	Very Low	Low	BPL
	Unfiltered Release	Low (Aging Waste)	Low	BPH
	HEPA Breached	Low	Low	HEPAL
	HEPA Breached	High	Low (H_2 Burn)	HEPAH
	Dome Collapsed	High (101-SY Waste)	H_2 Burn	DCH
	Dome Collapsed	High (Wet Waste)	High (Aircrash & Fire)	DCVH
	Dome Collapsed	High (Dry, Inert Waste)	High (Aircrash & Fire)	DCVHI
	Dome Collapsed	High (Dry, FECN Waste)	High (Aircrash & Fire)	DCVHF
	Dome Collapsed	High (Organic Waste)	High (Aircrash & Fire)	DCVHO
To Ground	Subterranean Leak	Small	N/A	SLK
	Subterranean Leak	Large, SST	N/A	LLKSST
	Subterranean Leak	Large, DST	N/A	LLKDST
To Atmosphere and Ground	Surface Spill	Small	N/A	SSP
	Surface Spill	Large	N/A	LSP
	Spray Leak	Small	Low	SSPRY
	Spray Leak	Large	Low	LSPRY
	Dome Collapse + Subterranean Leak	High	Moderate (H_2 Burn)	DCLLK
	Dome Collapse + Subterranean Leak	High (C-103 Waste Only)	High (H_2 Burn & Fire)	DCHORG
	Dome Collapse + Subterranean Leak	4 Tanks	Low (Seismic Event)	DCL4
	Dome Collapse + Subterranean Leak	12 Tanks	Low (Seismic Event)	DCL12
	Dome Collapse + Subterranean Leak	45 Tanks	Low (Seismic Event)	DCL45
	Dome Collapse + Subterranean Leak	122 Tanks	Low (Seismic Event)	DCL122

These segregated initiating-event types were used to develop accident sequences and quantify sequence likelihood. Event-sequence diagrams (ESDs) and event trees were developed for each initiator family. The ESDs document the subsequent system responses, phenomenological events, and mitigating actions that can occur in response to the initiator. The ESDs also specify the most appropriate release category describing the end state of each sequence. The release categories defined for the HTF PSA are presented in *Table 3.* A complete discussion of release category properties and the development of radionuclide source terms is presented in the source term. To quantify the accident-sequence frequencies for each initiator, event trees corresponding to the ESDs were developed and quantified. The results represent the total frequency per year of each release category for each initiator. Results from each initiator were summed to obtain the total release category frequencies over the entire tank farm.

3.1. FLAMMABLE GAS ACCUMULATION

The major concerns associated with flammable gas accumulation (FGA) and a subsequent burn in the tank and/or ventilation system are damage to the components and concurrent release of radioactive materials to the environment. In addition to the burn-induced aerosol release from the dome space (including material entrained from the crust liquid waste), a fraction of the radioactivity trapped within the exhaust system can be released. In addition, if the dome collapses into the tank, the surface of the waste can become exposed to the atmosphere, permitting entrainment of materials from the surface. Leaks also can develop from tank failure, resulting in a liquid pathway for the release of radioactive materials to the environment.

The accident sequences of concern contain the following elements.

- Concentrations of combustible gases exceed the lower flammability limit (LFL) in the tank dome space and/or ventilation system.
- An ignition source within the tank or ventilation system ignites the flammable gas.
- Combustion of the gases produces a pressure and temperature transient that results in a pressurized release of gases and entrained material into the environment. These transients could result in pressures sufficiently high to fail the high-efficiency particulate air (HEPA) filters, ventilation system, and even the tank.

Twenty-four tanks are included on the Flammable Gas Watch List at the present time. Six of these tanks are DSTs (such as Tank SY-101), and the remaining eighteen are SSTs. All of the DSTs on the Flammable Gas Watch List are actively ventilated. The SSTs on the watch list were placed there primarily because they exhibited increases in waste level without the addition of liquids. The Flammable Gas Watch List was used as the initial screening criterion for tanks that might be subject to burns. However, all of the 177 tanks were examined for their burn potential. In addition, double-contained receiving tanks (DCRTs) have been identified as having the potential for the accumulation of H_2 generated while the waste was held temporarily in these tanks. DCRTs are small holding tanks used for routing, sampling, and other operations in the course of on-site waste transfers.

In the waste tank environment, H_2 is believed to be produced by three mechanisms: (1) radiolysis, (2) thermolysis, and (3) corrosion. Radiolysis and thermolysis are estimated to produce approximately equal amounts of H_2 in Tank SY-101, and the contribution of corrosion is not negligible. Radiolysis and thermolysis occur in the liquid components of the tank waste. However, because γ– and β–radiation dominate the radiolysis component, the amount of H_2 produced by radiolysis is assumed to depend on the total radiolytic power of the tank.

3.2. INITIATING-EVENT FREQUENCY FOR DOUBLE-SHELL TANKS

As mentioned above, the DSTs on the Flammable Gas Watch List are all actively ventilated. Continuous operation of the ventilation system maintains flammable gas concentrations at levels well below the LFL. It is assumed that ventilation system failures are repaired promptly, long before appreciable concentrations can build up. Therefore, only episodic releases are of concern for the DSTs. The greatly reduced magnitude of burps in DSTs other than Tank SY-101 clearly indicates that the events associated with this latter tank dominate the flammability concern with DSTs.

Before the mixing pump was installed, Tank SY-101 experienced a spontaneous release (burp) of slurry gas approximately every 100 days. In developing the initiating-event frequency for burps in Tank SY-101, only events occurring after March 1989 were considered because the practice of air lancing in this tank was terminated at this time. Based on these considerations, the mean frequency of Tank SY-101 burps (initiator BURP) was estimated to be approximately 3.5 per year. A more detailed analysis of the release history yielded the following attributes for a probability distribution for the frequency of burps in Tank SY-101.

Statistical Group	*Events per Year*
5th Percentile	2.64
50th Percentile (Median)	3.44
Mean	3.51
95th Percentile	4.74

A review of the level drop information for the other DSTs indicates mean frequencies of release events from approximately 2.0 events/yr to approximately 3.2 events/yr (Tank SY-103). However, because the magnitudes of the burps in tanks other than Tank SY-101 have been relatively small and flammable gas concentrations in the dome and ventilation system have been only fractions of the LFL, only Tank SY-101 was considered in this evaluation. That is, the frequency of burns resulting from burps in other DSTs is negligible compared with that for Tank SY-101.

3.3. INITIATING-EVENT FREQUENCY FOR SINGLE-SHELL TANKS

The frequency of passively-vented SST failures because of burns was estimated probabilistically using two conservative assumptions:

- Each electrical spark from the level probe is capable of igniting the gas mixture when the LFL is exceeded. (The level probe is the only component inside the tank dome space that is deemed important as an ignition source in passively-vented tanks.)
- Each ignition/burn is capable of causing catastrophic failure (dome collapse and shell failure) of the tank.

Adiabatic burns (but not complete combustion) at relatively low H_2 concentrations produce a significant pressure rise in the tank. The presence of N_2O as one of the oxidizers increases the magnitude of the pressure rise for the H_2 concentrations of interest. The failure pressure of SSTs has been estimated to be only 11 psig ($7.6 \cdot 10^4$ Pa). Thus, there is a high probability that burns at low concentrations will fail an SST. Because there are no probability distributions for either the tank pressure loads induced by burns or the load capacities of the tank (except for Tank SY-101, which is a DST), it was assumed conservatively that each ignition/burn would cause a tank failure. This assumption does not appear to have a dramatic effect on tank failure frequency because sensitivity calculations performed for several of the tanks indicate that the frequency of ignition/burns at an assumed LFL of 5% is not significantly different from that for an assumed LFL of 4%. The failure pressure for SSTs is relatively low compared with the failure pressure calculated for Tank SY-101 and the other DSTs.

The H_2 gas released to the tank dome space by each of the three generation mechanisms (radiolysis, thermolysis, and corrosion) was represented by discrete distributions. The calculated values shown in *Table 4* were assigned a probability weight of 0.4. In recognition of the perceived conservative assumption that all of the gas generated is released continuously to the dome space, a gas-release value equal to 50% of the table value was assigned a probability of 0.2. The final point in the three-point distribution accounted for the fact that there is some probability that the H_2 release rate could be greater than the point estimates given in *Table 4*. This final point was assigned a value equal to 150% of the table value and was assigned a probability of 0.4.

3.4. SINGLE-SHELL TANK LEAK FREQUENCY

Initially, all tanks constructed at Hanford were SSTs. A total of 149 SSTs were built in various farms from 1944 to approximately 1964. A total of 68 SSTs are reported to be leakers in official Hanford records. Except for four tanks built in 1964, every group has tanks that developed leaks of various sizes, from a few hundred gallons (~ 100 m^3) to hundreds of thousands of gallons ($\sim 10^3$ m^3). When a leak is detected in any tank, efforts are made to minimise the continuing leak over the future years by removing the drainable liquid out of the tank. The time of detection of these leaks may not be precisely known because of the sparsity of leak detectors around the tanks and the difficulty in detecting leaks by observing changes in tank liquid level (1-in [2.5-cm] level change in a 75-ft [23-m] diameter tank = 2750 gallons [10.5 m^3]). Historically, both methods have suggested leaks.

The Tank Farm Surveillance Report provides tank data on the years the tanks were built, the years they began to leak, the quantity of various liquid and solid wastes they contain, and other relevant information.

Recent data analysis performed by Steve Agnew of Los Alamos National Laboratory revealed that at least 12 of the 68 reported tank leaks were attributable to overfilling and/or transfer-line leaks. Also, reported leak dates were revised to reflect the actual time of the leak rather than the declared dates. Therefore, the leak data included in the Surveillance Report were modified in light of this new information.

3.5. POSSIBLE ROOT CAUSES OF LEAKS

Two primary conclusions appear to be supportable from the above analysis of historical SST leak data. First, a dramatic change in SST leak frequency occurred when these were removed from active service and has continued to the present. Second, the patterns and timing of the recorded failures indicate that common-cause mechanisms are the primary causes of tank leaks. Physical evidence of failure causes is not readily available; however, several hypotheses have been suggested during our investigations that may help explain the observed data. These are discussed in the following paragraphs.

TABLE 4 Burn frequency for passively-vented tanks.

Tank	Mean Burn Frequency (events/yr)	Release Category Assignment
Flammable Gas Watch List Tanks:		
AX-101	1.40E-04	DCLLK
AX-103	0.00E+00	DCLLK
S-102	1.96E-04	DCLLK
S-111	4.52E-05	DCLLK
S-112	1.76E-04	DCLLK
T-110	0.00E+00	DCLLK
U-103	4.70E-07	DCLLK
U-105	1.86E-04	DCLLK
U-108	0.00E+00	DCLLK
U-109	0.00E+00	DCLLK
High Organic Watch List Tanks:		
S-102**	1.96E-04	DCLLK
B-103	0.00E+00	DCLLK
C-103	1.55E-04	DCHORG
TX-105	0.00E+00	DCLLK
TX-118	0.00E+00	DCLLK
U-106	3.65E-07	DCLLK
U-107	9.70E-05	DCLLK
U-111	0.00E+00	DCLLK
Other:		
A-102	3.60E-05	DCLLK
B-101	1.74E-05	DCLLK
B-203	8.69E-05	DCLLK
B-204	8.38E-05	DCLLK
BY-106	8.40E-05	DCLLK
S-101	1.91E-05	DCLLK
S-108	2.62E-06	DCLLK
S-110	1.10E-05	DCLLK
Total Burn Frequency (events/yr)	1.34E-03	

Application of the mean SST leak frequency to the binomial distribution for the recorded 379 tank-years of DST experience produces a probability of 36% that no failures would have been observed in this period. Thus, applying the SST leak frequency directly to the DSTs would not be completely unreasonable. A lower estimate of future DST leak frequency could be obtained through Bayesian updating of the SST leak projections with DST experience. However, this was not done because it was judged that it could produce non-conservative results for future performance. Although no leak has been detected for a total of 379 tank-service years, it is likely that some wearout has occurred and that the DSTs are operating in a somewhat degraded status. Also, independent of the issue of tank ageing, it is believed that most mechanisms leading to leaks are cumulative. Therefore, the likelihood of primary shell leakage is expected to increase with continued operation. Furthermore, the bulk of the SST tanks developed leaks after they were 20 years old, and only three DST tanks are older than 20 years. Therefore, the future leak frequency distribution developed for SSTs also applies to the primary shell of the DSTs.

3.6. SEISMIC RESPONSE ANALYSIS

A seismic response assessment was performed on a double-walled waste storage tank, the associated equipment pits, the tank gas exhaust ductwork system, and the support facilities; an assessment also was performed for a single-walled waste storage tank. The assessment consisted of the following:

- *Seismic Hazard Analysis.* Determination of the frequency of various potential peak ground accelerations at the site.
- *Fragility Analysis.* Determination of the seismic-initiated peak ground acceleration at which plant structures and components are predicted to fail.
- *Accident-Sequence Model.* Development of a logic model that depicts the potential component failure scenarios considering possible combinations of associated equipment or structure failures.
- *Preliminary Quantification and Results.* Assembly of seismic hazards, fragilities, and models and the quantification of the frequency of causing sufficient damage to release hazardous materials, as well as identification of dominant contributors.
- *Final Quantification.* Calculation of the uncertainty in the damage and release frequency. The seismic uncertainty analysis was not performed in this study because the contribution to total risk from seismic events was negligible compared with non-seismic-initiated events.

3.7. SEISMIC HAZARDS

The Woodward-Clyde Consultants (WCC) study performed in April 1989 predicted peak ground acceleration frequencies at six Hanford sites. As in other contemporary probabilistic hazard studies, the WCC results were based on (1) the location and geometry of earthquake sources relative to the site, (2) the recurrence of earthquakes of various magnitudes on the sources up to the maximum magnitudes for each source, and (3) the attenuation of ground motions from the sources to Site 1.

The source models and attenuation relationships that were developed earlier by WCC in the seismic hazard studies for the WNP-2 nuclear power plant, which is located in the region, and for the N Reactor studies performed earlier in 1987 were used in the WCC 1989 study. The uncertainty in the source models and attenuation relationships in these studies was embedded in the 1989 study.

The overall site plan indicates that the two sites (Sites 1 and N) are relatively close to each other. A comparison of the results for these two sites showed that there is a similarity in the annual probability of exceedance between the two sites. The hazard curves were extrapolated to an annual exceedance frequency of 10^{-7}. The upper bound at an acceleration of approximately 1.3 g is limited by the mean frequency of 10^{-7} a^{-1} because values lower than this are of little interest in the tank farm risk assessment.

4. Summary of Radiological Source Terms

The airborne release and liquid release estimates, in terms of kg material released, are summarised in *Table 5* for each release category. *Table 6* provides a snapshot of one measure of relative risk in terms of the quantity of material released to the atmosphere between the various release categories. The radiological consequences of these releases are presented in Section 6.

5. Source Terms

The mass and isotopic content of radioactive material released into the atmosphere and/or into the soil column is an important factor in determining consequences and risk. These releases, along with the timing of the release and the parameters that influence the dispersion of these materials in the environment (*e.g.* release height and the energy associated with airborne releases), are referred to as "source terms". Source terms were developed for each important release category. The airborne source terms, along with meteorological data and relevant demography, were input to atmospheric dispersion codes to determine the health effects associated with airborne releases. Liquid pathway source terms were input to codes that model ground transport and retention to determine long-term health effects resulting from releases into the ground. A similar approach was used for economic risk.

There is a basic difficulty in trying to combine the risks of population health consequences via the atmospheric and liquid pathways. Airborne releases typically involve transport times to receptors and exposure times to the passing cloud of minutes to hours. The population at risk and the exposure doses for various accident scenarios can be determined with reasonable accuracy based on dispersion modeling, current census data, and evacuation scenarios (if any). In contrast, the releases of liquids into the soil column can involve transport delay times to receptors of hundreds or even thousands of years. These delay times introduce substantial uncertainty in quantifying future health effects because they must be based on projections from current population distributions and lifestyles. In addition, there is considerable uncertainty involved with the current models for vadose zone and groundwater transport. This modeling is an essential element in the prediction of groundwater pathway health effects involving future generations.

Philosophical questions arise as well; *e.g.* how do you balance the importance of health effects thousands of years in the future against the priorities and concerns of today? The approach used here was to report the long-term doses and risks from the groundwater transport pathways separate from the doses and risks from the airborne transport pathway. Thus, two sets of source terms were calculated for

TABLE 5 Point estimates of risk based on total airborne releases.

Release Category	Release Category Freq (1/yr)	Mean Value of Total Release (kg)	Mean Risk (kg/yr)
DCL122	2.26E-06	2.98E+05	6.74E-01
DCL45	5.93E-06	4.21E+04	2.50E-01
DCLLK	1.19E-03	8.95E+01	1.06E-01
LSP	1.57E-01	6.10E-01	9.58E-02
HEPAL	2.89E-01	2.20E-01	6.36E-02
DCH	6.98E-04	8.95E+01	6.24E-02
LSPRY	4.52E-04	6.98E+01	3.16E-02
SSP	1.81E+00	1.70E-02	3.08E-02
DCHORG	1.55E-04	1.79E+02	2.77E-02
DCL12	7.91E-06	3.04E+03	2.40E-02
HEPAH	5.07E-03	2.90E+00	1.47E-02
BPH	8.52E-03	1.40E+00	1.19E-02
SSPRY	8.59E-03	1.26E+00	1.08E-02
BPL	7.04E+00	1.11E-03	7.81E-03
DCL4	1.03E-05	5.29E+02	5.45E-03
DCVHI	2.09E-07	3.63E+02	7.59E-05
DCVHO	1.54E-08	4.09E+03	6.29E-05
DCVH	2.71E-07	1.78E+02	4.81E-05
DCVHF	4.93E-08	3.63E+02	1.79E-05
SLK	5.33E-01	0.00E+00	0.00E+00
LLKSST	5.59E-02	0.00E+00	0.00E+00
LLKDST	1.10E-02	0.00E+00	0.00E+00

those release categories that involve both an airborne release and a liquid release into the soil column.

In addition to the frequency of the accident-sequence/release category (from the accident-sequence models section), it is important to know the location of the release to perform the consequence calculations described in the consequence assessment section. Because of the large number of tanks at the HTF and the impracticability of performing calculations for each individual tank, it was convenient to group the tanks by location for the source-term definition. As a result, four different source-term compositions were developed to be representative of the tanks in the NE, SE, SW, and NW quadrants of the 200 Area. The radiological source term for a given isotope in each quadrant is the product of the quantity of waste material released and the concentration (GBq kg^{-1}) of that isotope in that quadrant. This section describes the development of distributions for the quantities of material released to the environment. These distributions are appropriate for all tanks contributing to the particular release category. Details on the source term/dose modeling are shown in *Figure 3*.

Table 6. Total Quantities of Material Released to the Environment

Release Category	Airborne Releases (kg): Short-Term, Energetic Release: Low Value (10th %)	Mean Value	High Value (90th %)	Mean Release Duration	Short-Term, Ground-Level Release: Low Value (10th %)	Mean Value	High Value (90th %)	Mean Release Duration	Long-Term, Ground-Level Release: Low Value (10th %)	Mean Value	High Value (90th %)	Mean Release Duration	Liquid Release (kg): Low Value (10th %)	Mean Value	High Value (90th %)
BPL	--	--	--	--	8.95E-05	1.11E-03	2.21E-03	8 h	--	--	--	--	--	--	--
BPH	--	--	--	--	0.087	1.4	3.166	8 h	--	--	--	--	--	--	--
HEPAL	--	--	--	--	0.03	0.22	0.48	10 min	--	--	--	--	--	--	--
HEPAH	0.84	2.88	5.64	10 min	--	--	--	--	--	--	--	--	--	--	--
DCH	3.39	10.83	19.73	10 min	3.39	14.06	29.34	8 h	4.8	64.6	156.2	2 d	--	--	--
DCL4	--	--	--	--	10	59.84	131.52	8 h	30.48	469.52	1108	3 d	1.59E+05	1.06E+06	2.29E+06
DCL12	--	--	--	--	30.12	179.64	394.56	8 h	197.64	2856.24	6714.24	6 d	1.39E+06	3.18E+06	5.34E+06
DCL45	--	--	--	--	112.95	674.1	1479.6	8 h	3001.5	41450.4	97206.3	23 d	8.18E+06	1.20E+07	1.60E+07
DCL122	--	--	--	--	306.22	1827.56	4011.36	8 h	21740.4	296479.52	696105.16	60 d	2.76E+07	3.32E+07	3.80E+07
DCVH	26.87 (Cs Eh)	98.43(Cs Eh)	198.69 (Cs Eh)	30 min	3.4	14.06	29.48	8 h	1.42	64.98	158.65	2 d	--	--	--
DCVHI	137.73 (Cs Eh)	283.77 (Cs Eh)	447.01 (Cs Eh)	30 min	3.4	14.06	29.45	8 h	1.42	64.96	158.35	2 d	--	--	--
DCVHF	137.73 (Cs Eh)	283.77 (Cs Eh)	447.01 (Cs Eh)	30 min	3.4	14.06	29.45	8 h	1.42	64.96	158.35	2 d	--	--	--
DCVHO	761.27 (Cs Eh)	4007.6 (Cs Eh)	8068.9 (Cs Eh)	30 min	3.4	13.99	29.34	8 h	3.51	64.38	154.5	2 d	--	--	--
DCLLK	3.39	10.83	19.73	10 min	3.39	14.06	29.34	8 h	11.58	89.46	205.28	2 d	3.41E+04	9.51E+05	1.41E+06
DCHORG	10.55(Cs Eh)	98.08(Cs Eh)	229.87(Cs Eh)	30 min	2.22	14.26	32	8 h	0.8	66.18	163.57	2 d	6.80E+05	7.55E+05	8.31E+05
SSPRY	--	--	--	--	0.07	1.26	2.85	15 min	--	--	--	--	1.98E+01	9.82E+01	2.08E+02
LSPRY	--	--	--	--	4.12	69.82	149.49	8 h	--	--	--	--	5.80E+02	5.46E+03	1.22E+04
SLK	--	--	--	--	--	--	--	--	--	--	--	--	1.14E+04	3.42E+04	5.68E+04
LLKSST	--	--	--	--	--	--	--	--	--	--	--	--	8.52E+04	2.04E+05	3.12E+05
LLKDST	--	--	--	--	--	--	--	--	--	--	--	--	4.09E+05	1.11E+06	2.50E+06
SSP	--	--	--	--	5.70E-03	0.02	0.03	8 hr	--	--	--	--	5.70E+02	1.70E+03	2.84E+03
LSP	--	--	--	--	0.21	0.61	1.15	8 hr	--	--	--	--	2.07E+04	6.11E+04	1.15E+05

Notes: The total release of each radionuclide is equal to the product of the numbers listed in this table and the concentration (Ci/kg) of radionuclide in the released material.
Amounts shown for Release Categories DCL4, DCL12, DCL45, and DCL122 are the totals for all tanks.
Cs Eh = Cesium enhanced

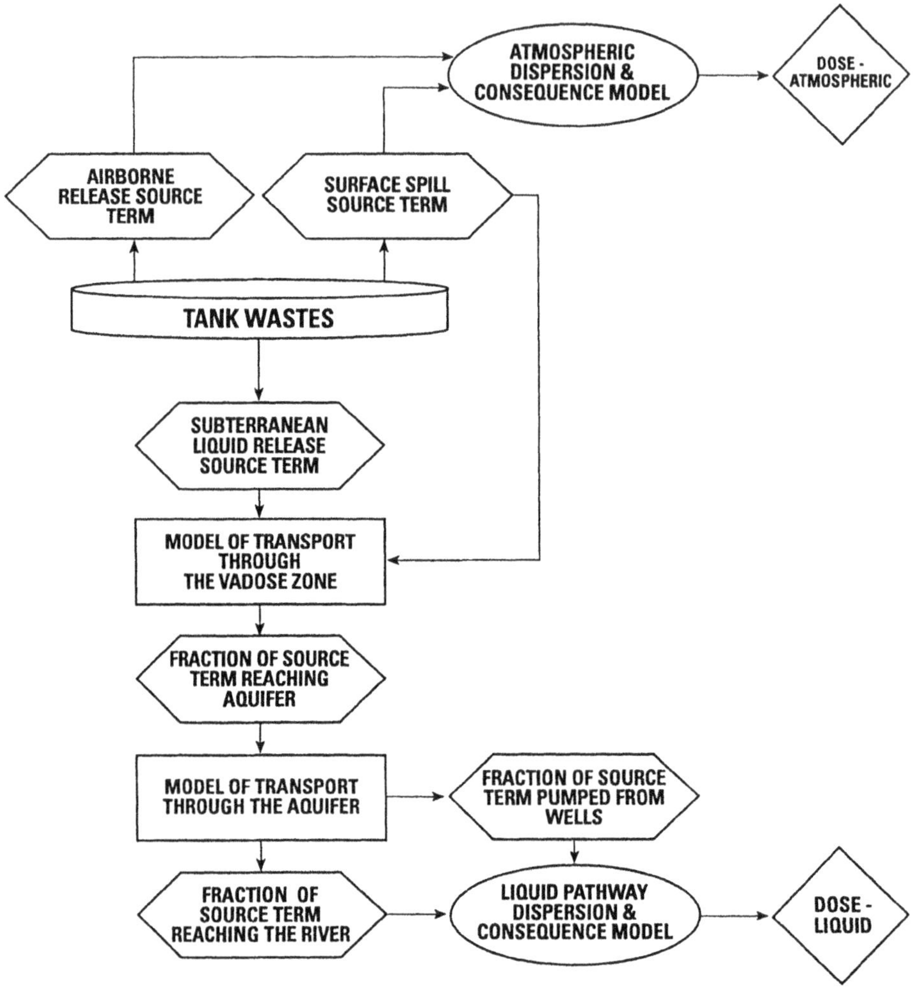

Figure 3. Hanford Tank Farm source term and consequence model assessment.

6. Consequence Assessment

This section presents the basis, and the results of dose calculations, for the source terms given in the previous chapter. For airborne releases, dose estimates were made for the on-site, co-located worker, and off-site populations. In most PSA consequence analyses, only the off-site population (general public) is considered, but because of the large number of workers in the 200 East and 200 West Areas close to the tank farms, it was necessary to consider workers in adjacent facilities as well.

The much longer term liquid pathway doses apply *only* to the off-site populations located along the Columbia River, downstream from the Hanford Site. As will be shown, potential exposures to liquid pathway releases are delayed for hundreds of years for some isotopes and thousands of years for others, so using the present population distribution is clearly questionable. Because there is no rational basis on which to predict that far into the future, no attempt was made to do so. It also was assumed that no water for human consumption would be drawn from on-site wells in close proximity to the tank farm locations. Although this assumption may be indefensible over the very long term (many centuries), it does provide a consistent basis for performing the liquid pathway consequence assessment.

The dose calculation procedures used here are similar to those used in the Tank SY-101 PSA. The principal difference is that there are four different sets of radionuclide source-term compositions, corresponding to the contents of tanks in the four quadrants of the 200 Area. Although the primary interest here is in the person-rem (1 rem = 0.01 Sv) population doses and concomitant LCF health effects calculated by AP-RISK, we also have reported maximum individual doses calculated with AI-RISK at various distances.

6.1. ATMOSPHERIC DISPERSION AND DOSE ASSESSMENT MODELS

Two related computer codes, AI-RISK [2] and AP-RISK [3], were developed by Los Alamos to perform radiation risk calculations for individuals and for collective populations, respectively. The codes originally were developed to facilitate comprehensive analyses of health consequences, ground contamination, and cleanup associated with possible energetic chemical reactions in HLW tanks. Because tank farm accidents can release aerosols containing significant fractions of large particles, AI-RISK and AP-RISK have provisions for environmental transport and dosimetry models for all dispersible particle sizes. Both codes use a tilted Gaussian plume model that tracks the dispersion of particles in five separate size groups or 'bins'. To estimate the range of potential doses to an individual, a cumulative probability distribution of dose values is constructed by using the joint-frequency meteorological data for the site.

Inhalation dose factors were calculated based on values of committed dose equivalent factor per unit intake given in ICRP30 [4]. Particle-size-dependent and solubility-class-dependent dose factors for various radionuclides (based on ICRP30) were used to calculate the resulting effective dose equivalent (EDE). In addition to the acute accident dose, the codes calculate appropriate long-term off-site doses (50-year) resulting from continuous ingestion of contaminated food and water. The codes also calculate the long-term occupational dose (50-year) resulting from post-accident occupancy of a partially contaminated (assuming clean-up to a Protective

Action Guideline [PAG] level of 5 rem / 0.05 Sv over 50 years) work place, the dose contributors being ground shine and the inhalation of resuspended particulate contamination. This dose contribution is based on user-specified site occupancy factors and post-accident cleanup criteria. Both codes also estimate the potential health effects from calculated doses based on ICRP-recommended models and methodology. They are presented as lifetime probabilities of (1) latent cancer mortality, (2) genetic effects, or (3) acute mortality, depending on the dose level.

6.2. EQUIVALENT RELEASE QUANTITIES

As discussed in the source term section (Section 4) and shown in *Table 6*, airborne releases can be characterised by three timeframes: short-term energetic (occurring in less than 2 h), short-term (occurring in less than 8 h), and long-term (occurring from 8 h to 60 days). In AP-RISK, the long-term releases are simulated by a series of short-term (2-h) releases over the appropriate time span for the particular scenario. This implemented in the code by a Monte Carlo simulation in which the normalised 2-h dose *vs* probability curve, based on the meteorological data, is repeatedly sampled for the total release time to obtain the dose for the longer duration releases.

The long-term, ground-level releases were considered only in the evaluation of the off-site population doses because the on-site personnel would not be permitted to return if there was a release lasting for days.

6.3. MAXIMUM INDIVIDUAL DOSES

Although not directly used in the population dose consequence assessment, dose factors for individual on-site receptors at 100 m, 1 km, 5 km, 10 km, and 15 km are presented in *Table 7* for reference purposes. They are the acute doses, received *during cloud passage* at the plume centerline, corresponding to a release quantity of 1 kg. To obtain the maximum individual doses corresponding to the various release quantities, it is only necessary to obtain the product of the appropriate release quantity (*Table 6*) and the dose factor (*Table 7*). Note that the factors for the elevated (fire) release should be used for categories DCVH, DCCVHI, DCVHF, DCVHO, and DCHORG and that the releases for the earthquake categories (DCL4, DCL12, DCL45, and DCL122) are for multiple tanks rather than a single tank. The major contributor to the dose factors listed in *Table 7* comprises TRU isotopes via the inhalation pathway.

The highest individual doses result from the large earthquake (DCL122). None of the release categories would cause a prompt fatal dose at 100 m. Because all the co-located workers considered in the dose calculations are at distances greater than this and have the benefit of building shielding, no fatalities are expected in this group.

6.4. EXPOSURE TO HAZARDOUS CHEMICALS

In addition to radionuclides, the tanks contain a number of chemicals that are potentially harmful to humans. If these chemicals are released through the vadose zone and the underground aquifer streams, they would ultimately reach and contaminate the Columbia River water. This release of the chemicals to the river could

Table 7. Acute Individual Doses (1-kg Release)

Quadrant	Distance from Release Point (km)	Ground (rem)		Elevated, Fire (rem)	
NE		50%	95%	50%	95%
	0.1	1.35E-01	8.00E-01	2.15E-04	4.93E-03
	1	2.81E-03	2.81E-02	9.64E-05	1.30E-03
	5	2.44E-04	2.24E-03	1.38E-05	2.56E-04
	10	8.48E-05	8.19E-04	6.00E-06	1.15E-04
	15	4.94E-05	4.59E-04	3.72E-06	6.99E-05
SE					
	0.1	1.19E+00	7.07E+00	1.46E-03	3.59E-02
	1	2.49E-02	2.50E-01	5.48E-04	9.86E-03
	5	2.16E-03	2.00E-02	9.32E-05	1.99E-03
	10	7.51E-04	7.31E-03	4.36E-05	8.95E-04
	15	4.38E-04	4.06E-03	2.80E-05	5.46E-04
SW					
	0.1	3.66E-01	2.18E+00	5.26E-04	1.23E-02
	1	7.67E-03	7.72E-02	2.18E-04	3.32E-03
	5	6.67E-04	6.19E-03	3.32E-05	6.66E-04
	10	2.31E-04	2.26E-03	1.52E-05	3.00E-04
	15	1.35E-04	1.25E-03	9.52E-06	1.83E-04
NW					
	0.1	2.52E-02	1.50E-01	5.78E-05	1.23E-03
	1	5.26E-04	5.27E-03	2.92E-05	3.04E-04
	5	4.57E-05	4.21E-04	3.53E-06	5.98E-05
	10	1.59E-05	1.54E-04	1.49E-06	2.68E-05
	15	9.25E-06	8.52E-05	8.96E-07	1.63E-05

occur decades after the actual spill/release of the tank contents to the ground. The contaminated river water can lead to a toxicological dose if it exceeds the safe concentration levels and is consumed by the surrounding human population. Accordingly, a scoping hazard analysis was performed to evaluate the potential hazards of the potential chemical exposure. The purpose of the scoping analysis was to determine and identify if there is a need to perform a detailed chemical dispersion analysis and hazard evaluation for the tank.

Hazardous chemical contamination of the Hanford Site groundwater and the Columbia River was examined comprehensively in the 1986 Hanford Environmental Impact Statement (EIS), This study evaluated the effects of potential leakage from all Hanford SSTs and found that hazardous chemical concentrations in the Columbia River remained well below the U.S. Environmental Protection Agency (EPA) drinking water standards for all remediation scenarios under all site conditions. The EIS also showed that groundwater within the current site boundaries would, over time, show hazardous chemical concentrations well above EPA standards with no tank remediation.

Because large leaks from DSTs were not evaluated explicitly in the EIS, a scoping analysis of the potential for contamination of the Columbia River was performed for Tank SY-101. This analysis confirms that leaks from DSTs represent no threat to the Columbia River based on the U.S. EPA maximum contaminant levels for drinking water. Because site boundaries were assumed to remain as is for this evaluation (as discussed in the initiating events section, Section 2), no further analysis of hazardous chemical risk was performed, and no health effects from hazardous chemical releases are included in the results reported here.

7. Results and Conclusions

The final results of the PSA performed for the HTF are the unconditional risk curves. Risk curves present the relationship between the frequency of occurrence of radionuclide release events and the level of damage sustained as a result of the release; they are presented as complementary cumulative distribution functions (CCDFs) for total health and economic consequences and for those release categories that contribute significantly to these risk indices. In addition, an uncertainty quantification was performed to generate risk bands (*i.e.* percentile curves) for the total unconditional health effects for both on-site and off-site receptors.

The health risk curve assembly process involved integrating the release category frequency distributions and the conditional risk curves corresponding to each release category. The release category frequency distributions were generated using the RISKMAN© personal computer software package. The mean value and 5th, 50th, and 95th percentiles of the release category distributions were used in the development of risk curves for health and economic consequences.

7.1. HEALTH RISK RESULTS

7.1.1. *Airborne Releases*

The consequences of airborne releases were quantified for two receptor populations: on-site personnel (*i.e.* co-located workers), and the off-site public. The risk curve assembly process outlined here is applicable to both population groups.

To facilitate the calculation of consequences, the release quantity distributions determined in the consequence assessment section for the various release categories were replaced with three-point discrete probability distributions. The quantities used for the discrete distributions were the 10th, mean, and 90th percentile of the more detailed distributions calculated in the consequence assessment section. The probability weights assigned to these three values were:

Release Quantity	Probability Weight
10th Percentile	0.2
Mean	0.6
90th Percentile	0.2

The radioisotope concentration in a release was based on the location of the release within the tank farm. The site was divided into quadrants: NW, NE, SW, and SE. As described in Section 4, representative radioisotope concentrations and doses were generated for each quadrant. *Table 8* shows the fraction of the release category frequency that results from tanks in each of the four quadrants.

Possible releases of material from four locations (with unique release concentrations), with three possible release quantities at each location, led to the development of 12 conditional consequence curves for each release category. The probability weight associated with each of the 12 curves is a function of the probability weight of the release quantity and the conditional probability that the event occurred in the given quadrant. For example, the list below *Table 8* shows the probability weightings associated with each of the 12 conditional consequence curves developed for release category BPL.

Not all release categories had frequency distributed throughout all four quadrants of the tank farm. For example, the frequency of Release Category DCH comes entirely from BURP events occurring in Tank SY-101, which is located in the SW quadrant. Therefore, conditional consequence curves for DCH were needed for the SW quadrant only.

The calculation of a probability distribution for the exceedance frequency of a given damage (person-rem) level requires the merging of the 12 weighted conditional consequence curves and the release category frequency distribution for each release category. The distribution for the unconditional exceedance frequency per year of damage level x for Release Category A is calculated as the product of the frequency distribution for Release Category A and the conditional exceedance frequency distribution at damage level x. The conditional exceedance frequency distribution for damage level x is a 12-bin discrete distribution (one bin for each of the conditional consequence curves), where the bin probabilities correspond to the probability weight assigned to each curve. To generate the mean risk curve with uncertainty bands for Release Category A over all damage levels, probability distributions for the exceedance frequency must be calculated at a number of damage levels to sufficiently define the shape of the curve.

The CCDFs for total off-site and on-site risk were calculated by summing the appropriate probability distributions for exceedance frequencies at corresponding damage levels for all contributing release categories. The 5th, 50th, and 95th percentile curves and mean risk curves were calculated for total off-site and total on-site consequences. *Figure 4* presents the mean curves for total off-site consequences and for the seven most risk-significant release categories. *Figure 5* shows the 5th, 50th, and 95th percentile total off-site consequence curves. *Figure 6* presents the mean curves for total on-site consequences and for the seven most risk-significant release categories. The 5th, 50th, and 95th percentile total off-site consequence curves are presented in *Figure 7*.

TABLE 8 Percentage Contributions to Release Categories by Quadrant

Release Category	Frequency (per yr)	Per Cent Contribution by Quadrant			
		Northeast	Southeast	Southwest	Northwest
BPL	7.04E+0	5.83%	58.06%	30.28%	5.83%
SSP	1.81E+0	10.96%	48.91%	35.14%	4.98%
HEPAL	2.89E-1	0.00%	61.63%	38.37%	0.00%
SLK	5.34E-1	38.18%	12.68%	17.58%	31.55%
LSP	1.57E-1	13.92%	44.30%	35.44%	6.33%
LLKDST	1.09E-2	0.00%	89.29%	10.71%	0.00%
LLKSST	5.59E-2	19.64%	21.43%	39.29%	19.64%
SSPRY	8.59E-3	13.92%	44.30%	35.44%	6.33%
BPH	8.52E-3	0.00%	100.00%	0.00%	0.00%
HEPAH	5.07E-3	0.00%	0.00%	100.00%	0.00%
DCH	6.98E-4	0.00%	0.00%	100.00%	0.00%
LSPRY	4.52E-4	13.92%	44.30%	35.44%	6.33%
DCLLK	1.18E-3	23.02%	14.89%	62.09%	0.00%
DCL4	1.03E-5	26.85%	17.45%	28.86%	26.85%
DCL12	7.91E-6	26.85%	17.45%	28.86%	26.85%
DCL45	5.93E-6	26.85%	17.45%	28.86%	26.85%
DCHORG	1.55E-4	0.00%	100.00%	0.00%	0.00%
DCL122	2.26E-6	26.85%	17.45%	28.86%	26.85%
DCVH	2.71E-7	21.59%	35.23%	27.27%	15.91%
DCVHI	2.09E-7	13.24%	23.53%	30.88%	32.35%
DCVHF	4.93E-8	68.75%	18.75%	0.00%	12.50%
DCVHO	1.54E-8	20.00%	20.00%	20.00%	40.00%

Probability Weights for Release Category BPL Conditional Consequence Curves				
Release Quantity (probability)	Tank Farm Quadrant (and Conditional Likelihood)			
	NE (0.06)	SE (0.58)	SW (0.30)	NW (0.06)
10th percentile (0.2)	0.012	0.116	0.06	0.012
Mean (0.6)	0.036	0.348	0.18	0.036
90th percentile (0.2)	0.012	0.116	0.06	0.012

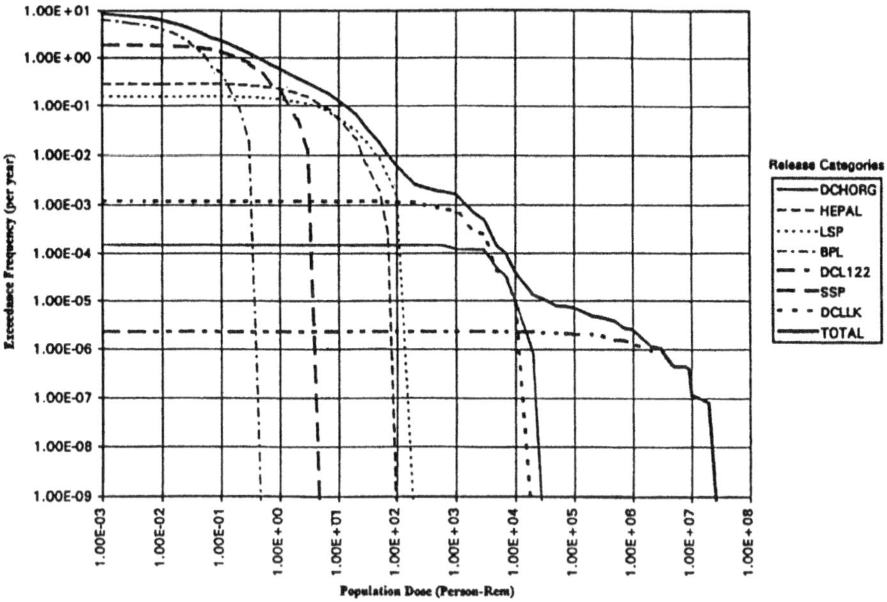

Fig. 4. Offsite total mean unconditional consequences.

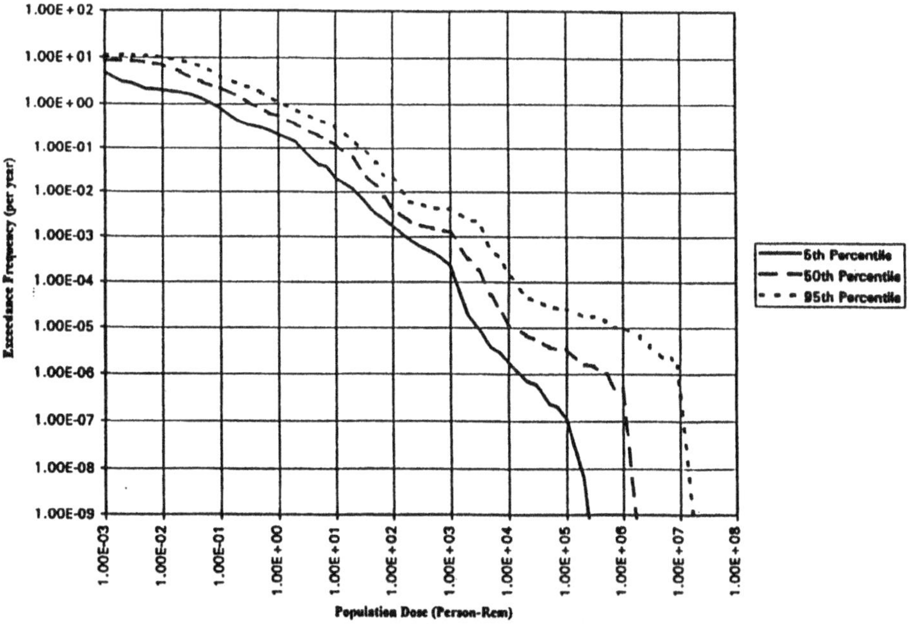

Fig. 5. Offsite total unconditional consequences.

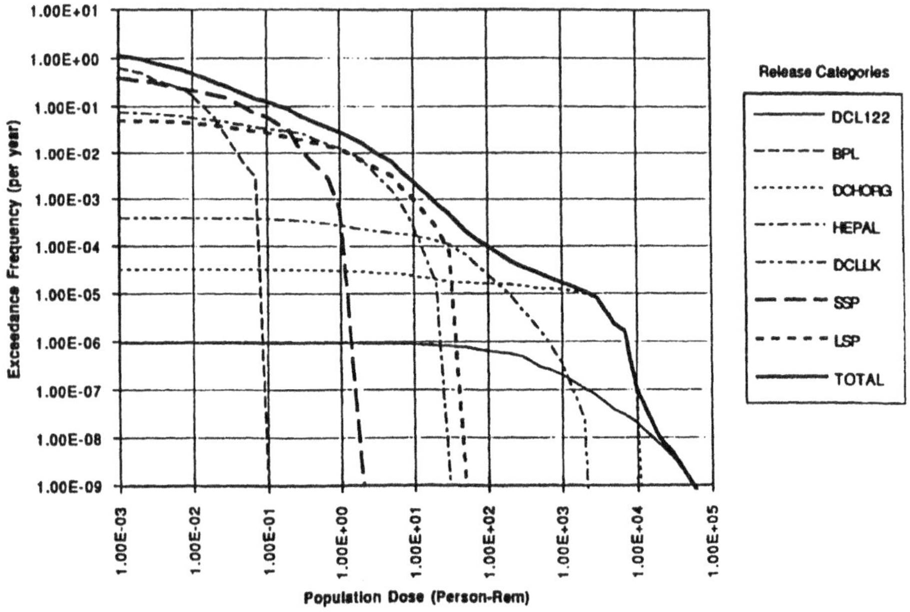

Fig. 6. Onsite total mean unconditional consequences.

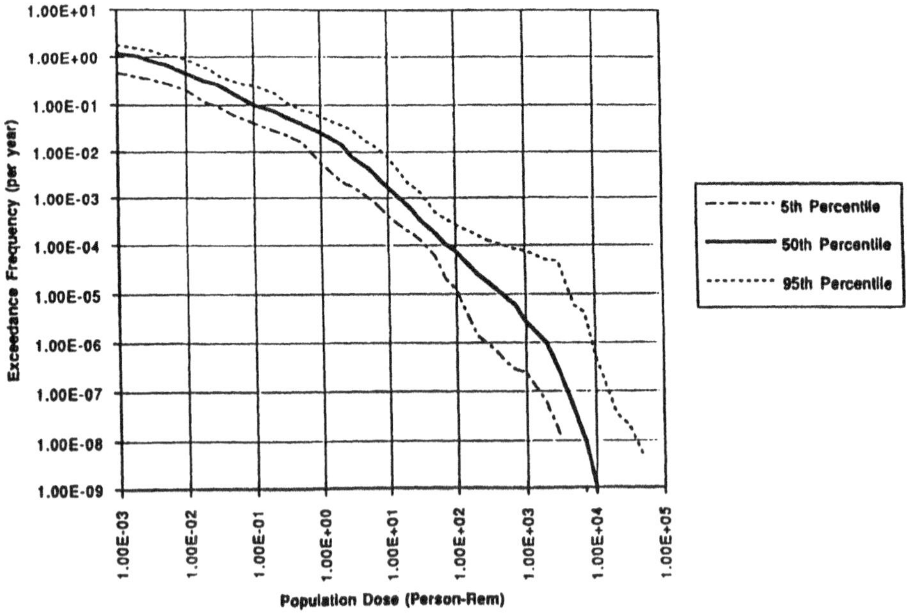

Fig. 7. Onsite total unconditional consequences.

7.1.2. *Liquid Releases*

The procedure described above for airborne releases also was used to develop health effect consequence curves for liquid releases. Because there are no on-site health effects associated with subterranean releases, results are presented for off-site consequences only. The mean risk was calculated for each release category involving a liquid release. Some of these release categories, such as DCLLK, cause an airborne release as well as a subterranean release. Risks as a result of the airborne component of these release categories were calculated. As shown, the radiation exposure risks posed by subterranean liquid releases are significantly lower than those for airborne releases. They also are delayed significantly in time compared with doses from airborne releases. *Figure 8* presents mean curves for total off-site consequences as a result of subterranean releases and for the seven most risk-significant liquid release categories. The 5th, 50th, and 95th percentile total off-site consequence curves for liquid releases are presented in *Figure 9.*

7.1.3. *Accident Sequences Important to Health Risk*

The mean total off-site health risk of 20.2 person-rem a^{-1} can be divided into 5 general accident-sequence groups.

1. Structural failures of SSTs caused by seismic events contribute 52% of the total risk. These events are characterised as medium- to high-intensity, low-frequency seismic events that fail multiple SSTs simultaneously.

2. Accidents during waste transfer operations account for about 18% of the total risk. The conditional consequences of these scenarios are generally low, but their frequency of occurrence is relatively high.

3. Another 18% of the total is a result of flammable gas combustion events. These include the BURP event for Tank SY-101 and H_2 combustion in the dome spaces of a number of SSTs as discussed in Section 4.2.1. These events are predicted to occur relatively infrequently, but their consequences can be severe.

4. Failures in ventilation systems contribute another 11% of the total risk. These events include loss of active ventilation, HEPA filter failures, and ventilation line breaches in actively- or passively-ventilated tanks. These are relatively high-frequency events with minor consequences.

5. Risks from an aircraft crashing into one of the 177 tanks contribute less than 0.1% of the total risk of off-site health effects. Releases from these events have very high conditional consequences, but their frequency of occurrence is extremely low.

Although the frequency of seismic events is very low, the involvement of multiple tanks and the potential for extended releases until recovery actions can be completed, caused the calculated consequences to be sufficiently large to rank first for overall health risk. Two points are important to remember in comparing risks from low-frequency seismic events with higher frequency operational events. Firstly, the frequency of large earthquakes is, by definition, difficult to predict. Also, the phenomena modeled in the source term analyses for these releases exhibit wide ranges of possible outcomes. Secondly, the damage postulated from

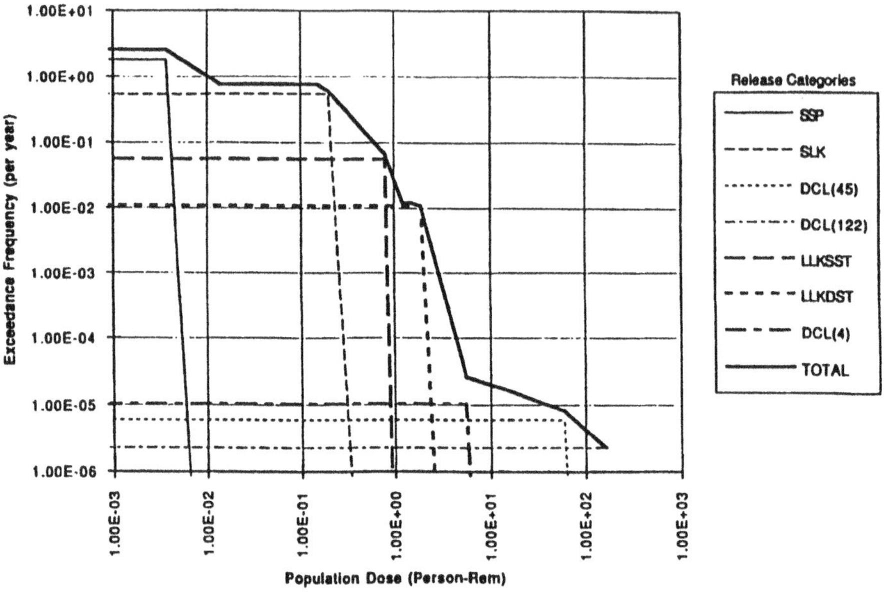

Fig. 8. Liquid total mean unconditional consequences.

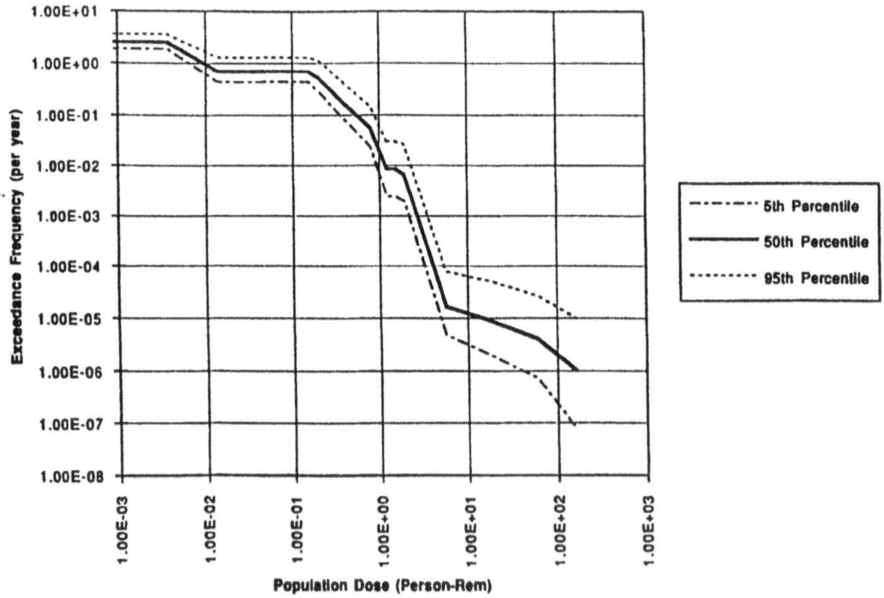

Fig. 9. Liquid total unconditional consequences.

radionuclide releases following a seismic event pale in comparison to the direct damage that would be caused by a large earthquake in the Hanford area.

7.1.4. *Tank Importance Based on Health Risk.*

Based on the relative contributions of each tank to each release category, the mean release quantity for each release category together with the average radioisotope concentrations for each quadrant of the tank farm, the mean risk associated with each of the 177 HLW tanks can be calculated. Rankings of the most risk-significant tanks and groups of tanks contributing to off-site and on-site health effects are presented in this section.

By way of example, *Table 9* presents the first 38 of 177 waste tanks sorted by mean off-site health consequences. Of the 177 waste tanks, only 35 tanks individually contribute more than 1% to the total off-site health risk. These 35 tanks constitute approximately 58% of the total risk. The leading tank relating to off-site risk (at 5.5% of the total) is Tank SY-101. The relatively high risk-rank results of Tank SY-101 are primarily because of the formation and periodic release of H_2 gas (*i.e.* the BURP initiating event). Of the 1.12 person-rem a^{-1} attributable to Tank SY-101, 1.03 person-rem a^{-1} (93%) is caused by the BURP initiating event. Operation of the mixing pump installed in Tank 101-SY in July 1993 appears to have precluded such BURP events. Without the BURP event, the mean annual risk for Tank SY-101 would drop to 0.107 person-rem a^{-1}, and the tank ranking would drop to number 78 in the list.

The second-ranked tank with respect to off-site risk is Tank C-103, contributing 4.8% of the total for off-site health effects. The dominant risk scenario for Tank C-103 is the generation of H_2 gas in the waste, with the gas accumulating to a concentration greater than the LFL in the dome space and igniting. The H_2 burn is assumed to ignite the floating organic layer known to exist in the tank. The tank is assumed to fail as a result of the H_2 combustion. The release category and source terms for this scenario (DCHORG) account for the increase in the airborne release resulting from the pool fire. This scenario accounts for 0.669 of the 0.960 person-rem a^{-1} consequences (70%) attributable to Tank C-103.

Tank AX-101 ranks third with respect to off-site risk with consequences of 0.875 person-rem/yr (4.3% of the total off-site consequences). As was the case with Tanks SY-101 and C-103, the dominant cause of releases from Tank AX-101 is the generation, accumulation, and ignition of H_2 within the tank leading to dome collapse. This scenario accounts for 67% of the risk associated with Tank AX-101.

SSTs A-102, S-102, U-105, and S-112 are listed in *Table 9* as the fourth, fifth, sixth, and seventh tanks with respect to off-site health risks. These tanks, as a group, account for 8% of the total off-site health consequences. H_2 generation, accumulation, and ignition scenarios causing dome collapse are the leading contributors to off-site consequences for all four of these tanks, accounting for 61% of the consequences attributable to these four tanks. The remaining 170 of the 177 tanks contribute 77% of the off-site health risk.

Summing the risk by tank type yields a total of 4.85 person-rem/yr for 28 DSTs and 15.4 person-rem/yr for 149 SSTs. Thus, even though the total risk from DSTs is about one-third of that for SSTs, the per-tank risk for a DST is about 70% higher than the risk from an SST. One of the reasons for this is that all of the DSTs are active tanks and, as such, are susceptible to releases that occur

TABLE 9 Tank ranking by off-site hazard.

Group Rank	Tank Rank	Tank ID	Tank Type	Quadrant	Health risk (p-rem a^{-1})*	% **
1	1	SY-101	DST	SW	1.12E00	5.53
2	2	C-103	100	SE	9.60E-01	4.76
3	3	AX-101	100	SE	8.75E-01	4.34
4	4	A-102	100	SE	4.41E-01	2.19
5	5	S-102	100	SW	4.05E-01	2.01
6	6	U-105	100	SW	3.90E-01	1.94
7	7	S-112	100	SW	3.75E-01	1.86
8	8	C-105	100	SE	3.45E-01	1.71
8	9	C-106	100	SE	3.45E-01	1.71
9	10	C-112	100	SE	2.91E-01	1.44
9	11	A-101	100	SE	2.91E-01	1.44
9	12	C-107	100	SE	2.91E-01	1.44
9	13	C-110	100	SE	2.91E-01	1.44
9	14	C-102	100	SE	2.91E-01	1.44
10	15	A-104	100	SE	2.71E-01	1.34
10	16	A-105	100	SE	2.71E-01	1.34
10	17	A-106	100	SE	2.71E-01	1.34
10	18	C-104	100	SE	2.71E-01	1.34
11	19	U-107	100	SW	2.57E-01	1.27
12	20	AY-101	DST	SE	2.36E-01	1.17
12	21	AY-102	DST	SE	2.36E-01	1.17
12	22	AZ-101	DST	SE	2.36E-01	1.17
12	23	AZ-102	DST	SE	2.36E-01	1.17
13	24	C-108	100	SE	2.17E-01	1.08
13	25	C-111	100	SE	2.17E-01	1.08
13	26	AX-103	100	SE	2.17E-01	1.08
13	27	AX-104	100	SE	2.17E-01	1.08
13	28	C-101	100	SE	2.17E-01	1.08
13	29	C-201	200	SE	2.17E-01	1.08
13	30	C-202	200	SE	2.17E-01	1.08
13	31	C-203	200	SE	2.17E-01	1.08
13	32	C-204	200	SE	2.17E-01	1.08
13	33	A-103	100	SE	2.17E-01	1.08
13	34	AX-102	100	SE	2.17E-01	1.08
13	35	C-109	100	SE	2.17E-01	1.08
14	36	S-111	100	SW	1.79E-01	0.89
15	37	S-101	100	SW	1.40E-01	0.69
16	38	SX-102	100	SW	1.31E-01	0.65

*: person-rem. 1 person-rem = 0.01 person-Sv.

**: percentage of total off-site health risk.

as a result of waste transfer activities, such as large and small spill and spray events. Only about one-third of the SSTs are active. Therefore, a disproportionate percentage of the risk associated with waste transfer accidents is included with the DSTs. Also, most of the DSTs are located in the SE quadrant of the tank farm, where the average radioisotope concentrations and worker density used in the consequence assessment are the most severe from a dose standpoint.

8. Insights and Conclusions

The results of this study are particular to the Hanford Site, and they should not be applied to other sites having waste tanks.

Health risks to co-located workers (on site) and the public (off site) from airborne releases in the Hanford HLW tanks are very low, about 0.014% and 0.027% of the exposure from background radiation for the average on-site co-located worker and the public, respectively.

The highest ranking class of accidents was found to be seismic events, contributing about half of the total mean value off-site risk. Although the frequency of these events is very low, the involvement of multiple tanks and the potential for extended releases until recovery actions can be completed, caused the calculated consequences to be sufficiently large to rank first for overall health risk.

After seismic events, scenarios contributing to health risk are high-frequency, low-consequence events, especially the airborne component of surface spills during waste transfers. Although there may be conservatism in the modeling of transfer spill events, improvements in data and modeling would not be expected to change the scenarios' relative position as important risk contributors. This finding implies that the potential risks of sluicing operations should be examined carefully as part of the preparations for waste retrieval.

Flammable gas accumulation and ignition are important risk contributors, even with elimination of BURP events in Tank SY-101. The results of this analysis indicate that flammable gas accumulation and ignition in passively-ventilated tanks is the most likely scenario capable of producing significant tank damage and a significant radionuclide release.

Risks from liquid pathway releases were shown to be significantly lower than those for airborne releases. This assessment was made assuming that the Hanford Site remains a restricted area. This means that only off-site consequences from the long-term transport of radionuclides to the Columbia River were evaluated. The conclusion that liquid pathway risks are very small is expected to remain valid for other site-use scenarios as long as the direct pumping and consumption of contaminated groundwater is prevented.

References

1. MacFarlane, D. R. et al. (1993) Risk assessment for Hanford high-level waste tank 241-SY-101, *Los Alamos National Laboratory Draft Report*, July, 1993, LANL, New Mexico.
2. Yuan, Y.C. and MacFarlane, D.R. (1992) AI-RISK: A computer program for calculating doses and health risks from accidental release of radioactive materials, *Los Alamos National Laboratory Document* LA-UR-92-2636, LANL, New Mexico.
3. Yuan, Y.C. and MacFarlane, D.R. (1993) AP-RISK: A computer program for calculating population doses and health risks from accidental release of radioactive

materials, *Los Alamos National Laboratory Document* LA-UR-93-3825, LANL, New Mexico.

4. ICRP (1979) ICRP30, International Commission on Radiation Protection Limits for Intakes of Radionuclides by Workers, *ICRP Publication* 30, Pergamon Press, Oxford.

UNDERGROUND AUTOCATALYTIC-CRITICALITY POTENTIAL AND ITS IMPLICATIONS TO WEAPONS FISSILE-MATERIAL DISPOSITION

JOR-SHAN CHOI
Science Fellow, 1995 -1996
Center for International Security and
Arms Control (CISAC), Stanford University
Stanford, CA 94305-6165, U.S.A.
Lawrence Livermore National Laboratory,
Livermore, CA 94550, U.S.A.

Abstract.

Several options for weapons fissile-material disposition, such as once-through mixed-oxide (MOX) fuel in reactors or immobilisation in waste glass, would result in end products requiring geologic disposal. The criticality potential of the fissile end products containing U-235 and Pu-239 and the associated consequences in a geologic setting are important considerations for the final disposal of these materials. The possibility of underground criticality, and especially autocatalytic criticality, is affected by (1) groundwater leaking into a failed waste container, (2) preferential leaching of neutron absorbers or of fissile material from a failed container, and (3) preferential deposition of fissile material in the surrounding rock.

Bowman and Venneri [1] have pointed out that fissile material mixed with varying compositions of water and silica can undergo a nuclear chain reaction. Some configurations can become autocatalytically supercritical resulting in considerable energy release, terminated finally by disassembly.

Some reviews [2] rejected the Bowman and Venneri warning as implausible because of low probabilities of scenarios that could lead to such configurations. Sanchez *et al.* [3] reported possible supercritical conditions in systems of Pu-SiO_2-H_2O and Pu-tuff-H_2O but concluded that the probability of forming such combinations is extremely low.

Kastenberg *et al.* [4] studied the potential for autocatalytic criticality of plutonium or highly enriched uranium in the proposed Yucca Mountain geologic repository. They concluded that plutonium or uranium could, theoretically, become supercritical, but that such criticality is unlikely given the hydrology, geology and geochemistry of the Yucca Mountain site.

These studies are not definitive. The possibility of criticality exists. Detailed mechanisms have not been sufficiently studied for clear conclusions on the proba-

M. J. Stenhouse and V. I. Kirko (eds.),
Defence Nuclear Waste Disposal in Russia: International Perspective, 263–278.

bilities of occurrence. More technical analysis is needed to understand the potential for underground autocatalytic criticality and the associated consequences. Here we present our calculations of possible conditions and a mechanism for criticality in a container containing fissile material in various different chemical forms, including conditions for autocatalytic criticality. We discuss criteria for designing waste solids for the disposition of weapons fissile material so that long-term dissolution of a neutron-absorber or of fissile material would not cause criticality. These results for simplified geometries outline the conditions for a more detailed analysis of practical systems. The potential for underground criticality is also studied for other fissile waste forms, including spent fuel from commercial reactors, research reactors, and naval reactors, all of which may be destined for geologic disposal.

If not carefully resolved on scientific grounds, underground criticality could potentially become the 'Achilles heel' of programmes to dispose of weapons materials and nuclear wastes[1]

1. Introduction

Several kinds of radioactive wastes now destined for geologic disposal can, theoretically, reach nuclear criticality in a geologic medium. These are surplus weapons-grade plutonium from dismantled nuclear weapons, spent nuclear fuel from commercial light-water reactors, and DOE-owned (U.S. Department of Energy) spent nuclear fuel from naval vessels, research reactors, and production reactors. One option for disposing of weapons-grade plutonium involves incorporating the plutonium into borosilicate glass logs, along with sufficient fission products, so that the resulting waste will be as self protecting as spent fuel when placed in a geologic repository.

Another option is to mix the weapons-grade plutonium with natural uranium, to form MOX fuel for light-water reactors. The discharged MOX fuel, containing a few percent of fissile U-235 and Pu-239, would subsequently be placed in a geologic repository. About 50 tonnes of weapons-grade plutonium might be disposed of in one of these two ways. Spent fuel from commercial reactors contains a few percent of fissile uranium and plutonium. About 63,000 tonnes of commercial spent fuel, containing about 950 tonnes of fissile material, are now destined for geologic disposal in the proposed Yucca Mountain repository. DOE-owned spent fuel contains a wide range of fissile components, including some highly enriched U-235. The amounts of Pu-239 and U-235 in typical waste containers planned for geologic disposal in the U.S.A. are listed in *Table 1*.

Possible mechanisms of reaching criticality in a geologic repository, after groundwater has leaked into a failed waste container, include:

1. preferential leaching of neutron absorbers from a failed waste container, and

2. preferential leaching and deposition of plutonium or fissile uranium in surrounding rock.

[1]This paper describes an independent study by the author and is not associated with his affiliation.

TABLE 1. Amounts of fissile material (Pu-239, U-235) in waste containers

Waste Type	Amount (kg)		
	wt. %	small container [a]	MPC [b]
Commercial spent fuel			
UO_2 spent fuel	1.5	20	143
Surplus weapons-grade plutonium			
MOX spent fuel	~ 4	54	378
Glass waste	~ 5	84	336
Ceramic waste	~ 12	80	320
DOE-owned spent fuel [5]			
N-reactor	1.4	74 [c]	495 [d]
ATR	30.4	94	664
Shippingport	60.9	44	262
Graphite	3.1	1.5	10

[a]: Small container holds 3 UO_2 or MOX PWR spent fuel assemblies, or 1 glass or ceramic waste form.

[b]: Multi-Purpose Container (MPC) holds 21 UO_2 or MOX spent PWR fuel assemblies, or 4 glass or ceramic waste forms.

[c]: Fissile content in small container designed for light-weight truck transport.

[d]: Fissile content in MOC designed for rail transport.

Bowman and Venneri [1] have pointed out that fissile material mixed with varying compositions of water and silica can undergo a nuclear chain reaction. Some configurations can become autocatalytically supercritical resulting in considerable energy release, terminated finally by disassembly.

Some reviews [2] have rejected the Bowman and Venneri warning as being implausible because of the low probabilities of scenarios that could lead to such configurations. Sanchez *et al.* [3] reported possible supercritical conditions in systems of Pu-SiO_2-H_2O and Pu-tuff-H_2O, but concluded that the probability of forming such combinations is extremely low.

Kastenberg *et al.* [4] studied the potential for autocatalytic criticality of plutonium or highly enriched uranium in the proposed Yucca Mountain geologic repository, focusing on the possibility of accumulating a critical assembly of migrating fissile material in the far field. They concluded that plutonium or uranium could, theoretically, become supercritical, although the critical mass required would be in the range of hundreds of kilograms. The requirement of such incredibly large critical masses for an underground criticality, apparently the result of the use of a one-dimensional slab model in their analysis, again led to the conclusion that underground criticality was unlikely given the hydrology, geology and geochemistry of the Yucca Mountain site.

These studies are not definitive, The possibility of criticality exists. Detailed mechanisms have not been sufficiently studied for clear conclusions on the probabilities of occurrence. More technical analysis is needed to understand the potential for underground autocatalytic criticality and the associated consequences. Here we present our calculations of possible conditions and mechanisms for criticality in a container holding fissile-material in various different chemical forms, including conditions for autocatalytic criticality. We discuss the criteria for designing waste solids for the disposition of weapons fissile material so that long-term dissolution of a neutron-absorber or of fissile material, would not cause criticality. The results for simplified geometries outline the conditions for a more detailed analysis of practical systems. The potential for underground criticality is also studied for other fissile waste forms, including spent fuel from commercial reactors, research reactors, and naval reactors, all of which may be destined for geologic disposal. The calculations have been made using the Monte-Carlo neutron-transport MCNP code [6].

2. Analysis

2.1. PU-239, SIO_2, H_2O SYSTEM

The effective multiplication factors, k_{eff}, for homogeneous systems involving Pu-239, SiO_2 and H_2O, reflected by an infinite reflector of silica, are shown in *Figures 1*, *2*, and *3* for mole ratios of SiO_2 to H_2O of 0, 0.67, and 2.70, respectively. The resultant curves could simulate a porous solid of silica and plutonium, with water occupying the pore volume. Each curve corresponds to a given ratio of silica volume to pore volume. We assume that the boron neutron absorber originally in the glass has been leached out preferentially. These curves begin at a water density of 1 g cm^{-3} within the pores, implying a water-saturated system. The temperature is assumed to be constant at the boiling point of water. The results indicate that the larger the system volume, the higher the potential for autocatalytic criticality. The calculations also indicate that the higher the SiO_2:H_2O molar ratio, implying less H_2O moderation, the slower the rise in k_{eff} as the water density decreases. Calculations for a homogeneous spherical 3-component system of Pu-239, SiO_2, and H_2O, were also performed by others [1, 3] and the results are compared in *Figure 4*.

Figure 5 shows the calculated critical mass of Pu-239 in spherical volumes of various radii, surrounded by SiO_2. For each volume ratio, the critical mass is calculated as a function of the effective density of water in the pores.

Figure 1. Homogeneous spherical system of Pu-239 and H_2O (no SiO_2 present).

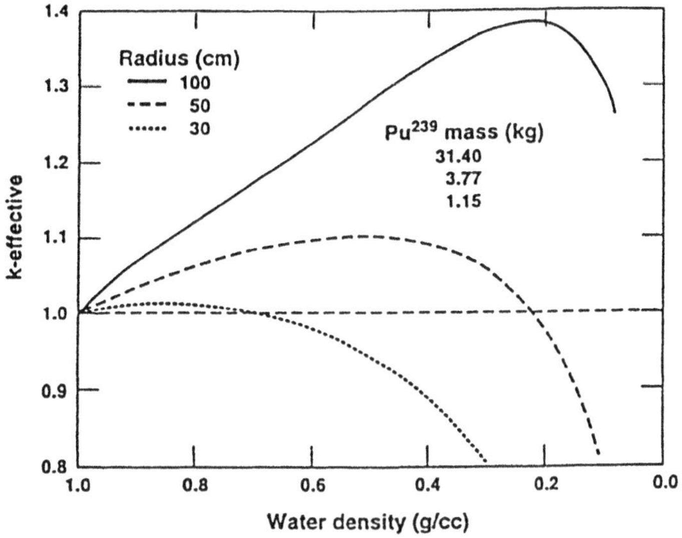

Figure 2. Homogeneous spherical system of Pu-239, SiO_2 and H_2O; mole ratio of SiO_2:H_2O of 0.67.

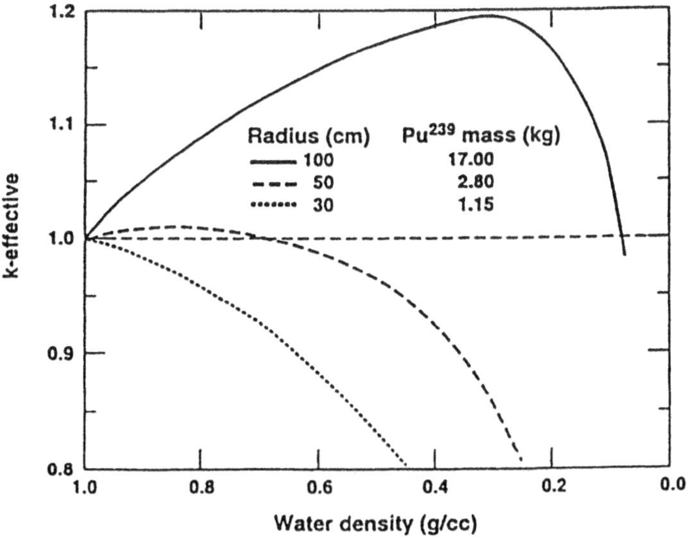

Figure 3. Homogeneous spherical system of Pu-239, SiO_2 and H_2O; mole ratio of SiO_2:H_2O of 2.70.

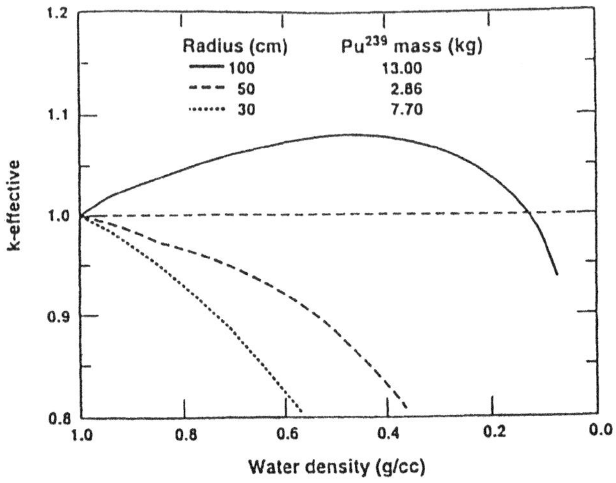

Figure 4. Comparison of k_{eff} results for systems of Pu-239, SiO_2 and H_2O

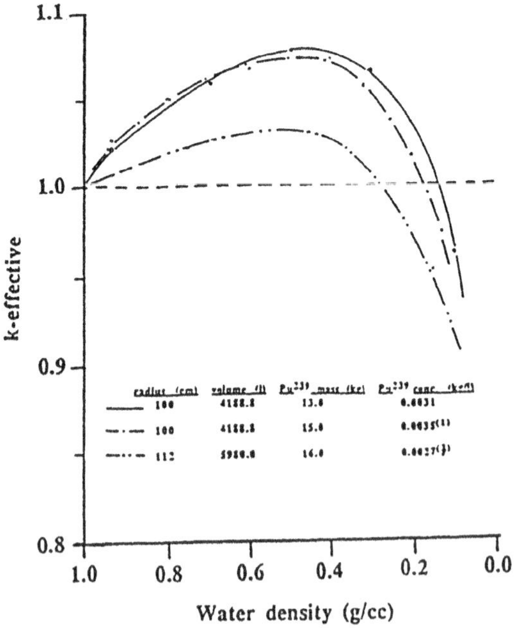

Figure 5. Spherical critical mass of Pu-239, SiO_2 and H_2O.

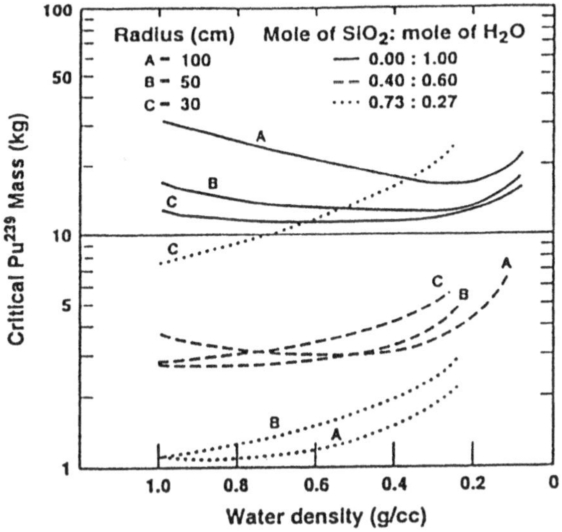

The curves proceed to lower water densities, implying that some liquid water may have been converted to steam by fission heating. Autocatalytic criticality would occur where a curve has a negative slope, because boiling the water would require less fissile mass for criticality. Where the slope is positive, the system is self regulating, because fission heating would increase the required critical mass, thereby causing the system to become subcritical. Here, autocatalytic criticality occurs when the system is over-moderated. Since more than the critically-optimum amount of water is present, water acts more as a neutron absorber, causing a positive water-void coefficient of reactivity. Boiling the water removes this neutron absorbing function and makes the system supercritical.

Autocatalytic criticality occurs more readily for large-volume systems. For smaller systems, neutron leakage becomes more important. Lower density of water increases neutron leakage and adds self-regulating feedback.

The lowest critical mass occurs for the curves of lower void fraction and larger volume, showing that there is an optimum composition and size for lowest critical mass. If plutonium were dissolved in the water instead of being a solid, boiling the plutonium-water solution could cause the egress of some plutonium and water, and the curves would depart significantly from those shown in *Figure 5*.

The effect of Yucca Mountain tuff on a change of system k_{eff} as a function of H_2O density is shown in *Figure 6*. A typical material composition

Figure 6. Homogeneous spherical system of Pu-239, tuff and H_2O.

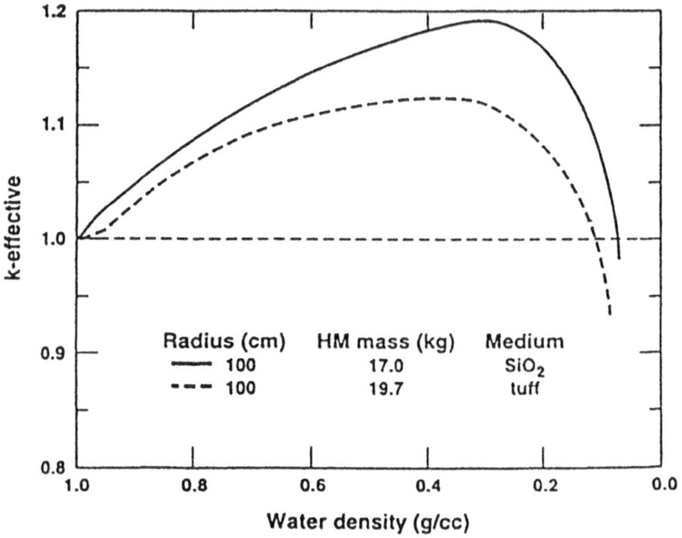

TABLE 2. Composition of Yucca Mountain tuff

Component	Percentage
SiO_2	71.6
Al_2O_3	12.1
H_2O	4.0
K_2O	3.5
CaO	2.4
MgO	trace
FeO	trace
CO_2	trace

of tuff is shown in *Table 2.*

Figure 6 shows the potential for autocatalytic criticality for a 3-component (Pu-239, tuff and H_2O) system, reflected by an infinite medium of tuff. The diagram also indicates that the tuff system yields a lower k_{eff} than the SiO_2 system, due mainly to neutron absorption by components in tuff other than SiO_2.

Figure 7. Homogeneous spherical system of Pu-239 (5 wt.%), U-238 and H_2O.

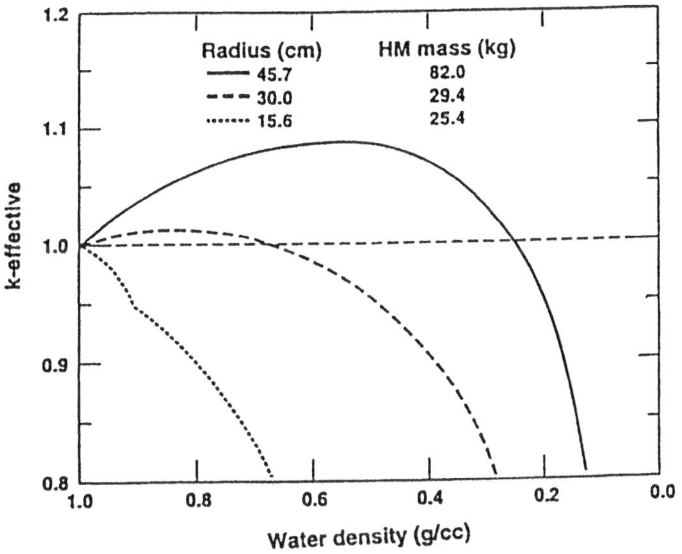

2.2. PU-239, U-238, H_2O SYSTEM

Figure 7 shows the criticality conditions presented in terms of the effective multiplication factor, k_{eff}, for reflected spherical homogeneous mixtures of Pu-239 (5 wt.%), U-238, and water, roughly simulating fresh MOX fuel assemblies in water. Other neutron absorbers are neglected. Here, a curve of positive slope is autocatalytic with respect to reduced water density, the maxima of curves for radii of 45.7 cm and 30.0 cm occur when enough water has been boiled away that further loss of water reduces k_{eff} because of less neutron moderation and increased leakage. The transition from autocatalytic to self-regulating criticality occurs at a higher water density than that in *Figure 5* because of the self-regulating effect of the absorption resonances of U-238.

Figure 8 shows the k_{eff} factor for reflected spherical homogenous mixtures of 10-year-old spent MOX fuel and water. Fission products (*i.e.* Sm-149, Gd-155, Rh-103 *etc.*) are included in the fuel mixtures. These neutron-absorbing isotopes significantly reduce the criticality potential of the spent fuel-water mixture. To achieve the initial criticality, a system with radius of 45.7 cm would require almost twice the mass, compared to the fresh MOX fuel of *Figure 7*. The transition from autocatalytic to self-regulating criticality also occurs at a much higher water density than that in *Figure 7*. For a radius of 30 cm, the system requires more than 2 times the mass

Figure 8. Homogeneous spherical system of spent MOX fuel H_2O.

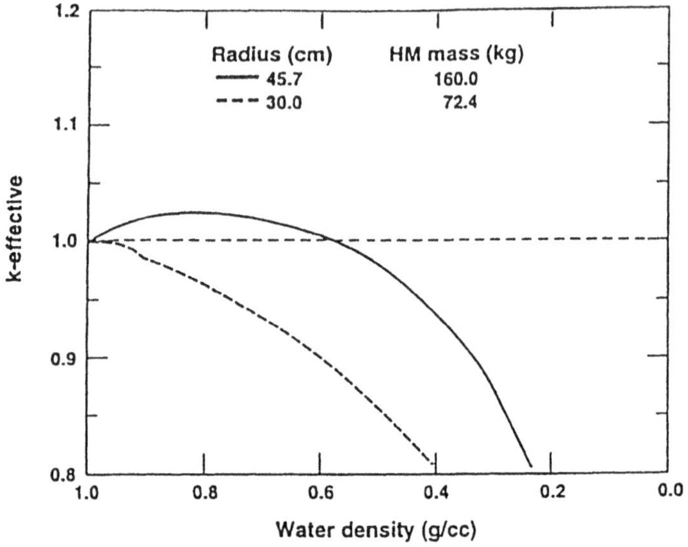

as that in *Figure 7*, and does not become autocatalytically critical.

2.3. U-235, U-238, H_2O SYSTEM

Figure 9 shows similar reactivity curves of k_{eff} for homogeneous spherical mixtures of solid enriched uranium (5 wt.% U-235) and water. More U-235 is required for criticality and there is less tendency towards autocatalytic criticality than for the Pu-239 system of *Figure 7*.

Figure 10 shows k_{eff} curves for homogeneous spherical mixtures of 10-year-old LWR spent fuel in water. Fission products (*i.e.* Sm-149, Gd-155, Rh-103 *etc.*) are included in the spent-fuel containing approximately 1.5% fissile content. These neutron-absorbing isotopes significantly reduce the criticality potential of the spent fuel-water mixture. Similar to the spent MOX fuel and water system, more mass is required for criticality, even for a larger system volume (radius of 50 cm instead of 45.7 cm), and the transition to self-regulating criticality occurs at a higher water density than that in *Figure 9*.

2.4. U-235,238 SPENT FUEL IN A MULTI-PURPOSE CANISTER

The k_{eff} curve for low-enrichment uranium fuel discharged from a pressurized light-water reactor (PWR) is shown in *Figure 11*, for a burnup of

Figure 9. Homogeneous spherical system of U-235 (5 wt.%), U-238 and H_2O.

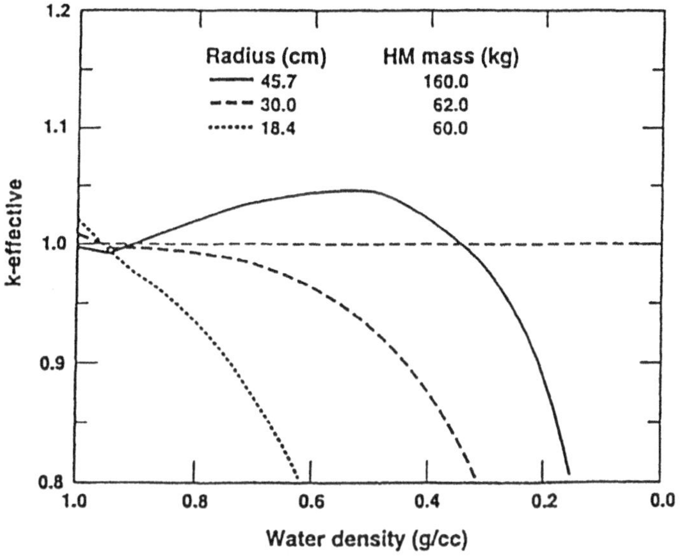

Figure 10. Homogeneous spherical system of spent UO_2 fuel plus H_2O

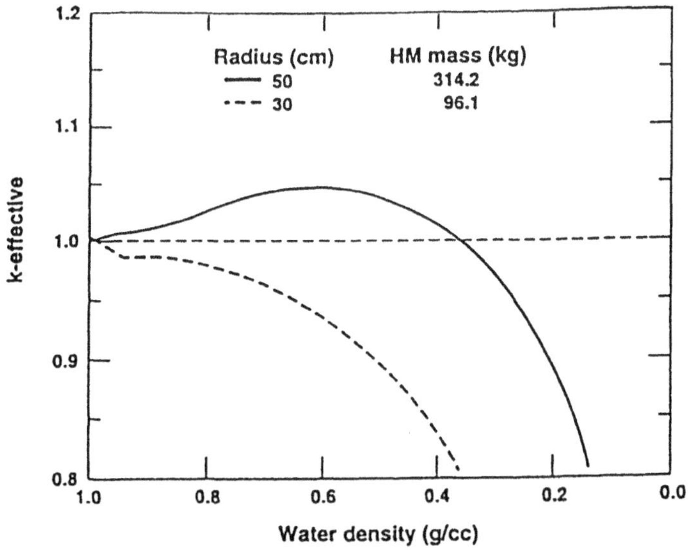

16 GWd Mg^{-1}, selected to optimise the criticality potential. The diagram also shows the k_{eff} curve for MOX spent fuel discharged with a burnup of 42.6 GWd Mg^{-1}. Fresh UO_2 fuel with an enrichment of 4.5% U-235 is included for comparison. The k_{eff} results are presented as a function of the mole ratio of water and heavy metal in the fuel assemblies, simulating the volumetric changes of internal materials due to the degradation of the waste container and possibly preferential leaching of internal structural materials.

The multi-purpose canister (MPC) contains 21 fuel assemblies. All three of the fuel assemblies considered here can become critical if the canister becomes flooded, as shown for the condition for a mole ratio of 7.1, which represents a water density of 1 g cm^{-3}. Additional neutron absorbers must be added for safety. If the system does become critical, reactivity will decrease if some water is boiled away by fission heating, as shown for decreasing mole ratio, so the system is self regulating as long as this fuel-rod spacing is preserved.

However, the waste container would be degraded over the long time periods typical of underground disposal, subjected to various hydrogeological and geochemical conditions resulting in waste-package corrosion and groundwater intrusion. Inside the canister, the fuel assemblies and the structural support material would also undergo chemical transformations upon contact with air and groundwater. These reactions could potentially alter the original fuel-rod spacing and subsequently change the volumetric distribution of the water and fuel mixture.

The reactivity effect due to changing mole ratio of water and heavy metal in spent fuel is shown in *Figure 11.* The region of mole ratio greater than 20 indicates conditions of autocatalytic criticality concern if the heavy metal in the MOX spent fuel could be dispersed in a volume of water greater than is available initially. The system could become autocatalytic because it is over-moderated and the fission heat would boil some water away. For spent UO_2 fuel, the transition occurs at a mole ratio of about 15. It is important that the waste-package design does not allow sufficient voids initially for water-filled criticality even with dispersed fuel, and that it precludes enough water for autocatalytic criticality.

2.5. NEUTRON ABSORBERS AND URANIUM DILUENTS TO AVOID CRITICALITY

Simply adding enough neutron poison, such as boron, to avoid criticality may not be an adequate solution. Since boron readily forms soluble compounds, the long-term preferential leaching of boron can result eventually in a critical system. If there is enough water for eventual autocatalytic criticality, approaching criticality by boron dissolution can result in auto-

Figure 11. k_{eff} of LWR fuel assemblies in a multi-purpose canister (MPC) containing 21 PWR assemblies, 0.12 wt.% boron, void filled with water.

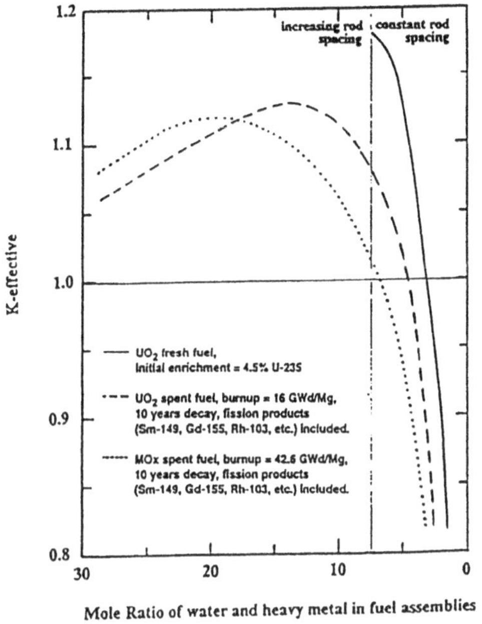

catalytic energy release. The use of gadolinium as a neutron poison may also be problematic due to its significant decrease in neutron absorption at elevated neutron energy (above 10 eV). More stable neutron absorbers, such as erbium, should be considered.

Another solution may involve the use of depleted uranium as filler material in the waste container. The depleted uranium would mix with U-235 as the uranium degraded and leached out from the fuel assemblies, thereby diluting the overall enrichment level required for criticality. The challenge is how to fill the depleted uranium in between the fuel assemblies to make it an effective diluent for U-235.

2.6. CRITICALITY FROM U-233, A DECAY DAUGHTER OF NP-237

The amount of U-233 in the spent fuel from low-enriched uranium LWRs, originally very small, increases as its parent Np-237, a radionuclide with half-life of $2.1 \cdot 10^6$ years, decays. U-233 generated from the decay of Np-237 in the fuel matrix, could dissolve and leach together with other uranium isotopes. If depleted uranium is used as filler material, the overall uranium enrichment level provided by U-233 and U-235 may still be maintained below the level required for criticality. However, if neptunium, an element with chemical behaviour different from that of uranium, dissolves and leaches out faster than uranium, U-233 could accumulate in the geologic medium from decay of Np-237 as the latter nuclide migrates. If sufficient U-233 accumulates in an underground volume of water and rock which exhibits an over-moderating condition, autocatalytic criticality could occur.

To prevent criticality from dissolution/migration of Np-237 and reconcentration of U-233 in the geologic medium, the geochemical conditions of the waste package and the repository should be "reducing" to minimise the dissolution of neptunium, and the waste package could be filled with depleted uranium used as diluent to reduce the overall uranium enrichment.

2.7. PU-SIO_2-H_2O SYSTEM WITH HEATING

Previous figures have assumed water boiling at constant temperature. However, if the system is autocatalytic, appreciable fission heating can occur. *Figure 12* presents a recent calculation by Bowman and Venneri [1], who have kindly agreed to its use in the present paper.

The three-dimensional (3D) surface shows k_{eff} as affected by the system temperature and by the relative amount of water in the Pu-SiO_2-H_2O mixture. In the left corner of the diagram, at boiling temperature and a water-saturated porosity of 35%, the system is slightly supercritical. If water can escape easily when it boils, the effective volume of liquid water in the pores (calculated at normal boiling temperature) will decrease and the reactivity increase, similar to one of the curves in *Figure 2*.

However, if the pores that contain water are of sufficiently small diameter, the permeability of the porous solid can be quite low. The system will heat further until the vapour pressure of water is sufficient to force water from the pores. The kinetic properties of reactivity feedback and fluid dynamics can generate a new pathway along the 3D surface. High temperatures are included, corresponding to vapour pressures high enough that the water becomes a supercritical fluid. In the limit of very low permeabilities, the pathway will lie on a path of nearly constant water density. Very large pressures could be generated.

Along any of the pathways that involve heating, the reactivity first increases because the average thermal energy of the neutrons increases, and the neutron thermal energy spectrum overlaps more of the large low-lying 0.3-ev absorption resonance of Pu-239, thereby increasing the effective fission cross section of Pu-239. This temperature effect is essentially absent for U-235 systems. For MOX fuel, Doppler broadening of the U-238 absorption resonances will reduce the net autocatalytic effect from Pu-239.

Figure 12. Criticality for the Pu-SiO_2-H_2O system; sphere of 200 cm diameter and SiO_2 reflected (from [1]).

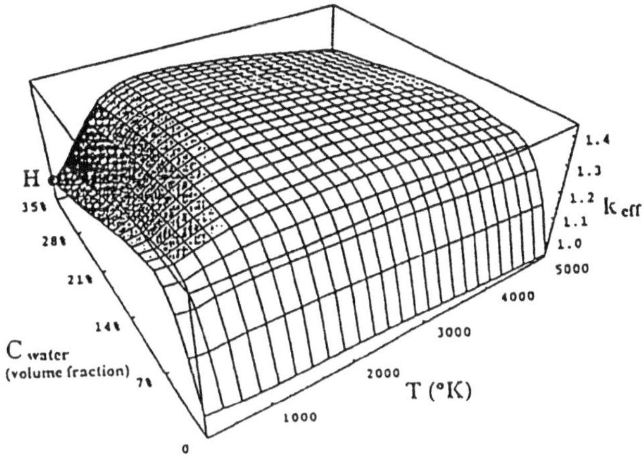

3. Conclusions

The following conclusions may be made from this work:

- Criticality and autocriticality are theoretically possible in failed waste canisters for systems involving spent fuel and water, MOX fuel and water, and plutonium in glass (*i.e.* SiO_2) and water, and possibly elsewhere involving plutonium in silica or tuff and water.
- There are regions of composition and volume to be avoided, as indicated by calculations of idealised systems, as well as the need for a more detailed treatment of geometry and composition in regions where criticality can potentially occur.
- Criticality and autocatalytic conditions in waste containers can be avoided by careful design, such as low-release neutron absorbers or uranium diluents (*e.g.* erbium, depleted uranium, or massive iron), and by limiting the amount of voids that could be filled with water.
- To prevent criticality from U-233 concentrated in the geologic medium from decay of Np-237 as neptunium dissolves and migrates through the medium, the geochemical conditions of the waste package and the repository should be made "reducing" to minimise the dissolution of neptunium.

- There is a theoretical potential for appreciable energy release from autocatalytic criticality. The extent and consequences of such an effect are not yet resolved. Hence, additional technical analysis is required.

References

1. BOWMAN C.D. AND VENNERI F. (1995) Autocatalytic criticality from plutonium and other fissile material, *Los Alamos National Laboratory Report* LA-UR-94-4022, LANL, New Mexico, also published in *Science & Global Security*, **5**, 1996.
2. VAN KOYNENBURG R.A. (Compiler) (1996) Comments on the draft paper "Underground supercriticality from plutonium and other fissile material", written by C. D. Bowman and F. Venneri (LANL), *Science & Global Security*, **5**, 303-322.
3. SANCHEZ R., MYERS W., HAYES D., KIMPLAND R., JAEGERS P., PATERNOSTER R., ROJAS S., ANERDSON R., AND STRATTON W. (1995) Criticality characteristics of mixtures of plutonium, silicon dioxide, Nevada tuff, and water, *Los Alamos National Laboratory Report* LA-UR-95-21130, LANL, New Mexico.
4. KASTENBERG W.E. *et. al.* (1996) On the potential of autocatalytic criticality of fissile materials in geologic repositories, Department of Nuclear Engineering Report, University of California, Berkeley.
5. RECHARD R.P. *et. al.* (1995) Performance assessment of the direct disposal in unsaturated tuff of spent nuclear fuel and high-level waste owned by U.S. Department of Energy, *Sandia National Laboratories Report* SAND94-2565, SNL, New Mexico.
6. MCNP (1991) MCNP – a general Monte Carlo code for neutron and photon transport, *Los Alamos National Laboratory Report* LA-7396-M Revision 2, LANL, New Mexico.

GEOLOGICAL HISTORY OF YUCCA MOUNTAIN (NEVADA) AND THE PROBLEM OF A HIGH-LEVEL NUCLEAR WASTE REPOSITORY

Y.V. DUBLYANSKY
Institute of Mineralogy and Petrography (IMP)
Russian Academy of Sciences (RAS), Siberian Branch
3 University Ave., Novosibirsk 630090, RUSSIA

J.S. SZYMANSKI
Technology and Resources Assessment Corporation (TRAC)
3800 Arapaho Avenue, Suite 225
Boulder, Colorado 80303, U.S.A.

A.V. CHEPIZHKO
Odessa State University
2 Shampanski per., Odessa 270058, UKRAINE

AND

B.N. LAPIN, V.N. REUTSKY
Institute of Geology
Russian Academy of Sciences, Siberian Branch
3 University Ave., Novosibirsk 630090, RUSSIA

Abstract.
Three types of deposits are widely distributed throughout the Yucca Mountain unsaturated zone: mosaic breccias, quartz-calcite veinlets, and micritic opal-calcite veins and associated slope deposits were found to be of hydrothermal origin. The inference is based on petrographic, accessory mineral, fluid inclusion, and stable isotopic data. The overall conclusion is that recurrent low-temperature hydrothermal (epithermal) activity occurred at Yucca Mountain in the geologically recent past (from *ca.* 8–10 Ma to as recently as 25 ka). New data require the currently-accepted concept of the hydrogeological system at Yucca Mountain to be reconsidered. The subject relates to the suitability of Yucca Mountain as a high-level nuclear waste site.

M. J. Stenhouse and V. I. Kirko (eds.),
Defence Nuclear Waste Disposal in Russia: International Perspective, 279–292.

1. Introduction

Yucca Mountain, located in south Nevada, has been designated for study as a potential national repository for high-level nuclear waste (HLW). As a result of the Nuclear Waste Policy Amendments Act of 1987 (Public Law 100-203), the mountain has been singled-out as the sole candidate site. However, evidence obtained during the last few years implies that the mountain may not be technically suitable to host the facility.

Yucca Mountain is a north-trending range composed of a sequence of Tertiary ash-flow tuffs flanked by broad alluvial basins from west and east (*Figure 1*). Two geological features make the mountain appear suitable for hosting the repository:

1. extensive unsaturated zone (the water table of the regional aquifer is located at depth 400–500 m); and
2. zeolitization of tuffs starting from depth 300–400 m.

The proposed permanent (non-retrievable) geological repository would be constructed within the unsaturated zone about 300 m above the water table. According to the safety regulations, the feasibility studies must demonstrate that the proposed repository would provide a geological barrier against any migration of the radionuclides and their penetration into the accessible environment for the period of 10,000 years. Hence, geologic issues of primary concern are: (1) for how long does the unsaturated zone exist; and (2) how probable is the inundation of the repository zone by upward movement of the groundwater in future. In this respect, any traces of the presence of water within the currently unsaturated zone in the geological past would indicate a potential hazard in future.

A number of geologic features suggests palæo-hydrothermal activity at Yucca Mountain, *viz.*:

- fault-associated mosaic breccias cemented by opal and calcite;
- quartz-opal-calcite veinlets encountered at depth 30 to 400 m;
- micritic wedge-shaped opal-calcite veins in fault gashes; and
- slope-parallel zones of opal-calcite cementation in alluvium and colluvium, developed downdip from vein-hosting faults.

These opal-calcite deposits are interpreted by the Yucca Mountain Project scientists as being pedogenic (*i.e.* precipitated from rainwaters that have percolated along interconnected rock fractures carrying dissolved carbonate from the overlying soil environment; see, for example, [1]).

We have evaluated the results produced by the U.S. Department of Energy in the course of site characterisation activities [2] and have performed a set of original field and laboratory studies. We have found that the currently accepted pedogenic interpretation is in conflict with a large

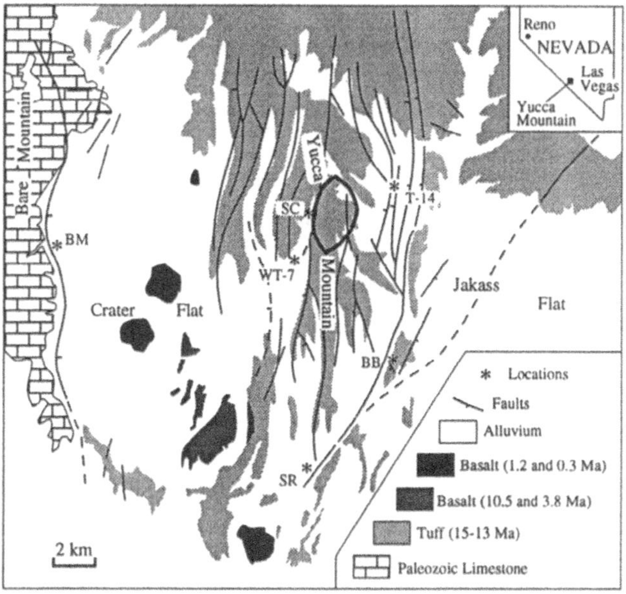

Figure 1. Geologic map of Yucca Mountain and its vicinity. Locations discussed in text. BM = Bare Mountain; WT-7, T-14 = Trench 14; BB = Busted Butte; SC = Solitario Canyon; SR = Stagecoach Road fault. Bold contour shows the extent of the proposed repository.

body of data. The following text presents evidence of a hydrothermal origin for three types of deposits widely distributed throughout the unsaturated zone and at the surface of Yucca Mountain: (1) mosaic breccias, (2) quartz-opal-calcite veinlets, and (3) micritic opal-calcite veins and associated slope deposits.

2. Mosaic Breccias

2.1. FIELD OCCURRENCE AND PETROGRAPHIC FEATURES

The so-called AMC (authigenic-mineral-cemented) breccias form dike-shaped bodies 30-70 cm wide along fault scarps. Some of these breccias are formed by epigenetic cementation of the CTM (crushed-tuff-matrix) breccias developed along faults [3]. In some localities the AMC breccias form nearly isometric (in plan) bodies a few metres in diameter. The latter are also associated with fault zones.

The AMC breccias revealed a number of petrographic features typical of epithermal deposits, such as granoblastic and crustification textures, multiple episodes of cementation; recrystallisation of cement, *etc.*

2.2. MAJOR ELEMENT RELATIONS

At Yucca Mountain, the welded tuffs reveal substantial enrichment in CaO (0.5–4.5%), Al_2O_3 (0.5–5.0%), Fe_2O_3 (0.5–4.5%), MnO (0.25–1.75%), TiO_2 (0.1–0.5%) relative to the unchanged volcanic glass, which is believed to represent the composition of the parent melt [4, 5]. At the same time, the tuffs are depleted in SiO_2 (4–10%), K_2O (0.5–3.0%), and Na_2O (0.5–2.5%). This feature has been attributed to a large-scale alteration by epigenetic hydrothermal processes which occurred between 10 and 2 Ma ago. Our studies have shown that during the formation of the AMC breccias, there have been further significant inputs of MgO (40–45%), CaO (20%), TiO_2 (12–14%), SiO_2 (10%), and P_2O_5 (8%) and removal of Al_2O_3 (34%), K_2O (40%), Na_2O (30%), and Fe_2O_3 (10%) (relative to the bulk bedrock tuff composition; [3]). One breccia from eastern Crater Flat has revealed an extreme enrichment in CaO (> 1200%) and MgO (540%).

2.3. ACCESSORY ZIRCON

Accessory zircon has provided further insight into the origin of the AMC breccias. This mineral has been extracted from 7 large samples (5–15 kg) of the bedrock tuffs (including welded, moderately- and non-welded varieties) and 11 samples of the AMC breccias. The Yucca Mountain accessory zircon was found to comprise two genetic types. **Relict-detrital** zircon occurs as rounded, fractured, and in some instances metamict grains. It is interpreted as xenogenic, derived from sedimentary rocks of the basement. In contrast, crystals of **epigenetic hydrothermal** zircon have perfect crystallographic shape, and morphologic features, typical of hydrothermal zircons from elsewhere. They often contain fluid inclusions. This zircon is interpreted as being formed *in situ*, which is implied by the presence of delicate (submicron size) growth sculpture on crystal surfaces and intergrowth relationships with other minerals (*e.g.* calcite and magnetite). Such fragile features would not have survived any significant transport.

Welded varieties of tuffs at Yucca Mountain hosting breccia bodies typically contain both relict-detrital (mean content 1.6 ppm) and epigenetic hydrothermal (mean content 1.2 ppm) zircons. The presence of the latter might be indicative either of large-scale epigenetic hydrothermal alteration, or of local bedrock alteration by younger fault-based hydrothermal activity. The AMC breccias from three sites displayed elevated contents of epigenetic hydrothermal zircon (Trench 14 – 1.82 ppm, Busted Butte – 3.01 ppm, and Solitario Canyon – 11.90 ppm), which suggests a hydrothermal origin for those geologic bodies. Two more accessory minerals determined in the AMC breccias, which possess characteristic features of the hydrothermal origin, are apatite and titanite [6].

2.4. ORIGIN OF THE AMC BRECCIAS

The AMC breccias at Yucca Mountain had been interpreted earlier as being formed *"by varying combinations of surface erosion and deposition ... and recent near-surface fault movement"* with *"most of the authigenic mineral cementation ... related to surficial pedogenic process"* [7] or even as *"colluvial breccia or simple alluvium cemented by authigenic minerals"* [8].

In contrast, our studies have shown that the AMC breccias have the preponderance of petrographic features typical of the low-temperature hydrothermal breccias form elsewhere (multiple episodes of cementation, presence of several mineral sequences, *etc.*) and should, therefore, be considered as epithermal deposits. This inference is reinforced by major-element geochemistry of the breccias, revealing significant input of some elements and removal of others during breccia formation, as well as by the presence of epigenetic hydrothermal accessory minerals such as zircon, apatite, and titanite.

3. Quartz-Opal-Calcite Veinlets

The minerals found in fractures within the Yucca Mountain unsaturated zone at depth 40–400 m are referred to as: secondary calcite, calcite-opal association, or micritic calcite-opal-sepiolite association. These minerals are often interpreted as follows: *"The unsaturated zone calcite appears to have precipitated from rainwaters that have descended along interconnected fractures carrying dissolved carbonate from the overlying soil environment..."* [9]. This interpretation, however, is at odds with a large body of data (*e.g.* petrography, fluid inclusions, ESR, ^{18}O in calcite-opal pairs, $\Delta^{18}O/\Delta z$ gradients in calcite from drill holes) implying elevated temperatures and geothermal gradients in the Yucca Mountain subsurface in the past [2]. Here we will discuss some new data obtained in the course of our studies of samples collected from the first 400 m of the exploratory tunnel under Yucca Mountain.

3.1. PETROGRAPHIC FEATURES

Petrographic studies revealed features which are not compatible with postulated pedogenic origin but are rather indicative of the low-temperature hydrothermal setting [3], *viz.*

1. Euhedral crystals of zeolite up to 0.8 mm in size pre-date deposition of calcite. Electron microprobe analysis yielded composition typical

of hydrothermal heulandite.
2. Calcite reveals up to 4 major episodes of deposition with hiatuses marked by corroded surfaces, entrapment of solid impurities, and crystallisation of quartz on the calcite surface. In many instances, several generations of calcite were found to be re-crystallised to form optically continuous crystal.
3. Opal (chalcedony) often grades into the euhedral quartz crystals. Quartz formed between different stages of calcite deposition displays features of competitive growth and/or free growth (growth into the void) with crystal sizes reaching 1.0–1.5 mm.

3.2. FLUID INCLUSIONS

Only 7 reliable homogenisation temperatures have been obtained by calcite from the Yucca Mountain subsurface in the course of the DOE characterisation activities ([10]; drilled cores from depths 178–386 m). The data, though restricted in number, suggest a palæothermal gradient as high as 150–160 °C km^{-1} [2]. Inclusions, suitable for instrumental studies were observed by Roedder and coworkers [9]; no numerical data have been reported, however.

In several calcite samples removed from the Yucca Mountain exploratory tunnel, we have observed four major types of inclusions [11]:

1. all-liquid;
2. gas-liquid with low vapour-to-liquid ratio (*ca.* 0.1 and less);
3. gas-rich with variable, but generally high, gas-to-liquid ratio (*ca.*0.4–0.8);
4. all-gas (may contain some water, though it is not always distinguishable optically).

All-liquid and gas-liquid inclusions are similar morphologically; they often occur in clusters, and represent, most probably, the same fluid. Normally, only one out of 150–200 inclusions contains a bubble. Most homogenisation temperatures measured in these inclusions fall in the range 35 to 55 °C (*Figure 2*). Such extremely low and consistent homogenisation temperatures provide strong evidence against any stretching or leakage. The new data are in good agreement with the palæo-geothermal gradient inferred from earlier fluid inclusion results ([2], *Figure 3*).

All but three of the inclusions subjected to the freezing experiments (39) contained diluted water. It should be noted that the term "diluted water", as used in fluid inclusion studies, describes water with a concentration of less than *approximately* 0.5–1.0 wt.% (at such low concentrations the freezing method rarely yields reliable data due to metastability problems). These concentrations, however, lie well within the 'hydrogeological'

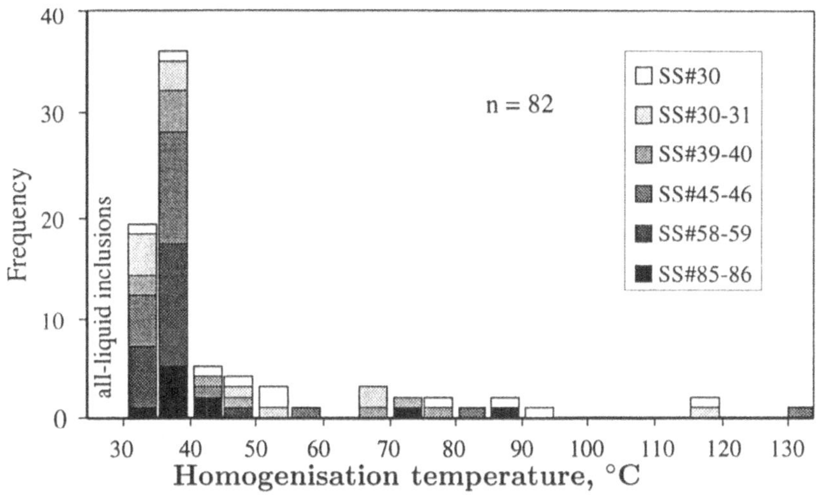

Figure 2. Homogenisation temperatures measured in calcites from exploratory tunnel.

range of salinities (1 wt.% = 10 g l^{-1}). Three all-liquid inclusions from one sample yielded concentration of 11 wt.% of $MgCl_2$.

Gas-rich and all-gas inclusions are rarer. They are larger than all-liquid and gas-liquid inclusions and are restricted to the earliest generations of calcite. On crushing, the bubbles from three gas-rich inclusions decreased in volume by a factor of 30 to 120; hence, they could not have been trapped at atmospheric pressure characteristic of the vadose setting. Semi-quantitative selective absorption volumetric analysis of individual bubbles revealed the following gases: methane + other hydrocarbons (80–84 vol.%); nitrogen+noble gases (16–18 vol.%); carbon dioxide (traces). Oxygen was not detected. Bubbles in four all-gas inclusions decreased by factors of 20 to 27 on crushing. They revealed similar gas composition, except three of them contained more carbon dioxide (10–15 vol.%). A bubble from one all-gas inclusion showed a slight increase (factor 1.8) on crushing while maintaining gas chemistry typical of the gas-rich inclusions (methane + hydrocarbons + nitrogen). The presence of methane (1–5 ppm) and trace amounts of heavy and light hydrocarbons was confirmed by bulk gas chromatographic analysis.

The data on gas-rich and all-gas inclusions may be interpreted as follows. These inclusions were trapped from a fluid carrying both dissolved and gaseous methane. An inclusion trapped as one-phase liquid followed the isochore and, once it was cooled enough, the 'normal' shrinkage bubble appeared (since the fluid sealed in the inclusion contained dissolved

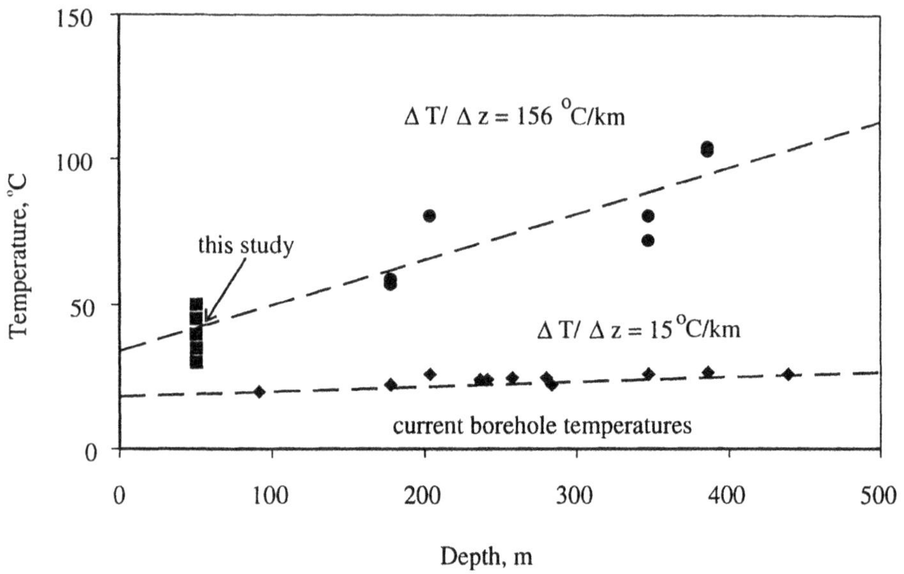

Figure 3. Temperatures of mineral deposition by fluid inclusions (data from [10] and this study), current temperatures measured in boreholes (data from [20]), and implied geothermal gradients.

methane, the latter appeared in a shrinkage bubble). At room temperature and pressure, the internal pressure in such an inclusion will be less than 1 atm. (10^5 Pa); hence, its bubble will decrease on crushing. An inclusion trapped as two-phase (liquid + gas) fluid will maintain a pressure corresponding to the entrapment pressure. An increase in bubble volume on crushing indicates that the latter was higher than 1 atm.

3.3. FORMATION ENVIRONMENT EVIDENCED BY FLUID INCLUSIONS

In earlier studies, the gas-liquid inclusions in the Yucca Mountain calcite have been interpreted as being indicative of depositional temperatures *"comparable to modern ambient temperatures"*, and large bubbles, occupying 40–80% of vacuoles, have been interpreted as the *"air-water-carbon dioxide phase"* trapped during the growth of calcite from films of percolating rainwater in the vadose zone [9]. In contrast, the results presented above imply the formation of calcite and other minerals of the assemblage at temperatures higher that those expected in the vadose zone, *i.e.* from epithermal (low-temperature hydrothermal) fluids. The presence of different types of inclusions (all-liquid+gas-liquid and gas-rich+all-gas) indicates that there has been more than one stage of hydrothermal activity.

4. Micritic Opal-Calcite Veins and Slope Deposits

4.1. OCCURRENCE AND TEXTURES

Opal-calcite veins are associated with faults, bordering the tectonic block of Yucca Mountain and adjacent tectonic depression of Crater Flat (*Figure 1*). The veins are usually wedge-shaped: their thickness at the topographic surface is 1.0–2.5 m; while at a depth of 1–3 m, they narrow to several tens of cm. The veins are composed of micritic calcite (25–55%) and opal. The content of detrital material ranges from 0 to 15%. One of the studied veins (Busted Butte) is set in a non-lithified aeolean sand ramp. Its position corresponds precisely to the Paintbrush Fault beneath the ramp. The vein is composed of numerous thin veinlets (few mm to few cm wide) and grades into the slope-parallel deposit at the ramp surface.

Slope carbonate deposits are the horizons of soil, colluvium, and alluvium cemented by calcite (35–65%) and, to a lesser extent, by opal. Slope-parallel bodies extend from faults downdip; some of them can be traced over a distance of several km. Their thickness does not normally exceed 1.0–1.5 m. The petrographic textures of the slope deposits change in a regular manner with increasing distance away from faults. The M-texture ('matrix'; calcite and opal that contain little or no detrital material) is characteristic of the wedge-shaped veins and slope deposits immediately at the veins (suspected spring orifices). The F-texture ("floating"; detrital clasts "float" in the calcite-opal matrix, the fragments do not touch) typically occurs within the first 400–800 m from orifices. In more distal lithofacies, the F-textured deposits grade into the GS-textured ones ("grain-supported"; detrital fragments are in contact, cement fills the space between clasts). Such textural patterns imply a decrease in intensity of mineral deposition with increasing distance away from the faults.

4.2. STABLE ISOTOPIC PATTERNS

Our stable isotopic profiling has demonstrated that in the calcite vein at Busted Butte, $\delta^{18}O$ increases from depth toward the Earth's surface ($\Delta\delta^{18}O/\Delta z$ = -1.97 ‰ per 100 m). This is expected for calcite deposited from ascending thermal waters. In slope deposits, both $\delta^{18}O$ and $\delta^{13}C$ increase with increasing distance downdip away from veins (*Figure 4*). Similar patterns have been recorded in four profiles: at Bare Mountain, site WT-7, Busted Butte, and Stagecoach Road over distances of 300 to 3,000 m ($\Delta\delta^{18}O/\Delta x$ = 0.03–0.90 ‰ per 100 m; $\Delta\delta^{13}C/\Delta x$ = 0.06–0.82 ‰ per 100 m). We interpret these lateral gradients as a consequence of isotopic fractionation caused by: (a) cooling of fluids; (b) CO_2-degassing; and (c) evaporation of mineral-forming fluids [12].

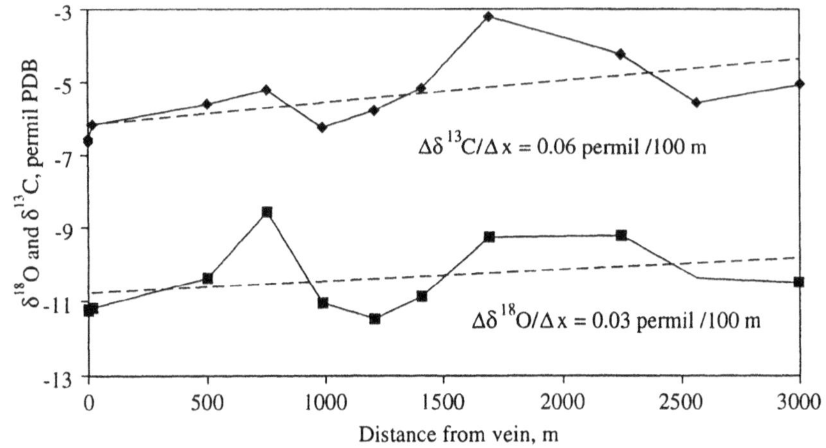

Figure 4. An example of a stable isotopic profile at Bare Mountain. Dashed lines show the linear fit to the data.

4.3. ORIGIN OF VEIN CALCITE AND SLOPE CARBONATES

According to the officially accepted **pedogenic concept** [1, 8], the Yucca Mountain carbonate deposits were formed as a result of "soil-forming" processes. This concept claims that the calcium carbonate was brought to the surface of the mountain by wind (as dust) and, in part, by rainwaters. The latter, percolating through soil, dissolved carbonate dust and carried it down. On evaporation, the water precipitated calcite and formed the slope-parallel carbonatized layers, calcretes. Micritic veins were deposited in places where the waters penetrated into fissures. The water, which infiltrated along fissures down to a significant depth (hundreds of metres) deposited crystalline calcite (veinlets).

An alternative **hypogene concept** [13, 2] considers carbonate deposits at Yucca Mountain as a result of the low-temperature hydrothermal (epithermal) activity. Hot waters upwelled along faults and transported calcium carbonate, dissolved from the Palæozoic limestones underlying Tertiary tuffs. At depths in the range 30–400 m, where hydro- and thermodynamic conditions were quiet, crystalline calcite was formed. Near the surface, in a more unstable environment, calcite was deposited as micritic masses filling fault gouges. Slope-parallel deposits were formed by the waters that discharged on the surface and flowed downslope.

Systematic changes of textures and stable isotopic properties of calcite with distance away from suspected palæo-orifices cannot be explained within the pedogenic concept. The only large-scale isotopic trend known for pedogenic carbonates is the change of both δ^{13}C and δ^{18}O with altitude [14]; no lateral isotopic trends have been reported to date. At the same time, isotopic trends analogous to those determined at Yucca Mountain (*i.e.* increase of $\delta^{13}C$ and $\delta^{18}O$ with distance away from orifices) were reported for modern hot-spring travertine deposits in western U.S.A. and elsewhere [15]. Importantly, the ambient-temperature springs do not show such a pattern.

The inference about the hydrothermal origin of micritic veins at Yucca Mountain is reinforced by the finding of elevated concentrations (up to 7 ppm) of epigenetic hydrothermal accessory zircon (vein at Solitario Canyon; [6]).

5. The Age of Hydrothermal Deposits at Yucca Mountain

The data discussed above give strong and unequivocal evidence of hydrothermal activity at Yucca Mountain in the past. In the context of concerns over the site's suitability as a repository, it is important to establish the age of this hydrothermal activity.

Mosaic breccias at Yucca Mountain are normally truncated by surface erosion and do not provide geologic evidence of a young age. An attempt to apply the fission-track method to date accessory zircon from two breccia bodies has been done by Levy and Naeser [7]. Part of the zircon crystals yielded ages similar to those of the bedrock tuffs (*ca.* 13.0–13.5 Ma). These crystals might be interpreted as relict-detrital zircons, whose old ages were annealed by heating when the crystals were incorporated into the ascending acidic magma. Zircons with ages younger than the bedrock (5–8 Ma) could be epigenetic hydrothermal, formed during later episodes of hydrothermal activity. The presence of individual grains with ages much older than the age of the bedrock tuffs suggests detrital contamination at temperatures lower than temperatures of fission track annealing (*ca.* 250–300 °C for zircon). In turn, this is indicative of a shallow (near-surface) setting during the formation of breccias.

Field relationships imply that the micritic opal-calcite veins and associated slope carbonates at Yucca Mountain are geologically young. Many occurrences of slope deposits are conformable with the tectonic and erosional relief of the mountain. At Busted Butte, both calcite vein and slope deposits are developed in a non-solidified aeolean sand ramp. A summary of the U-series ages obtained from carbonate deposits in the Yucca Mountain vicinity is given in *Figure 5.* It is apparent from this figure that the hydrothermal system was recurrently active during the last 450 ka (conventional limit for the U-series method), and probably more.

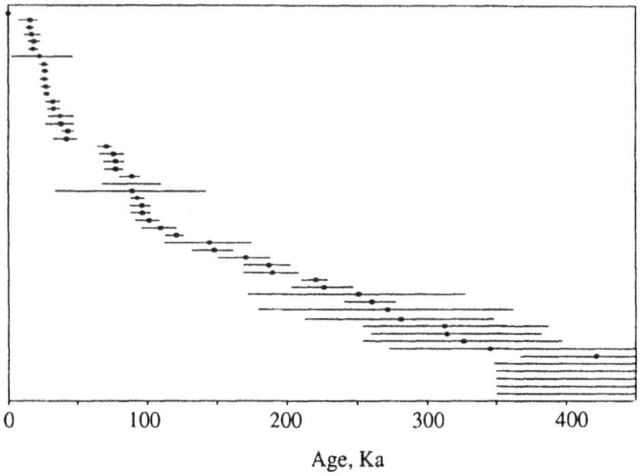

Figure 5: Age of calcite veinlets, micritic veins, slope deposits and spring deposits in the Yucca Mountain area (U-series; data from [21]); dot = determined age, line = instrumental error.

6. Discussion and Conclusions

The AMC breccias, subsurface veinlets, micritic opal-calcite veins and associated slope deposits at Yucca Mountain, reveal clear indications of a hydrothermal origin. At a number of sites, there is an apparent legacy of episodes of hydrothermal activity. For instance, at Trench 14, Busted Butte, site WT-7, and Solitario Canyon, hydrothermal cementation of older AMC breccias was followed by deposition of younger calcite-opal veins and slope calcretes, also from hydrothermal fluids.

Petrographic textures of subsurface quartz-calcite veinlets reveal up to 4 depositional episodes within individual specimens. Fluid inclusion studies determined at least two contrasting types of fluids taking part in calcite deposition. Field and petrographic relations, together with the U-series ages, reveal up to 6 episodes of deposition for different micritic veins and slope-parallel deposits. These results imply recurrent hydrothermal activity within a time span ranging from 8–10 Ma to as recently as 25 ka, and probably less. During pulses of this activity, waters carrying calcite dissolved from Palæozoic carbonates of the basement, upwelled along faults and discharged to the topographic surface. The time series of hydrothermal episodes (*Figure 5*) suggests that more pulses of hydrothermal activity may be expected within the future 10,000-year period covered by safety regulations.

The possible inundation of the repository zone, which upon filling would host up to 70,000 tonnes HLW, would be extremely hazardous. It may not only cause the migration of radionuclides and their penetration into the accessible environment, but also may lead to the underground autocatalytic criticality as suggested by Bowman and Venneri [16]. The latter concept has been criticised and discarded as implausible by several teams of reviewers [17,18]. The critique was based on the claim that the scenario of underground super-criticality is highly unlikely *given the geology, hydrogeology, and geochemistry of the Yucca Mountain site* [19]. However, since the whole concept of the geological and hydrogeological history and, in turn, the geological and hydrogeological future of Yucca Mountain, must be reconsidered in light of the new data presented, autocatalytic criticality may become an issue of paramount concern for the assessment of the site's suitability.

The new evidence for recurrent invasion of Yucca Mountain by hydrothermal fluids in the geologically recent past must be taken into consideration when assessing the suitability of the mountain as a site for a HLW repository.

Acknowledgments

The research was carried out by a joint team of the Technology and Resource Assessment Corporation, Nevada (TRAC-NA) and the Institute of Mineralogy and Petrography (IMP) of the Russian Academy of Sciences (RAS) under a contract agreement with the State of Nevada No. 94/96.003. Petrographic studies were performed in the Laboratory of Magmatic Formations, Institute of Geology, Siberian Branch (SB) of RAS, and fluid inclusion studies were carried out in the Laboratory of Fluid Inclusions, Institute of Mineralogy and Petrography, SB RAS, Novosibirsk, Russia. Accessory minerals were studied in the Laboratory of Accessory Minerals, Odessa State University, Odessa, Ukraine. Isotopic analyses were performed at McMaster University, Hamilton, Ontario, Canada (Laboratory of Prof. H. Schwarcz) and at the University of Arizona in Tucson, U.S.A. (Laboratory of Prof. A. Long).

References

1. NAS/NRC (1992) *Ground Water at Yucca Mountain: How High Can It Rise?* National Research Council, National Academy Press, Washington D.C., 231 pp.
2. Hill, C., Dublyansky, Y., Harmon, R., and Schluter, C. (1995) Overview of calcite/opal deposits at or near the proposed high-level nuclear waste site, Yucca Mountain, Nevada: pedogenic, hypogene, or both? *Environmental Geology* **26** (1), 69-88.
3. Dublyansky, Y. and Lapin, B. (1995) Bedrock tuffs, mosaic breccias, and young volcanic rocks at Yucca Mountain: field observations, petrography, and chemistry, *Technology and Resource Assessment Corporation Report* to the Nuclear Waste Project Office, State of Nevada, Carson City, 175 pp.
4. Sommerville, M.R., J.S., Frazier, G.A., and Archambeau, C.B. (1992) *Yucca Mountain Hydrothermal History*, Technology and Resource Assessment Corporation, Boulder, Colorado.
5. Livingston, D.E. (1993) A review of the major element geochemistry of Yucca Mountain, Nye County, Nevada, *Technology and Resource Assessment Corporation Report* to the Nuclear Waste Project Office, State of Nevada, Carson City, 48 pp.

6. CHEPIZHKO, A. AND DUBLYANSKY, Y. (1995) Accessory minerals from bedrock tuffs, mosaic breccias, and calcite/opal veins at Yucca Mountain (Feasibility Study), *Technology and Resource Assessment Corporation Report* to the Nuclear Waste Project Office, State of Nevada, Carson City, 139 pp.
7. LEVY, S.S. AND NAESER, C.W. (1991) Bedrock breccias along fault zone near Yucca Mountain, Nevada, Los Alamos National Laboratory Draft Report LA-UR-91, LANL, New Mexico.
8. Department of Energy (1993) Report on the origin of calcite-silica deposits at Trench 14 and Busted Butte and methodologies used to determine their origin, U.S. DOE, *Yucca Mountain Site Characterization Project Report YMP193-11-R*, 64 pp.
9. ROEDDER, E., WHELAN, J.F., AND VANIMAN, D.T. (1994) Fluid inclusion studies of calcite veins from Yucca Mountain, Nevada; tuffs: environment of formation, *Proc. 5th. International High Level Radioactive Waste Management Conf.*, Las Vegas, Nevada, pp. 1854-1860.
10. YMP (1993) Data released by the Yucca Mountain characterization Project Office, Carson City, Nevada.
11. DUBLYANSKY, Y., REUTSKY, V., AND SHUGUROVA, N. (1996) Fluid inclusion in calcite from the Yucca Mountain exploratory tunnel, *Proc. PACROFI-VI Conf.*, Madison, WI, U.S.A., May-June 1996, pp. 38-39.
12. DUBLYANSKY, Y. AND SZYMANSKI, J. (1996) Carbonate deposits at Yucca Mountain (Nevada, U.S.A.) and the problem of high-level nuclear waste disposal, *Chemistry for Sustainable Development* **4**, 149-161.
13. SZYMANSKI, J.S. (1989) Conceptual consideration of the Yucca Mountain groundwater system with special emphasis on the adequacy of this system to accommodate a high-level nuclear waste repository, U.S. DOE Internal Report, US Department of Energy, Las Vegas, Nevada.
14. QUADE, J., CERLING, T.E., AND BOWMAN, J.R. (1989) Systematic variations in the carbon and oxygen isotopic composition of pedogenic carbonates along elevation transects in the Southern Great Basin, *U.S. Geol. Soc. Am. Bull.* **101**, 464-475.
15. CHAFETZ, H.S. AND LAWRENCE, J.R. (1994) Stable isotopic variability within modern travertines, *Géographie physique et Quaternaire* **48** (3), 257-273.
16. BOWMAN, C.D. AND VENNERI, F. (1994) Underground autocatalytic criticality from plutonium and other fissile material, *Los Alamos National Laboratory Report* LA-UR-94-4022, LANL, New Mexico.
17. SANCHEZ, R., MYERS, W., HAYES, D., KIMPLAND, R., JAEGERS, P., PATERNOSTER, R., ROJAS, S., ANERDSON, R., AND STRATTON, W. (1995) Criticality characteristics of mixtures of plutonium, silicon dioxide, Nevada tuff, and water, *Los Alamos National Laboratory Report* LA-UR-95-21130, LANL, New Mexico.
18. KASTENBERG, W.E., PETERSON, P.F., AHN, J., BURCH, J., CASHER, G., CHAMBRÉ, P., GREENSPAN, E., OLANDER, D.R., VUJIC, J., BESSINGER, B., COOK, N.G.W., DOYLE, F.M., AND HILBERT, B. (1996) Considerations of autocatalytic criticality of fissile materials in geologic repositories, Research Preprint, Dept. of Nuclear Engineering, University of California, Berkeley
19. CHOI, J.-S. (1997) Underground autocatalytic-criticality potential and its implications to weapons fissile-material disposition, in M.J. Stenhouse and V.I. Kirko (eds.), *Defence Nuclear Waste Disposal in Russia*, Proc. NATO Advanced Research Workshop, Krasnoyarsk, June 24-26, 1996, Kluwer Academic Publishers, Dordrecht (this volume).
20. SASS, J.H., LACHENBRUCH, A.H., DUDLEY, W.W., PRIEST, S.S. AND MUNROE, R.J. (1987) Temperature, thermal conductivity, and heat flow near Yucca Mountain, Nevada: some tectonic and hydrologic implications, *U.S. Geol. Soc. Open-File Report* 87-649, U.S.G.S., Denver, Colorado.
21. HILL, C.A. AND SCHLUTER, C.M. (1994) *Data Chart, Yucca Mountain and Vicinity: Updated June 1994*, Technology and Resource Assessment Corporation Report, Boulder, Colorado, 60 pp.

CHEMICAL ASPECTS ASSOCIATED WITH THE ASSESSMENT OF RADIONUCLIDE MIGRATION THROUGH NATURAL GEOCHEMICAL BARRIERS

G.R.KOLONIN
United Institute of Geology, Geophysics and Mineralogy
Siberian Branch of Russian Academy of Sciences
3, Universitetsky Ave, Novosibirsk, 630090 Russia

Abstract.
The migration and subsequent release of radionuclides from a repository to the accessible environment is an important part of the assessment of natural barrier systems. The speciation of key radionuclides in natural waters as well as the solubility limits for the principal solid phases are prime problems for physico-chemical modelling. A simple and effective method for calculating equilibrium actinide concentrations at the interface between solid phases and solution was achieved using the dissolution of PuO_2 as an example. Complex formation processes were taken into account using the Froneaus function, which reflects the extent of complexing and depends on ligand concentrations. Possible increases in Pu concentrations of 6 to 8 orders of magnitude were predicted for groundwaters of Böttstein granites and Hanford basalts, respectively, owing to plutonium-fluoride complexing. The increasing influence of oxalate complexes on the dissolution of Pu-containing solid phases can be expected at oxalate concentrations $> 10^6$ M.

1. Introduction

In accordance with both the safety principles and criteria of IAEA [1] and the scientific basis of natural barrier system (NBS) [2,3,4], geochemical barriers are a very important part of the general concept of the geological disposal of radioactive waste and nuclear waste management. In particular, information about the maximum cumulative releases of radionuclides from a repository to the accessible environment is not only desirable but necessary in some countries [5,6].

Two decades ago, model calculations of five hypothetical hydrogeolog-

M. J. Stenhouse and V. I. Kirko (eds.),
Defence Nuclear Waste Disposal in Russia: International Perspective, 293–302.

ical situations were carried out by de Marsily *et al.* [7] yielding the main conclusion that retardation by chemical reactions was most effective in slowing radionuclide migration. Four important components of such migration and associated processes were also defined by this author in another paper, cited by Patera *et al.* [8], *viz.*

1. the rate of release of radionuclides, depending on their initial speciation and on the solubility limits of the complexes formed;
2. the pathways to the accessible environment;
3. the driving forces along the pathways (groundwater flow, molecular diffusion, thermal diffusion, osmosis, *etc.*);
4. the reactions occurring along the pathways, including changes in partitioning between solid and aqueous phases (resulting from changes in chemical composition, temperature, pressure, *etc.*).

Many of the above components of a complex physico-chemical model of radionuclide migration have been discussed by Patera *et al.* [8] in the context of the Yucca Mountain project. In general, sufficient progress has been made in understanding and modelling the dynamic and kinetic aspects of mass transfer across possible geochemical barriers [4,9]. On the other hand, the chemical details of migration processes are far from being understood at a quantitative level. As a contribution towards a solution to this problem, this paper briefly reviews the most probable aqueous radionuclide species and the main types of solubility reactions for PuO_2 as solid phase.

2. Scope

In the cited paper [8] there is a useful discussion about the theoretical and experimental approaches towards investigating radionuclide migration in connection with the proposed high-level radioactive waste (HLW) repository in the Nevada tuffs as host-rock (Yucca Mountain Project). This discussion is worth extending here given that other hard rock environments are being considered for disposal sites in Canada, Scandinavia and other European countries (France, Spain, Switzerland, Russia *etc.*) [10].

For this reason, data are presented in *Table 1*, extracted from [4], which summarise the properties of various groundwaters from different types of rocks in terms of temperature, pH, their anionic component and total mineralisation. These data suggest that the reactions of nuclides with carbonate, sulphate, chloride and fluoride components are expected for the majority of the geochemical systems under discussion. However, the hydrolysis of actinides as well as the possibility of organic complex formation should also be given attention. In this context, oxalic acid and oxalates are convenient chelating agents for modeling the possible participation of a variety of carboxylate complexes during radionuclide migration. Plutonium dioxide

TABLE 1. Anionic component (mg l^{-1}) and other properties of groundwaters in fractured rock, tuff and clay utilised in radioactive waste performance assessment studies (from [4]).

Rock type Location Country	**Granite** Finnsjön Sweden	**Granite** Böttstein Switzerland	**Basalt** Hanford U.S.A	**Tuff** Yucca Mnt. U.S.A.	**Clay** Boom Clay Belgium
Temp. (°C)	25	55	60	25	25
pH	6.9	6.8	9.4	7.6	9.2
HCO_3^-	220	117	46	125	250
Cl^-	61	6620	148	6.9	5.1
SO_4^{2-}	8.9	1560	108	18.7	2.0
F^-	0.6	3.6	37	2.2	3.0
Total mineralis.	420	13274	712	275	390

(PuO_2), an artificial analogue of natural uraninite (UO_2), is selected as the model for the immobilised form (solid phase) of actinides.

3. Radionuclide speciation in aquatic systems

Siborg and Katz [11] first compiled comprehensive information about the fundamental solution chemistry of actinides in a series of volumes. Some 25 years later, the chemistry of plutonium was updated in [12]. Both hydrolysis and complex-forming processes are the essence of radionuclide behaviour in aqueous systems. The co-existence of variable valency states is also characteristic of these elements.

According to [11], the following important trends have been observed:

1. The tendency to undergo hydrolysis should follow the trend: $Pu^{4+} > Np^{4+} > U^{4+} > Th^{4+}$.
2. The complexing properties of Pu(IV) with anions shows the following trend with monovalent anions:
   $F^- > NO_3^- > Cl^- > ClO_4$
   and with bivalent anions:
   $CO_3^{2-} > SO_3^{2-} > C_2O_4^{2-} > SO_4^{2-}$
   ("the weaker the acid...the stronger the complexing properties of the anion").
3. Disproportionation mechanisms, *e.g.*
   $3Pu(IV) = 2Pu(III) + Pu(VI)$ and
   $2Pu(V) = Pu(IV) + Pu(VI)$

are possible in solution and the participation of the redox couples Fe(II)/Fe(III) or Ce(III)/Ce(IV) should be taken into account.

In general, actinides should strongly complex with "hard" (ionic) oxygen-donor ligands, thereby extending significantly the solubility limits of actinide-immobilising mineral forms.

The well-known similarity in chemical properties of actinides and lanthanides can help considerably in assessing the performance of possible solid phases, owing to the availability of more complete sets of stability constants for rare-earth elements (REE), as discussed by Wood [13,14]. Such data suggest that carbonate complexes are most likely to form from the interaction of groundwaters with nuclear waste nuclides.

Fluoride complexes are also of great importance, especially in comparison with chloride complexes, since F^- is a "harder" ligand than Cl as well as the other halogen anions and sulphur-containing ligands. Phosphate species could be of potential interest with regard to the importance of solid phases, *e.g.* monacite – (Ce,La,Nd)PO_4, xenotime – YPO_4 and, by analogy, phosphates of plutonium and other transuranic elements. However, such compounds are outside the scope of this paper.

The initial compilation of the vast quantity of information concerning the stability constants of inorganic and organic complexes of actinides and lanthanides was subsequently revised and extended by Smith and Martell in their six-volume series [15]. Recently, a special volume was created under IAEA sponsorship, which contains a compilation and discussion of the chemical thermodynamics of aqueous inorganic complexes of actinides [16]. A rather brief but concise theoretical analysis of the simplest soluble forms of actinides was presented by Faure [17], who provided calculated data [18] combined with valuable geochemical information.

According to the conclusions of many researchers, including Patera *et al.* [8], it is futile to expect that such simple aqueous actinide species will exist under natural conditions, owing to the complexity of real geochemical processes governing migration and water-rock interactions, which involve coupling of reactions involving hydrolysis and complex formation (including multiple-ligand, polymerised and poly-nuclear forms) against a background of redox equilibria involving the important actinides.

4. Estimation of radionuclide migration

The simplified chemical scheme in this section permits estimation of the ability of soluble actinides to migrate depending on their equilibrium concentrations at the interface between solid and liquid phases. Such an approach is reasonable not only because values of concentration or equivalent parameter feature in kinetic and mass transfer equations, *e.g.* see [19], but also owing to the relative ease of calculating equilibrium concentrations as theoretical maxima.

In such cases, it is conventional to subdivide the dissolution process into two consecutive stages:

1. Dissolution into simple ionic forms, such as UO_2^{4+}(aq), Pu^{4+}(aq) or $Pu(OH)_5$(aq) below.
2. Complex formation (include hydrolysis, if necessary).

For example, uraninite dissolution in a multi-ligand aqueous system under low redox (reducing) conditions with respect to U(IV) stability, may be expressed:

$$UO_2 + 4H^+ = U_{aq}^{4+} + 2H_2O \tag{1}$$

$$U_{aq}^{4+} + nH_2O + mCl^- + pF^- = U(OH)_n^{(4-n)+} + UCl_m^{(4-m)+} + UF_p^{(4-p)+} \tag{2}$$

The following conventional subdivision of reaction 2 into three or more simple reactions is possible. Subsequently, manual calculations are sufficient to obtain some auxiliary function f, representing the sum of simpler so-called Froneaus functions [20] which reflect the extent of complexing with each ligand separately, *viz.*

$$f = [U_{tot}]/[U_{aq}^{4+}] = 1 + \Sigma\beta_n^{OH}[OH^-]^n + \Sigma\beta_m^{Cl}[Cl^-]^m + \Sigma\beta_p^F[F^-]^p \tag{3}$$

where $[U_{tot}]$ and $[U_{aq}^{4+}]$ are the total and purely ionic uranium concentrations, respectively, and coefficients $\beta_{n,m,p}$ signify the sets of stability constants of uranium complexes with ligands identified by the upper indexes. Such "visualisation" of dissolution reactions which are naturally complicated, is considered rather useful for understanding water-rock interactions and migration processes and as a means of obtaining rough estimates of equilibrium actinide concentrations quickly. Numerous routine computer codes exist for minimising the free energies of physico-chemical systems for a more exhaustive modelling.

The initial thermodynamic constants for plutonium compounds are rather limited. However, the use of information from [17] (appendix) allowed calculation of the equilibrium constant for the basic dissolution of PuO_2 at 25 °C under intermediate redox conditions, in which Pu(IV) is stable:

$$PuO_2 + 4H^+ = Pu_{aq}^{4+} + 2H_2O \quad logK_4 = 7.4 \tag{4}$$

It is then possible to modify this basic plutonium hydrolysis reaction for alkaline solution by calculating an equilibrium constant for the following reaction:

$$Pu_{aq}^{4+} + 5H_2O = Pu(OH)_5 + 5H^+ \quad logK_5 = 15.0 \tag{5}$$

TABLE 2. Results of calculations of theoretical concentrations of plutonium in aqueous systems in equilibrium with PuO_2 at 25 °C: intermediate redox conditions and in the absence of complex forming processes.

pH	0	1	2	3	4	6	8	10	12	14
log [Pu^{4+}_{aq}]	-7.4	-11.4	– -15.4	-19.4	-23.4			insignificant		
log [$Pu(OH)_5$]	-22.4	-21.4	– -20.4	-19.4	-18.4	-16.4	-14.4	-12.4	-10.4	-8.4

Simple addition of equations (4) and (5) and their equilibrium constants yields the equilibrium constant for reaction (6), which is more appropriate for acidic solutions:

$$PuO_2 + 3H_2O = Pu(OH)_5 + H^+ \quad logK_6 = 22.4 \qquad (6)$$

It is now easy to estimate the probable concentrations of plutonium over a wide pH range as a result of the equilibrium dissolution of PuO_2. The resultant data (*Table 2*) present theoretical concentrations of dissolved plutonium as a function of pH.

Concentrations lower than 10^{-24} M (*i.e.* Pu^{4+}_{aq} concentrations at pH>4) are deemed insignificant and do not appear in *Table 2*. Other data, namely for the pH intervals 0–4 for reaction (4) and 4–14 for reaction (6), can be considered as generating a basic curve of theoretical maximum possible plutonium concentrations in the absence of the more soluble solid phase $Pu(OH)_4$, as well as of complex-forming processes. It should be appreciated, however, that in natural solutions at pH>3, plutonium hydroxo complexes are more important than the simple aqueous ion Pu^{4+}. According to reaction (5), both soluble plutonium species should exist in equal concentrations at pH=3. On the other hand, only very acidic or alkaline solutions (see *Table 2*) could result in potentially hazardous plutonium concentrations if they are in equilibrium with the environment, especially if no complexing ligands are present.

Tentative estimates have been determined of the influence of important ligands in increasing the solubility limits for Pu-containing solid phases. The first example illustrates the possible influence of fluoride on complexes of Pu(IV) owing to the presence of fluoride which can be appreciable in groundwaters of Böttstein granites as well as in groundwaters of Hanford basalts (see *Table 1*). Calculated values of the auxiliary function, f, of the type characterised by reaction (3), were obtained at a total fluoride concentration (F_{tot} = [HF°] + [HF^-]) of $1.9 \cdot 10^{-4}$ mol l^{-1} (M) (3.6 mg l^{-1}) for

TABLE 3. Initial stability constants of plutonium fluoride and oxalate complexes, used in calculations.

Reaction	T °C	μ medium	log β_1	log β_2	Ref.
$Pu^{4+}+nHF=PuF_n^{(4-n)+}+nH^+$	25	$4(H,Na)NO_3$	4.47*	7.46*	[16]
$Pu^{4+}+nC_2O_4^{2-}=Pu[C_2O_4]_n^{(4-2n)+}$	25	$1(HNO_3)$	8.75	16.9	[22]
	25	$4(H,Na)NO_3$	8.30	14.9	[23]

*: In this paper, recalculated values of log β_1=7.67 and log β_2=13.9 for reaction $Pu^{4+} + nF = PuF_n^{(4-n)+}$, have been used.

TABLE 4. Estimates of the influence of fluoride complexes on the solubility of PuO_2 in acid solutions (values in the last two rows are for $\log[Pu]_{tot}$, M).

pH	0	1	2	3	4
$\log[Pu^{4+}]$	-7.4	-11.4	-15.4	-19.4	-23.4
$\log[F^-]/[F]_{tot}$	-3.2	-2.2	-1.2	-0.2	0
Böttstein granite	-4.2	-7.2	-10.2	-13.2	-17.0
Hanford basalt	-2.2	-5.2	-8.2	-11.2	-15.0

granite and $1.9 \cdot 10^{-3}$ M (36 mg l^{-1}) for basalt. Using the stability constants of fluoride complexes from *Table 3*, log f_1 = 6.4 in the first case and log f_2 = 8.4 in the second case.

An additional correction connected with the association of F^- ion with HF° is necessary for acidic solutions of fluoride. For this reason, the term $\log[F^-]/F_{tot}$ was included in *Table 4*. The results of the final stage of calculation, *i.e.* summation of this fluoride term with each auxiliary function and the initial Pu_{aq}^{4+} concentrations, yield values of $\log[Pu]_{tot}$, shown in the last two rows of *Table 4*. A significant increase in total plutonium concentration in solution is apparent as well as an increase in pH range over which potentially hazardous concentrations occur, especially for the Hanford basalt underground waters.

The second example addresses the situation where organic ligands act as complex-forming agents. Means *et al.* [21] emphasised the role of technological ligands (*e.g.* EDTA, used in decontamination operations) in the migration of radionuclides from the original disposal sites at Oak Ridge National Laboratory (U.S.A). In addition, these authors noted the exis-

TABLE 5. Estimate of the correction to the solubility of PuO_2 in acid solution, owing to oxalate complex formation.

pH	0	1	2	3	4
$\log[Pu^{4+}]$	-7.4	-11.4	-15.4	-19.4	-23.4
$\log[C_2O_4^{2-}]/[H_2C_2O_4]_{tot}$	-5.6	-3.7	-2.4	-1.3	-0.5
$[Pu]_{tot}$ M	-1.9	-4.1	-6.7	-9.6	-12.2

tence of mobile Pu-contaminated leachates in waste pits at the Hanford, West Valley and Maxey Flats facilities, as well as the presence, or suspected presence, of waste-complexing agents at Chalk River, West Valley and Maxey Flats (U.S.A).

Our modeling example concerning the influence of oxalate-ion on the solubility limit of PuO_2 was carried out since oxalates are used in the Purex process [23] and in other plutonium technologies. An additional basis for performing the calculations was the availability of reliable stability constants of Pu oxalate in both the American and Russian literature (*Table 3*). At a total oxalate concentration of 10^{-2} M and using stepwise constants from [23], the extent of complex formation was obtained as $\log f = 10.9$. After calculation of the correction term:

$$log \frac{[C_2O_4^{2-}]}{[H_2C_2O_4]_{tot}} = log \frac{K_1 K_2}{[H^+]^2 + [H^+] \cdot K_1 + K_1 K_2} \tag{7}$$

where $K_1 = 10^{-1.28}$ and $K_2 = 10^{-4.28}$ are the first and second dissociation constants of oxalic acid [24], the required correction for oxalate-ion hydrolysis was calculated and summarised in *Table 5* together with the initial PuO_2 solubilities and using $\log f = 10.9$. It is worth noting that a change in total oxalate concentration by one order of magnitude will increase or decrease values of the function f by 2 logarithmic units over the wide range of oxalate concentrations up to 2.9 at $[H_2C_2O_4]_{tot} = 10^{-6}$ M.

5. Conclusions

Given the widespread use of concentrations of chemical constituents in any kinetic or mass transfer equations involving physico-chemical simulations, a simple theoretical approach can be useful where a rapid, preliminary assessment of possible radionuclide migration and precipitation at geological barriers is required, for example, when discussing the potential of some sites for geological disposal of radioactive waste. Two problems are associated with the estimation of solubility limits of solid actinide phases, *viz.* the participation of complex-forming processes and the relatively complicated composition of groundwaters (with the possible addition of technological ligands).

The well-known "extent of complexing" (Froneaus function) offers a suitable way of solving such problems. Examples of calculations of the possible influence of fluoride complexes of plutonium on the solubility limits of PuO_2 in both Böttstein granite and Hanford basalt groundwaters has been demonstrated. An increase in Pu_{total} concentration in these aqueous systems by up to 8 orders of magnitude is predicted. The situation involving organic complexing ligands was discussed using oxalate complexes as an example, when a similar increase in the solubility limit of PuO2 was obtained. A similar approach can also be adopted to simulate the effects of coupled complex-forming reactions in multi-ligand aqueous systems, but individual Froneaus functions are combined.

Two final comments are worthwhile. The first concerns the somewhat arbitrary nature of all calculations discussed above, since the relatively small influence of ionic strength on solution equilibria was neglected. Secondly, it is important to emphasise the basic dependence of actinide concentrations on which phases are considered as equilibrium solid phases. In the case of plutonium, if PuO_2 is converted to plutonium hydroxide $Pu(OH)_4$, this change of equilibrium solid phase will result in an increase in plutonium concentration in the above examples, as well as in any other environmental geochemical situation, by about 7 orders of magnitude. Of course, imperfections such as the two identified above can be taken into account using more sophisticated computer modelling.

References

1. IAEA (1983) Criteria for Underground Disposal of Solid Radioactive Wastes, *IAEA Safety Series* **No 60**, IAEA, Vienna.
2. APTED, M. (1995) Repository and barrier concept, in D. Savage (ed.), *The Scientific and Regulatory Basis for the Geological Disposal of Radioactive Waste*, John Wiley & Sons, Chichester, New-York, pp. 51-73.
3. APTED, M., GRINDROD, P., AND SAVAGE, D. (1995) The near-field, *ibid.*, pp. 75-118.
4. GRINDROD, P., MCEWEN, T., ROBINSON, P. *et al.* (1995) The far-field, *ibid.*, pp. 119-183.
5. NRC (1987) *U.S. Nuclear Regulatory Commission Code of Federal Regulations*, Title 10, Part 60, p. 627.
6. EPA (1986) *U.S. Environmental Protection Agency Code of Federal Regulations*, Title 40, Part 191, p. 7.
7. DE MARSILY, G., LEDOUX, E., BARBREAN, A., AND MARGAT, J. (1977) Nuclear waste disposal: can the geologists guarantee isolation? *Science* **197**, 519-527.
8. PATERA, E.S., HOBART, D.E., MEIJER, A., AND RUNDBERG, R.S. (1990) Chemical and physical processes of radionuclide migration at Yucca Mountain, Nevada, *J. of Radioanalytical and Nuclear Chem.*, **Articles 142**, 331-347.

9. ROCHELLE, C.A., BATEMAN, K., MILODOWSKY, A.E. *et al.* (1992) Reactions of cement pore fluids with rock: implications for the migration of radionuclides, in Y.K. Kharaka and A.S. Maest (eds.), *Water-Rock Interaction*, Balkema, Rotterdam, pp. 423-426.
10. TSYPLENKOV, V.S. (1997) Geological disposal of high level radioactive waste, in P.A. Sterne, A. Gonis, A.A. Borovoi (eds), *Actinides and the Environment*, Kluwer Academic Publishers, Dordrecht (in press).
11. KATZ, J.J. AND SEABORG. J.T. (1957) *The Chemistry of the Actinide Elements*, Methuen & Co Ltd, London.
12. W.T. CARNALL AND G.R. CHOPPIN (1983) (eds.) Plutonium Chemistry, *ACS Symposium Series* **216**.
13. WOOD, S.A. (1990) The aqueous geochemistry of the rare-earth elements and yttrium, 1. Review of available low-temperature data for inorganic complexes and the inorganic REE speciation of natural waters, *Chemical Geology* **82**, 159-186.
14. WOOD, S.A. (1990) The aqueous geochemistry of the rare-earth elements and yttrium, 2. Theoretical predictions of speciation in hydrothermal solutions to 350 °C at saturation water vapour pressure, *Chemical Geology* **88**, 99-125.
15. SMITH, R.M. AND MARTELL, A.E. (1974-1989) *Critical Stability Constants*, Volumes 1-6, Plenum Press, New-York, London.
16. FUGER, J., KHODAKOVSKY, I.L., SERGEEVA, E.I., *et al.* (1992) The Actinide Aqueous Inorganic Complexes, The Chemical Thermodynamics of Actinide Elements and Compounds, Part 12, IAEA, Vienna.
17. FAURE, G. (1991) *Principles and Applications of Inorganic Chemistry*, MacMillan Publishing Company, New-York.
18. BROOKINS, D.G. (1988) *Eh-pH Diagrams for Geochemistry*, Springer-Verlag, Berlin.
19. SHELL, W.R., LINKOV I., MYTTENAERE C., AND MOREL B. (1996) A dynamic model for evaluating radionuclide distributions in forest from nuclear accidents, *Health Physics* **70** (3), 318-335.
20. HARTLEY, F.R., BURGESS, G., AND ALCOCK, R.M. (1980) *Solution Equilibria*, John Willey & Sons, Chichester, New-York.
21. MEANS, J.L., CRERAR, D.A., AND DUGUID, J.O. (1978) Migration of radioactive wastes: radionuclide mobilization by complexing agents, *Science* **200**, 1477-1481.
22. CHOPPIN, G.R., BOKELUND, H., AND VALKERS, S. (1983) Oxalate complexation in dissolved carbide systems, *Radiochimica Acta* **33**, 229-232.
23. MOSKVIN, A.I. AND GEL'MAN, A.D. (1958) Determination of composition and instability constants of oxalate and carbonate complexes of plutonium(IV), *Russian J. Inorg. Chem.* **3**, 962-974.
24. KETTLER, R.M., PALMER, D.A., AND WESOLOWSKY, D.J. (1991) Dissociation constants of oxalic acid in aqueous sodium chloride media to 175 °C, *J. Solution Chem.* **20**, 905-927.

APPLICATION OF BIOTECHNOLOGY TO THE PROBLEM OF THE DISPOSAL OF RADIOACTIVE WASTES

A.V.BELYI, YU.L.GUREVICH, P.P.PUSTOSHILOV
& M.G.SADOVSKY
Laboratory of the Ecological Biotechnology
Institute of Biophysics of Siberian Division of
Russian Academy of Sciences
Krasnoyarsk, 660036 RUSSIA

Abstract.
Artificial biological systems are proposed for use in the treatment of low- and intermediate-level radioactive wastes. The microbial conversion of organic and inorganic substances under continuous conditions, involving the turnover of matter, will allow a significant reduction in volume of radioactive wastes. Such specialised systems can act as a biogeochemical barrier for radionuclide transport.

1. Introduction

Concerns regarding the long-term safety of nuclear waste involves difficulties associated with the safe storage of both high- and low-activity radioactive wastes. The methods used to treat highly-active wastes differ from those for low-activity wastes. Low-level waste (LLW) has a smaller total activity, although a much greater volume, whereas high-level waste (HLW) has a relatively small volume but significantly greater total activity. This difference in activity does not necessarily mean, however, that the hazard from LLW is less in comparison with that from HLW, or that the corresponding disposal requirements need not be as stringent. Since the large volumes of LLW do not allow the application of rigorous isolation methods similar to those developed for HLW, there may be a possible reduction in the long-term reliability of waste immobilisation and, therefore, an increase in the cost of measures necessary to ensure the safety of these wastes.

Clearly, the reliability of long-term isolation will increase, and the cost correspondingly less, as the volume of wastes for disposal decreases. To ensure the safe disposal of radioactive wastes, wastes are usually emplaced

M. J. Stenhouse and V. I. Kirko (eds.),
Defence Nuclear Waste Disposal in Russia: International Perspective, 303–308.

in geological formations (*e.g.* clay layers, granitic rock *etc.*), to guarantee their long-term isolation. Besides the geological barriers, use is frequently made of geochemical barriers, one concept being based on the use of artificial minerals (*e.g.* zircon *etc.*) which incorporate radionuclides by stable chemical bonding [1-3].

2. Biological methods for radioactive waste treatment

The search for a suitable solution to the problem of safe radioactive waste disposal requires an extension of existing methods and techniques used to concentrate and immobilise radionuclides. In this context, one can develop artificial biological systems which would perform an isolation function for toxic elements, *i.e.* act as a biological barrier. Such an approach is relatively exotic, but has potential from our point of view. It should be noted that the development of natural geochemical barriers often involves biological reactions, thereby providing a form of biogeochemical barrier. The biological components of systems of this type already fulfil such a function, such as the utilisation of biomass and/or biopolymers of various organisms, *e.g.* plants, yeast, fungi, bacteria [4, 5]. One can also treat organic wastes (spirit, oil, packing plastics, filtering materials) via biological degradation (bio-incineration under 'soft conditions') [6].

The above examples involving the biotechnological treatment of radioactive wastes can be extended to include methods which simulate the basic processes of mass transfer and concentration of substances and chemical elements in natural systems. Under such circumstances, the question arises whether it is possible to develop a biotechnological system which can act as a barrier for radionuclide transport (*i.e.* concentrate and immobilise radionuclides). Such bio-systems must be isolated from the environment and achieve a partial turnover of matter.

To address this question, one must consider three essential factors which are specific to the natural processes of matter turnover: (1) matter consumption itself, which results in (2) the concentration and (3) the immobilisation of a number of chemical elements. It should be stressed that the latter two phenomena result from the biological transformation of substances, which are subsequently involved in the geochemical reactions of both organic and inorganic species.

The development of specialised systems which simulate these properties results in the explicit manifestation of the characteristics of biological species which allow a barrier function to be established.

Table 1 shows the data on the concentration, by some biological species, of several radionuclides which are considered to be hazardous and characteristic components of nuclear wastes. It is readily apparent from the data

TABLE 1. Concentration Coefficients* for various elements

Element	Concentration Coefficient	Organic agent
Cs	5	ocean
Sr	20	phytoplankton
Ra	100	– " –
Co	60000	– " –
Zr	170000	– " –
Sr	96	benthos
Nb, Po	1000	– " –
Pu	2600	– " –
Zr	60000	– " –
Ce	90000	– " –
Ru	200000	– " –
Sr	30	bottom
Cs	1000	sediment
Co	3000	– " –
Ce	5000	– " –
Pu, Zr, Nb, Ru	10000	– " –

* equal to the ratio of element concentration in an individual component of the ecosystem to that in solution

that the 'concentration process' can be very efficient (by factors of 10^3 to 10^4).

The results shown in *Table 1* were obtained for natural ecosystems [8-10]. The *concentration coefficient* depends strongly on the individual component of the ecosystem that achieves the concentration effect, and may vary significantly. For example, the ratio of concentration coefficients can be seen to vary from 200 for caesium, to 20 for cobalt, and 6–17 for zirconium. It is important, therefore, to establish an appropriate structure and community composition of the biotechnical system.

We believe that the high bio-accumulation values observed in *Table 1* are determined by the properties of an *open* biological system, through which matter and energy are able to flow continuously. The transformation of substances via cyclic processes has a significant impact on the bio-accumulation and removal of many elements during the turnover of matter. For a continuous system, one could hardly expect to obtain concentration values close to those observed for regular methods of adsorption on a biomass.

It is evident that a bio-system designed to perform a barrier function must simulate some components of natural ecosystems. This problem is quite difficult, but achievable. At least one example exists of an artificial ecosystem which provides the complete turnover of oxygen and water [11].

Let us distinguish some types of biological processes which are relevant in this respect.

1. Mineralisation and deposition of organic matter during microbial destruction: Mineralisation during anaerobic oxidation processes results in the removal of a significant part of matter as carbon dioxide. The mass of organic matter decreases by a factor of approximately 5 after biological oxidation. The anaerobic reduction process yields an even more effective loss in mass - by a factor of up to 10. Heavy metals could be deposited out effectively via such processes.
2. Destruction of organic substances in artificial systems with trophic webs: A simple trophic 'bacteria-protozoa' system enhances the mineralisation of organics. The cyclic conversion of inorganic phosphate into organic and back to inorganic, specifically promotes the coagulation and sedimentation of dissolved and suspended matter.
3. Transformation of inorganic substances in chemosynthetic processes:

   – Chemo-autotrophic microorganisms usually inhabit the areas around geochemical barriers. Both chemo-autotrophic nitrification and denitrification, *viz.*

   $$NH_4 \longrightarrow NO_3 \longrightarrow NO_2 \longrightarrow N_2$$

   allow the elimination from the relevant wastes of bound nitrogen, which is extremely mobile. Consequently, the mobility of accompanying radionuclides is reduced. The occurrence of biosorption in this case is not known.

   – The cyclic transformation of sulphate and sulphide sulphur, *viz.*

   $$SO_4{}^{2-} \longleftrightarrow S^{2-} \longrightarrow S^0$$

   appears to be rather effective for removing heavy metals into sediments.

   – Processes involving the bacterial oxidation of iron are highly effective and productive. Ferric iron obtained from a bacterial fermentor shows an increased potential for flocculation and sorption compared with that which is derived by chemical means [13].

If inorganic substances are used by bacteria as an energy source, tens to hundreds of grams of an oxidising substance are necessary to synthesise one gram of bacterial biomass. Biomass accumulation is much less in comparison with heterotrophic processes.

3. Conclusions

Thus, given relatively moderate funding and careful control of the microbial transformation of organic and inorganic matter, the possibility exists for developing closed systems which enable matter turnover, with the potential for biological species to immobilise and concentrate radionuclides. Hence, such species can provide the basis for developing special biotechnological systems for the treatment of low- and intermediate-level radioactive wastes.

Bio-systems based on ecological principles should be readily adaptable to solving problems concerning the remediation of polluted territories or the creation of protective zones with an enhanced capacity for sorbing radionuclides around nuclear plants, reprocessing plants and disposal sites.

References

1. RINGWOOD, E. A. (1978) *Safe Disposal of High-level Nuclear Reactor Wastes: a New Strategy*, Canberra.
2. BURAKOV, B. E., ANDERSON, E. B., GALKIN, B. YA., STARCHENKO, V. A. AND VASILIEV, V. G. (1996) The crystalline host-phases for immobilization of weapons plutonium and waste actinides, in E.R. Merz and C.E. Walter (eds.), *Disposal of Weapon Plutonium*, pp.85-89 Kluwer Academic Publishers, Dordrecht,.
3. EWING, R. C., WEBER, W. J., AND LUTZE, W. (1996) Crystalline ceramics: waste forms for the disposal of weapons plutonium, in E.R. Merz and C.E. Walter (eds.), *Disposal of Weapon Plutonium*, pp.65-83 Kluwer Academic Publishers, Dordrecht, The Netherlands.
4. ASHLEY, N. V., POPE, N. R. AND ROACH, D. J. W. (1987) Feasibility study of the application of biotechnology to nuclear waste treatment: research reviews and patent abstract, *Biotechnol. Adv.* **7** (1), 118.
5. GUDIN, C. AND THEPEMER, C. (1991) Métaux lourds et radionucléides, *Biofutur.*, **106**, 74-75.
6. HAFEZ, M. B., HAFEZ, N. AND RAMADAN, Y. S. (1992) Uptake of cerium, cobalt and cesium by *Potamogeton crispus*, *J. Chem. Technol. and Biotechnol.* **54** (4), 333-340.
7. MACASKIE, L. E., JEONG, B. C., AND TOLLEY, M. R. (1994) Enzymatically accelerated biomineralization of heavy metals: application to the removal of americium and plutonium from aqueous flows: Paper presented at FEMS Symp. Metals and Microorganisms: Relationships and Appl., Metz, 6-7 May, 1993; *FEMS Microbiol. Rev.* **14** (4), 351-368.
8. TUSA ESKO (1992) IVO's resinating bacteria make light work of waste treatment. *Nucl. Eng. Int.* **37**, (451), 39.

9. ANCELLIN, J., GUEGUENIAT, P. AND GERMAIN, P. (1979) *Radioécologie marine; étude du devenir des radionucl{*eides rejetes en milieu marin et aplication á la radioprotection. Ed. Eyrolles, Paris.
10. BOWEN, H. J. M. (1979) *Environmental Chemistry of the Oceans*, Academic Press, London, New York.
11. *Radioactivity in the Marine Environment* (1971) Report prepared by the Panel on Radioactivity in the Marine Environment of the Committee on Oceanography, National Res. Council, Nat. Acad. Sci., Washington D.C.
12. GITELSON, J. J., KOVROV, B. G., LISOVSKY, G. M. ET AL. (1975) Experimental ecological systems containing man, *Problems of Space Biology.* Moscow, Nauka.
13. GUREVICH, YU. L., LADYGIN, V.P., AND TEREMOVA, M.I. (1975) Degradation of technogenic flows of organic matter by communities of microorganisms and ciliates, *Izv. RAN, Ser. Biol.* **2**, 226-230.
14. BELYI, A. V., GUREVICH, YU. L., GRASYKOVA, N. A., AND KADOCHNIKOVA, G. G. (1997) *Thiobacillus* in the biotechnology of waste water treatment, *Siberian Ecological Journal* (in press).

EVALUATION OF THE SAFE DISPOSAL OF RADIOACTIVE WASTE AND SPENT FUEL IN THE VICINITY OF BASHMACHNAYA BAY ON THE NOVAYA ZEMLYA ARCHIPELAGO

N.N. MELNIKOV, V.P. KONUKIRN, & V.N. KOMLEV
Mining Institute, Kola Science Centre, RAS
24 Fersman str., 184200 Apatity
Murmansk Region, RUSSIA

Abstract.

The problem of disposing of the radioactive waste and spent fuel accumulated in north-west Russia, in the Novaya Zemlya Archipelago, is discussed in this paper. An analysis of risk factors associated with the construction of a pilot facility is given. The investigations carried out are considered to be insufficient in the context of guaranteeing long-term isolation of radioactive waste and spent fuel in permafrost carbonate rocks within Bashmachnaya Bay. Recommendations are proposed concerning the scientific approach to solving this radioactive waste management problem.

1. Radioactive waste and spent fuel in the Murmansk and Arkhangelsk regions

The north-west (NW) region of the Russian Federation, including the Murmansk and Arkhangelsk provinces, is one of the most over-burdened areas in terms of accumulated radioactive waste and the associated radiation hazard. Russia's Northern Navy, having over 80 operating surface ships and submarines equipped with nuclear power units, is based in this area. In addition to these vessels, about 70 decommissioned nuclear submarines, still containing nuclear fuel, are on military marine bases or in shipyards.

The Murmansk Shipping Company has 7 nuclear-powered icebreakers and one nuclear-powered lighthouse vessel together with some special ships for collecting and storing radioactive waste and spent fuel.

M. J. Stenhouse and V. I. Kirko (eds.),
Defence Nuclear Waste Disposal in Russia: International Perspective. 309–315.

The Kola nuclear power plant (NPP), situated on the shore of Lake Imandra, has 4 VVER-440 nuclear reactors.

Over 100 industrial organisations and research institutes of the Murmansk and Arkhangelsk regions use ionizing radiation sources. In addition, dozens of lighthouses with RITEGs (nuclear-powered units) are located along the coast of the Barents Sea.

All of the above-mentioned military and civil organisations using nuclear and hazardous radiation facilities have, over several decades, produced radioactive waste which, apart from waste which has been disposed of in the sea, is currently concentrated in temporary on-land and floating storage facilities.

According to an assessment performed at the beginning of 1996, about 16,000 m^3 of solid and 6,000 m^3 of liquid radioactive waste have been accumulated in the region, as well as 5,100 m^3 of radioactive salt concentrate and about 1,600 m^3 radioactive spent resins. About 1,500 spent ionizing radiation sources are currently stored centrally in temporary *Radon* storage centres, or stored directly at the facilities of various companies, as well as medical and research institutes.

The most difficult and hazardous situation concerns the Northern Navy because the radioactive wastes arising during the operation of nuclear-powered ships had, until 1993, been dumped in the sea and, for this reason, the need for facilities for conditioning and safe intermediate storage was not foreseen. The total activity of all radioactive waste in the region is estimated to be about 2.3 PBq (1 PBq = 10^{15} Bq).

It should also be noted that the recent build-up of radioactive waste is expected to increase dramatically after 2003 as a result of the decommissioning of the Navy's submarines and surface ships (according to the requirements of the international treaty on nuclear weapon decommissioning) as well as the future dismantling of the reactors at Kola NPP.

It is also foreseen that the reprocessing of reactor cores from the icebreakers *Lenin*, *Arktika* and *Sibir*, and the floating maintenance base *Lepse*, will generate large quantities of radioactive waste for subsequent disposal.

According to a preliminary assessment performed by the Kola Science Centre Mining Institute (KSCMI), the predicted volume of solid radioactive waste arisings will be 42,000 m^3 by 2004, and 59,000 m^3 by the year 2010.

Of extreme concern to the NW region is the fate of spent fuel. By the end of 1996, the number of spent fuel assemblies in this area amounted to about 57,000. Although most of these assemblies are planned to be transported to the Mayak reprocessing plant in the Chelyabinsk province, defect spent fuel together with test fuel which has 'consumed' up to 5-10% of its total quantity, are to be disposed of in the region.

A study of the experience gained in tackling similar problems in western European countries, USA and Japan, shows that a decision concerning the management of radioactive waste arisings can be made only on the basis of detailed scientific and experimental investigations and on a comparison of possible alternatives for radioactive waste disposal sites as well as waste isolation technologies. A notable feature in all the attempts to solve the problem of radioactive waste and spent fuel in NW Russia is the fact that practically no comparative analyses have been performed on candidate sites.

Given the need for additional studies concerning potential sites for radioactive waste and spent fuel disposal, the authors consider it appropriate to focus additional scientific investigations on the island of Novaya Zemlya as well as other possible candidate sites.

2. Project Concerning the Isolation of Radioactive Waste and Defect Nuclear Fuel in Permafrost Carbonate Rocks on the Island of Novaya Zemlya (Bashmachnaya Bay)

A feasibility study (VNIPI Concept) for the construction of a pilot facility on Novaya Zemlya Island for the underground isolation of solid radioactive waste and defect fuel assemblies, was carried out by the VNIPI Promtekhnologiya Institute, under contract with the government of the Russian Federation [1].

In 1995, this project was included in the Federal Target-Oriented Programme "Management of Radioactive Waste and Spent Fuel Materials, Their Utilization and Disposal in 1996-2005". In the context of our independent study, the main principles of the VNIPI Concept are discussed below.

The initial postulate of the Concept states that permafrost is a natural geological formation having sub-zero temperatures throughout with all pores and cracks filled with ice formed *in situ*. Frozen rock zones potentially suited for radioactive waste isolation would consist of permafrost having a continuous distribution both laterally and vertically, and with rocks which are themselves stable and cryogenically 'impermeable' [2].

The permafrost rock mass in the south-west part of Yuzhny Island, to the north of Bashmachnaya Bay and 10 km from the Barents Sea, was selected as host rock for the pilot facility for radioactive waste and spent fuel disposal. The land around the site is relatively flat with some hills up to 130 m above sea level. The thickness of permafrost, which consists of limestone, marls and sandstone with layers of clayey shale (Upper Devonian and Coal Periods), is about 300 m. The depth of seasonal thawing is up to 2 m and the annual mean temperature of the rock is -4 °C.

Three types of excavations/disposal concepts are planned for (solid) radioactive waste and spent fuel disposal:

- 15 m deep trenches for radioactive waste with activity in the range Category 1–2 (*i.e.* low- to medium-activity),
- 300 m boreholes with diameter ~ 400 mm - for disposal of Category 2 (medium) activity radioactive waste, and
- up to 300 m deep boreholes of diameter 1000 mm - for Category 3 (high-level) activity radioactive waste (unprocessed spent fuel).

The estimated annual radioactive waste input to the facility is about 2,000 m^3 [3].

Disposal of fuel assemblies and spent fuel in permafrost is particularly problematic. In such cases, the temperature in the repository and in the surrounding rock mass is expected to reach, but not exceed, 100 °C [4]. In addition to the elevated temperature, the distribution of heat is likely to cause localised thawing in areas around highly radioactive materials. Even in these circumstances, however, it is stated (in the Concept) that a safe, cryogenic barrier will exist beyond the localised area of thawing which is expected to last for several decades.

The authors of the Concept consider the following principle to be the basis for the provision of ecological safety:

> *If the frozen host rock is cryogenically impermeable, and the radioactive waste repository is sufficiently deep that the rock is 'stable' from effects of seasonal thawing and frost-activated geological processes, then the only requirement for preventing radionuclide transport beyond the repository zone over a period long enough to isolate the potential hazard from radioactive waste, is that the engineered barrier continues to act efficiently beyond the time taken for the heat-impacted rock to be recover its original (cryogenically impermeable) state [1,4].*

3. Risk factors for the disposal of radioactive waste and spent fuel in the pilot facility

Firstly, it is appropriate to focus on the remarks made by the State Ecological Expertise Commission under the leadership of S.S. Belov, concerning the feasibility study for a pilot facility on the Novaya Zemlya archipelago in 1992 and on the comments of the Mining Institute.

During an analysis of climatic effects, as well as the engineering-geological and geo-cryological conditions for construction, the following factors emerged.

- The area for the proposed disposal consists mainly of carbonate rocks which can contain karst cavities. Fractures and karst cavities in rock masses are not completely filled with ice and can contain air. On thawing, these rocks can also become water permeable, especially those which are highly mineralised.

- Since the potential area for radioactive waste disposal is situated close to the sea, it is important that the degree of connectivity between groundwater and seawater be evaluated. It is possible that the frozen rock of the coastal plain below sea level contains frost cavities with trapped water and, therefore, is not continuously impermeable. Furthermore, if the boreholes containing nuclear fuel penetrate a zone containing highly 'aggressive' (saline) water, cryogenic isolation would not occur.
- There is no guarantee that the thickness of permafrost is 300 m according to the design requirements of the project. Examination of the topography and geological maps shows that the thickness of the cryogenic zone can vary significantly.

With regard to the problem of a possible change in climate in the Novaya Zemlya area, scientific opinion is divided. Experts from the Moscow State University consider that during the design life of the facility, no significant change will occur in either climate or permafrost behaviour. On the other hand, the conclusion of the GTI Department of Climate Research is that the annual temperature within the area may increase by 8–10 °C over the next 60 years owing to an increase in atmospheric CO_2 content. According to Swedish experts, the temperature in the Arctic has increased already by 3 °C over the last 20 years.

It is relatively straightforward to calculate that such a warming effect will degrade the permafrost zone, resulting in an increase in the depth at which the upper horizon appears. Such a downward displacement of the frozen zone would eventually represent a severe hazard in terms of a radioactive waste and spent fuel repository situated close to the surface.

The layer which undergoes seasonal thawing in the northern permafrost area is not expected to exceed 2.0 m. Although frost-rock weathering processes are restricted to this near-surface area, cases are known of rock uplift, frost fracturing and frost lifting during engineering construction under permafrost conditions, down to a depth of 20-30 m, *i.e.* greater than the depth of the seasonal thawing layer. Despite such known occurrences, no long-term evaluation of such processes has been considered for the current disposal situation.

It is also important to estimate the conditions under which permafrost will remain stable during climatic changes in the Arctic over periods covered by short-lived and long-lived radionuclide decay.

4. Conclusions

In addition to considering investigations of the permafrost geological formations in northern Russia with regard to their potential for radioactive waste isolation, we accept, in principle, the need to construct a radioactive waste disposal pilot facility on the Novaya Zemlya Archipelago to carry out research studies in permafrost. Nevertheless, we believe that the decision concerning the disposal of large volumes of radioactive waste and spent fuel on Novaya Zemlya Island (Bashmachnaya Bay) does not yet have a sufficient scientific basis and is, therefore, premature.

Before making a decision on selecting one or more sites for radioactive waste and spent fuel final disposal in the geological formations of NW Russia, a full-scale comparative feasibility analysis should be carried out taking into account all other alternative regions, in particular, the Kola peninsula with its deep crystalline basement.

It should also be noted that to reduce the radiation hazard in the region, it would be more appropriate to consider the disposal of high-level waste and unprocessed spent nuclear fuel at a later date when the Russian economy is stable. Rather, at present the main financial effort should be aimed at constructing safe facilities for the interim storage of hazardous radioactive materials and supplying these facilities with modern equipment as required.

The resultant interim period would allow time to carry out all the necessary theoretical and experimental research required to determine the optimum site(s) for a repository for the safe disposal of radioactive waste and spent fuel in NW Russia, as well as provide a sound scientific basis for all the technical factors associated with such disposal.

References

1. VNIPI (1992) Summary resolution of State Ecological Expertise Commission of the Russian Federation (RF) Ministry of Ecology and Natural Resources on "Pre-feasibility study for construction of pilot facility for underground isolation of low and intermediate level solid radioactive waste and defect spent fuel assemblies in permafrost" under the task of the RF Government N BM-1113-23853 of June 20, 1992, in *Underground Disposal of Radioactive Waste*, Annex 1, pp. 202-210, Russian Academic Publishers, Apatity.
2. LOPATIN, V.V. (1996) Underground isolation of radioactive waste in permafrost rocks of the Novaya Zemlya island, *Bulletin of the Centre for Public Information on Atomic Energy* **6**, 21-25.
3. LOPATIN V.V. *et al.* (1994) Principal technological solutions for underground isolation of radioactive waste at pilot facilities in permafrost, in *Underground Isolation of Radioactive Waste in Permafrost*, pp. 50-56, Minatom, Moscow.
4. KAZAKOV, A.N. *et al.* (1994) Thermo-physical aspects of the underground isolation of heat-emitting radioactive waste in permafrost, in *Underground Isolation of Radioactive Waste in Permafrost*, pp. 17-22. Minatom, Moscow.

5. ENGELMANN, H.J. *et al.* (1996) Radioactive waste management and a conceptual repository design for the Kola peninsula region of North-West Russia, Final Report of DBE and Mining Institute KSC RAS, Peine, Germany.

COMMENTS ON SEISMIC CONSIDERATIONS WITH RESPECT TO DEEP RADIOACTIVE WASTE DISPOSAL

G.G. SEMENOV
Krasnoyarskgeology
Marks Street
660049 Krasnoyarsk
Russia

The available seismic data disprove the concept that tectonic plates are tectonically passive regions. Massive rebound, and anomalously high formation pressures displayed in oil and gas fields, are undoubtedly connected with high tectonic stresses which must exceed lithostatic and hydrostatic stresses by a large factor. Within the plate regions, beyond activation zones (plate boundaries), considerable tectonic stresses associated with seismic zones are known practically all over the continents.

Occasionally, plate earthquakes are close in magnitude to seismically active regions. For instance, the New-Madrid earthquakes in 1811-1812 were of magnitudes 8.2, 7.8, 8.1 and 8.3. As a result of high tectonic stresses, earthquakes of the same magnitude are more destructive, cover larger areas, and require less stress for their initiation in platform regions than in seismically active regions.

An estimate made concerning the tectonic stresses required for the formation of one type of geological phenomena associated with platform regions – metamorphic domes, indicates that a stress of 5–8 kbar (1 bar = 10^5 Pa) is required in the upper parts of the Earth crust. The tectonic stresses of the New-Madrid seismic focal zone are close to this magnitude. Evidence from all continents indicates that compressive stresses exceed stresses associated with lithostatic load by a factor of 3–5 (10–20 locally). The nature of the forces of higher tectonic horizontal stresses and the role of platform tectonics are currently not understood.

There are twenty or more forms of these forces, for which we propose a unified basis. According to the tectonic plate model, there are a number of reasons for considering the Earth as a single geodynamic system, where the main tectonic factor is the asthenosphere.

M. J. Stenhouse and V. I. Kirko (eds.),
Defence Nuclear Waste Disposal in Russia: International Perspective, 317–319.

As a result of the analysis of depth, energy and mass associated with geological processes which occur initially over continents, we propose the following classifications of geodynamic conditions at various scales.

1. On a geoplanetary scale, geodynamic forces are associated with the general lithosphere-asthenosphere structure of the Earth. The depth of the asthenosphere is about 5–6 km under the mid-ocean ridges, and $\sim$ 200-250 km under the continental plates. Rock masses slipping from such a geoplanetary 'hill' create gigantic stress fields all over the Earth. In the central parts of the continents, these stresses are intensified, especially in the upper crust, by the synclinal structures under the continents at the lithosphere-asthenosphere boundary.
   According to our model, the horizontal tectonic stresses in the central parts of the continents may reach 5–10 kbar. The geoplanetary geodynamic state and the associated tectonic stresses are intimately connected for all continent blocks, in contrast to the smaller-scale forces (see below).
2. On a subglobal scale, geodynamic forces are associated with subduction zones and may be subdivided into two types: (a) true subduction and (b) collision. A dramatic example of the latter type involves the collision of the Hindustan and Eurasian continental plates which resulted in the formation of the spectacular, relatively young mountainous belt of the Alpine-Himalayan chain, covering an area up to 10^7 km^2.
3. On a subregional scale, the geodynamic setting depends on intensive, local variations in density of the asthenosphere mantle. An example of this type is the Baykal rift zone, where the underlying force is isostasy.
4. Regionally, the geodynamic state is a function of the Earth's crust under the fold mountains and platform regions.

Thus, given the above discussion, in addition to established methods of characterisation of sites for radioactive waste disposal, we propose additional types of investigation:

- Investigations of tectonic horizontal stresses in nature, with a view to discovering current tectonic stress fields.
- Geodynamic investigations of the lithosphere-asthenosphere level, K boundary, including special investigations of the asthenosphere (as the principal tectonic factor in the system lithosphere-asthenosphere) and the lower part of the 'granitic' crust (as a secondary factor). The purpose of such investigations is to explain the geodynamic nature of current areal tectonic stresses with a view to predicting future stress changes.

- With respect to neotectonic and seismic studies, the M boundary is non-informative and should be excluded from geophysical investigations.

SUMMARY OF WORKSHOP DISCUSSIONS

There were several opportunities for in-depth discussions over the course of the Workshop, both as comments on individual papers and, more generally, as contributions to open discussion periods. A summary of the main topics discussed is provided below.

Radioactive Waste Treatment

Routine solidification methods used in Europe were discussed at the Workshop and, based on the urgent need to stabilise existing wastes at defence facilities in Russia, *e.g.* wastes arising from reprocessing at the Krasnoyarsk-26 nuclear facility, participants concluded that the most straightforward and relative inexpensive conditioning methods should be used. Vitrification of (liquid) highly-active waste, as phosphate glass rather than borosilicate glass, was recommended since the technology for the latter is familiar and available within Russia at the present time. For intermediate- and low-level wastes, solidification in cement/concrete was recommended, in common with conditioning practices in almost all European countries.

Radioactive Waste Disposal – Site Selection

At the meeting, scientists heard how local geophysicists have started regional geophysical investigations of the granitic basement in the Krasnoyarsk Krai Territory, prior to more detailed characterisation work. A significant start has also been made on the identification of prospective disposal sites, with support from scientists from the V.G. Khlopin Radium Institute in Moscow.

Some members of the local scientific community considered that any site close to Krasnoyarsk-26 was not acceptable on geological/safety grounds, mainly due to concerns over future seismic events. In this context, it was pointed out that seismic effects, either current or future (predicted), are likely to be much more significant in other countries, *e.g.* Japan. Again, the *multi-barrier waste disposal principle* was emphasised, whereby both near-field and far-field barriers contribute to the overall safety of a disposal concept. In some disposal concepts, greater emphasis is placed on near-field barriers; in others, the far-field barrier is more important.

In discussions concerning site selection, the phrase 'best site' was mentioned frequently. However, it was pointed out that finding the 'best' site is not a viable proposition. Rather, what is required is an *acceptable* site, *i.e.* one for which the results of the asociated safety assessment performed for a particular disposal concept, indicate an assessment value (*e.g.* dose to the population) which is below the national (or international) authorisation safety limit (or guideline).

Several other site-selection issues were discussed, *viz.*

- With reference to an international methodology for site selection (IAEA, NEA), preliminary site selection should be based on a careful evaluation of all relevant factors and not just geological input. Thus, for example, distance from the waste storage location to a potential repository site is an important consideration and may well be a key factor towards the selection of a local site. The site selection methodology should also address site characterisation work (drilling programme, seismic investigations, geochemical sampling and analysis, and hydrogeological data collection) and the type of characterisation data which are necessary in order to be able to assess the suitability of previously identified sites.
- Site selection should be an open process and should be recognised as such by the local community. Western experience has shown (often the hard way!) that such a characteristic is best achieved by publishing technical reports on site characterisation progress in a timely manner and making them widely available.
- The need for a *multi-disciplinary approach* to site selection and characterisation was emphasised in discussions. No single scientific discipline can provide all the expertise necessary to support the decision on a suitable site. Thus, geological/hydrogeological input is certainly critical to the overall evaluation of a site's suitability, but other key inputs are hydrochemistry, geochemistry, transportation, as well as social and environmental (impact) issues. Coordination of such an integrated effort is essential. For the Krasnoyarsk region, local scientists (Krasnoyarsk State University) are well placed to coordinate such a multi-disciplinary approach, and similarly for other regions in Russia.

Deep Injection of Liquid Radioactive Waste

At Krasnoyarsk-26, liquid waste from reprocessing facilities has been disposed of by injection into relatively impermeable horizons close to the facility, similar to that which occurred at Chelyabinsk and, to a lesser extent, at Tomsk, and the fate of such radioactivity was one of the major topics discussed at the Workshop. It was noted that, based on the results of detailed safety assessments, authorisation (Disposal Licence) has been given for deep-well injection of liquid waste until 2001. This liquid waste comprises mainly Sr-90 and Cs-137 radionuclides, but has been treated to remove actinides prior to disposal by injection. After 2001, it was unaminously agreed that radioactive waste management should involve the disposal of solidified waste only, with appropriate solidification methods, as summarised above.

Independent assessment of the resultant radioactivity in geological formations at Tomsk had been performed by UK scientists as part of an EC study, and this work was reported at the meeting. The main conclusion of this assessment was that the resultant radiological impact (dose to the population) from low-level liquid waste disposed at Tomsk was several orders of magnitude below the typical national safety limit (0.1 mSv a^{-1}) in Western European countries. Similar studies addressing liquid waste disposal at Chelyabinsk and Krasnoyarsk-26, are being carried out at the present time and NATO delegates recommended that the results of this work be completed and published as quickly as possible, to allay any fears of the local population.

PERSPECTIVE OF RUSSIAN FEDERATION PARTICIPANTS

Representatives of various industrial and scientific organisations from Krasnoyarsk, Moscow, Tomsk, St. Petersburg, Novosibirsk, Apatity and Krasnodar in Russia, as well as from the U.K., Germany, U.S.A., Canada, Belgium, Bulgaria and Portugal, took part in the Workshop. More than twenty-five papers were presented and discussed over a three-day period. In addition, general discussion sessions, as well as a press conference, were held during the meeting.

A general conclusion from the Workshop is that the deep disposal of radioactive waste – the final stage of waste management, should be able to offer the necessary radiological protection to the general public and the environment around disposal sites, provided that the selection of the disposal site and disposal concept is carried out according to existing (international) guidelines and that certain requirements are fulfilled during the construction of the repository.

Extensive research and development of disposal technologies are underway in various scientific centres throughout the world and in some countries, industrial-scale radioactive waste disposal (surface) sites are in operation.

Workshop delegates heard that deep disposal by injection of liquid radioactive waste has occurred at the Chemical Mining Plant of Minatom (at Zheleznogorsk, Krasnoyarsk region) since 1967 and that research work is currently being carried out, the aim of which is to select and characterise potential sites for the disposal of solid radioactive waste in low-permeable geological formations.

Delegates at the Workshop were in favour of substantial international collaboration in several aspects of radioactive waste disposal, *viz.*

- *General problems*, specifically:
  - development of a methodology and the corresponding geodynamic and geological research technology for the construction of a deep repository;
  - development of methods addressing the safety of radioactive waste disposal systems, taking advantage of international databases and using predictive modelling combined with probabilistic risk assessment;
  - research and development of engineered and natural barriers, taking into account geomechanical, tectonic, physico-chemical and geochemical phenomena and processes;
  - study of operational radioactive waste repositories and examination of the historical background to their characterisation (including liquid radioactive waste repositories) in Russia;
  - study of natural analogues in order to make use of experience gained during the development of immobilisation technologies and validation of radioactive waste disposal involving different waste forms and different types of geological formations;
  - development of methods of geodynamic exploration, covering tectonic and seismic phenomena, and methods of non-destructive monitoring of mountain ranges via seismic, electromagnetic and acoustic measurements;
  - preparation and submission of proposals to appropriate funding bodies for conducting joint research projects in the field of radioactive waste disposal.
- *Radioactive waste disposal in the Krasnoyarsk region*
  - comprehensive research programme and additional investigation of the geology of the Mining Chemical Plant region in connection with both the planned construction of a deep repository for solid wastes in the Kansko-Agansky mountain range, and the ongoing liquid radioactive waste disposal at the *Severniy* site;
  - expert evaluation of the current and future environmental status of region around the *Severniy* site, using both Russian and foreign specialists (according to Russian Federation legislation);

There was a general consensus among the delegates for the establishment of an Overview Committee/Group to consider the problems of deep radioactive waste disposal in the Krasnoyarsk region. Such a committee would consist of representatives of the administrative and scientific organisations of the Krasnoyarsk region, Minatom and other interested environmental and public organisations as well as foreign specialists.

Finally, substantial enthusiasm was expressed for holding a follow-up Workshop in 1998.

Michael Stenhouse
Vladimir I. Kirko
Co-Directors, NATO Advanced
Research Workshop

INDEX

GPSR Compliance

The European Union's (EU) General Product Safety Regulation (GPSR) is a set of rules that requires consumer products to be safe and our obligations to ensure this.

If you have any concerns about our products, you can contact us on ProductSafety@springernature.com

In case Publisher is established outside the EU, the EU authorized representative is:

Springer Nature Customer Service Center GmbH
Europaplatz 3
69115 Heidelberg, Germany

Batch number: 10761697

Printed by Printforce, the Netherlands